ORIGIN OF COSMIC RAYS

INTERNATIONAL ASTRONOMICAL UNION
UNION ASTRONOMIQUE INTERNATIONALE

SYMPOSIUM No. 94

JOINTLY WITH INTERNATIONAL UNION OF
PURE AND APPLIED PHYSICS
HELD IN BOLOGNA, ITALY, JUNE 11–14, 1980

ORIGIN OF COSMIC RAYS

EDITED BY

GIANCARLO SETTI
University of Bologna and Istituto di Radioastronomia, CNR, Bologna, Italy

GIANFRANCO SPADA
Istituto Tecnologie e Studio Radiazioni Extraterrestri, CNR, Bologna, Italy

ARNOLD W. WOLFENDALE
Physics Department, University of Durham, U.K.

D. REIDEL PUBLISHING COMPANY
DORDRECHT : HOLLAND / BOSTON : U.S.A. / LONDON : ENGLAND

Library of Congress Cataloging in Publication Data

Main entry under title:

Origin of cosmic rays.

    At head of title: International astronomical union; Union astronomique internationale.
    1. Cosmic rays-Congresses. I. Setti, Giancarlo, 1935– II. Spada, Gianfranco. III. Wolfendale, A. W. IV. International astronomical union.
QC485.8.07074   523.01'97223   81-2574
ISBN 90-277-1271-9   AACR2
ISBN 90-277-1272-7 (pbk.)

Published on behalf of
the International Astronomical Union
by
D. Reidel Publishing Company, P.O. Box 17, 3300 AA Dordrecht, Holland

All Rights Reserved
Copyright © 1981 by the International Astronomical Union

Sold and distributed in the U.S.A. and Canada
by Kluwer Boston Inc.,
190 Old Derby Street, Hingham, MA 02043, U.S.A.

In all other countries, sold and distributed
by Kluwer Academic Publishers Group,
P.O. Box 322, 3300 AH Dordrecht, Holland

D. Reidel Publishing Company is a member of the Kluwer Group.

No part of the material protected by this copyright notice may be reproduced or utilized in any form or by any means, electronic or mechanical, including photocopying, recording or by any informational storage and retrieval system, without written permission from the publisher

Printed in The Netherlands

TABLE OF CONTENTS

PREFACE	ix
THE ORGANIZING COMMITTEES	xi
LIST OF PARTICIPANTS	xiii

V.L. GINZBURG / THE ORIGIN OF COSMIC RAYS (INTRODUCTORY REMARKS)	1
P. MEYER / REVIEW OF COSMIC RAYS	7
H. REEVES / ISOTOPES IN GALACTIC COSMIC RAYS	23
W.R. Webber / The Charge and Isotopic Composition of $Z \geq 10$ Nuclei in the Cosmic Ray Source	31
C.J. Waddington, P.S. Freier, R.K. Fickle and N.R. Brewster / Isotopes of Cosmic Ray Elements from Neon to Nickel	33
M. Cassé, J.A. Paul and J.P. Meyer / Wolf Rayet Stars and the Origin of the ^{22}Ne Excess in Cosmic Rays	35
Y.V. Rao, A. Davis, M.P. Hagan and R.C. Filz / CR-39 Plastic Track Detector Experiment for Measurement of Charge Composition of Primary Cosmic Rays	37
J.P. WEFEL / SUPERNOVA AND COSMIC RAYS	39
G.G.C. Palumbo and G. Cavallo / What can we learn about Cosmic Rays from the UV, Optical, Radio and X-Ray Observations of Supernova 1979c in M 100?	51
J. LINSLEY / VERY HIGH ENERGY COSMIC RAYS	53
M. Giler, J. Wdowczyk and A.W. Wolfendale / Diffusion of High Energy Cosmic Rays from the Virgo Cluster	69
T.K. Gaisser, T. Stanev, P. Freier and C.J. Waddington / On the Detection of Heavy Primaries above 10^{14} eV	71
A. Ferrari and A. Masani / Interstellar and Intracluster Tunnels and Acceleration of High-Energy Cosmic Rays	73
J. Nishimura / Features of the High Energy Electron Spectrum	75
P.H. FOWLER, M.R.W. MASHEDER, R.T. MOSES, R.N.F. WALKER and A. WORLEY / ULTRA HEAVY COSMIC RAYS	77
W.R. Binns, R. Fickle, T.L. Garrard, M.H. Israel, J. Klarmann, E.C. Stone and C.J. Waddington / The Heavy Nuclei Experiment on HEAO-3	91

R. COWSIK / PROPAGATION STUDIES RELATED TO THE ORIGIN OF COSMIC
   RAYS   93

R.J. Protheroe, J.F. Ormes and G.M. Comstock / Interpretation of
   Cosmic Ray Composition: the Pathlength Distribution   107
R.I. Epstein / Are Stellar Flares and the Galactic Cosmic Rays
   Related?   109

E.B. FOMALONT / EXTENDED RADIO SOURCES   111

I.I.K. PAULINY-TOTH / COMPACT RADIO SOURCES   127

M.J. REES / NUCLEI OF GALAXIES: THE ORIGIN OF PLASMA BEAMS   139

N. Panagia and K.W. Weiler / The Common Properties of Plerions
   and Active Galactic Nuclei   165
E. Hummel / Central Radio Sources in Galaxies   167
L. Maraschi, R. Roasio and A. Treves / A Self-Consistent Multiple
   Compton Scattering Model for the X and γ-Ray Emission from
   Active Galactic Nuclei   169
K.O. Thielheim / Symmetry Breaking and Invariant Mass Approach to
   the Spiral Structure of Galaxies   171
R. Silberberg and M.M. Shapiro / Neutrino Emission from Galaxies
   and Mechanisms for Producing Radio Lobes   173

J. ARONS / PARTICLE ACCELERATION BY PULSARS   175

M. Morini and A. Treves / γ-Ray Emission from Slow Pulsars   205
M.M. Shapiro and R. Silberberg / Distribution of Neutrino Fluxes
   from Pulsar Shells   207

R. SANCISI and P.C. VAN DER KRUIT / DISTRIBUTION OF NON-THERMAL
   EMISSION IN GALAXIES   209

G. Sironi and G. De Amici / Cosmic Rays and Galactic Radio Noise   215
C.G.T. Haslam, C.J. Salter and H. Stoffel / The All-Sky 408 MHz
   Survey   217
S. Kearsey, J.L. Osborne, S. Phillipps, C.G.T. Haslam,
   C.J. Salter and H. Stoffel / The Large-Scale Distribution of
   Synchrotron Emissivity in the Galaxy   223
R. Beck and U. Klein / Radio Emission from Nearby Galaxies at
   High Frequencies   225

W.B. BURTON and H.S. LISZT / HIGH-DENSITY, COOL REGIONS OF
   INTERSTELLAR MATTER IN THE GALAXY   227

C.J. Cesarsky and R.M. Kulsrud / Cosmic-Ray Self-Confinement in
   the Hot Phase of the Interstellar Medium   251

M. Lachièze-Rey / Rayleigh Taylor Instabilities in the
  Interstellar Medium   253
R.J. Stoneham / Nonlinear Landau Damping of Alfven Waves and the
  Production and Propagation of Cosmic Rays   255
T.K. Gaisser, A.J. Owens and G. Steigman / Cosmic Ray Antiprotons
  5-12 GeV   257
J. Szabelski, J. Wdowczyk and A.W. Wolfendale / Anti-Protons in
  the Primary Cosmic Radiation   259

J. TRÜMPER / THE X-RAY SKY   261

A. Kembhavi and A.C. Fabian / Quasar Contribution to the X-Ray
  Background   273
S. Ikeuchi and A. Habe / Dynamical Behaviour of Gaseous Halo in
  a Disk Galaxy   275
P. Giommi and G.F. Bignami / The Fluctuations of the Cosmic X-Ray
  Background as a Sensitive Tool to the Universal Source
  Distribution   277

L. SCARSI, R. BUCCHERI, G. GERARDI and B. SACCO / THE GAMMA-RAY
  SKY   279
A.W. WOLFENDALE / GAMMA RAYS FROM COSMIC RAYS   309

T. Montmerle, J.A. Paul and M. Cassé / Cosmic Rays from Regions
  of Star Formation - I. The Carina Complex   321
M. Cassé, T. Montmerle and J.A. Paul / Cosmic Rays from Regions
  of Star Formation - II. The OB Associations   323
J.A. Paul, M. Cassé and T. Montmerle / Cosmic Rays from Regions
  of Star Formation - III. The Role of T-Tauri Stars in the Rho
  Oph Cloud   325
T. Dzikowski, B. Grochalska, J. Gawin and J. Wdowczyk / High
  Energy γ-Rays from the Direction of the Crab Pulsar   327
G. Pizzichini / A New Kind of Gamma Ray Burst ?   329
M. Salvati, E. Massaro and N. Panagia / The Components of the
  Galactic γ-Ray Emission   331
M. Giler, J. Wdowczyk and A.W. Wolfendale / Gamma Rays from
  Galaxy Clusters   333
F. Giovannelli, S. Karakula and W. Tkaczyk / High Energy Gamma
  Rays from Accretion Disc   335
G.F. Bignami / Search for X-Ray Sources in the COS-B Gamma-Ray
  Error Boxes   337

W.I. AXFORD / THE ACCELERATION OF GALACTIC COSMIC RAYS   339

H.J. Völk, G.E. Morfill and M. Forman / Cosmic Ray Acceleration
  in the Presence of Losses   359
C.J. Cesarsky and J.P. Bibring / Cosmic-Ray Injection into
  Shock-Waves   361

L. O'C. Drury and H.J. Völk / Shock Structure including Cosmic
  Ray Acceleration 363
J. Pérez-Peraza and S.S. Trivedi / Selective Effects in Cosmic
  Rays Induced by Coulombian Interactions with Finite
  Temperature Plasmas 365

C. CHIUDERI / HIGH ENERGY PHENOMENA IN THE SUN 367

G. HAERENDEL / MAGNETOSPHERIC PROCESSES POSSIBLY RELATED TO THE
  ORIGIN OF COSMIC RAYS 373

L.G. Kocharov and G.E. Kocharov / On the Mechanism of Generation
  of Solar Cosmic Rays Enriched by Helium-3 and Heavy Elements 393
E. Möbius and D. Hovestadt / On ^{3}-He Rich Solar Particle Events 395
H.S. Ahluwalia / Cosmic Ray Evidence for the Magnetic
  Configuration of the Heliosphere 397

G. SETTI / CONCLUDING REMARKS 399

GENERAL DISCUSSION 405

PREFACE

The cosmic radiation was discovered by Hess in 1912 but its origin is still the subject of much controversy and considerable study. For several decades most workers in the cosmic ray field were interested in the Nuclear Physical aspect of the particle beam and many important discoveries were made, notably the identification of the positron, the muon, the pion and the strange particles. More recently however, emphasis has changed to the Astrophysical aspect both with regard to the origin of the radiation and to its relation with the other radiation fields.

Mindful of the increasing importance of the Astrophysical facets of the subject the Cosmic Ray Commission of IUPAP approached the High Energy Astrophysics Commission of the IAU with the suggestion of a joint Symposium on Cosmic Ray Origin. The plan was to bring together workers in all the various astronomical fields - from Radio, through Optical to Gamma Rays - with Cosmic Ray physicists and to fully explore the various interrelations. The approach was received with enthusiasm and this book contains the proceedings of the ensuing Symposium, (styled IUPAP/IAU Symposium No. 94) which was held in Bologna from 11th - 14th June 1980. Virtually all the papers presented are reproduced here.

The Scientific Organizing Committee chose the invited speakers whose papers formed the major scientific component of the meeting. Additional short contributed papers were also called for and very brief 2-page resumés (which have not been refereed) are given in the proceedings.

The Symposium was sponsored by IAU, IUPAP and the National Research Council of Italy and to these bodies the organisers express their grateful thanks. The organisers are also very indebted to the Mayor of Bologna for the warm reception offered to the participants and for permission to use the "Stabat Mater" Aula for the opening session of the Symposium. The organizers wish to express their gratitude to the secretarial staff, especially to Miss Pia Tamborrino, of the TE.S.R.E. Institute for their invaluable help offered in the organization of the meeting.

GIANCARLO SETTI
GIANFRANCO SPADA
ARNOLD W. WOLFENDALE

SCIENTIFIC ORGANIZING COMMITTEE

A.W. Wolfendale (Chairman), S. Hayakawa, K.I. Kellermann, F.B. McDonald,
F. Pacini, K. Pinkau, J.P. Wefel, G. Setti, I.S. Shklowsky

LOCAL ORGANIZING COMMITTEE

G. Setti (Chairman), G. Spada (Secretary),
G. Cavallo, S. Cecchini, M. Galli

LIST OF PARTICIPANTS

Ahluwalia, H.S., University of New Mexico, Albuquerque, U.S.A.
Arons, J., University of California, Berkeley, U.S.A.
Attolini, M.R., Istituto TE.S.R.E., Bologna, Italy
Axford, W.I., MPI für Aeronomie, Katlenburg-Lindau, W. Germany
Bartolini, C., University of Bologna, Italy
Beck, R., MPI für Radioastronomie, Bonn, W. Germany
Belli, B.M., Istituto Astrofisica Spaziale, Frascati, Italy
Bignami, G.F., Istituto Fisica Cosmica, Milano, Italy
Bijleveld, W., Sterrewacht Leiden, The Netherlands
Blake, P.R., University of Nottingham, England
Bloemen, J.B.G.M., Huygens Laboratorium, Leiden, The Netherlands
Bradt, H., MIT, Cambridge, U.S.A.
Brini, D., University of Bologna, Italy
Burton, W.B., University of Minnesota, Minneapolis, U.S.A.
Camenzind, M., Universität Zürich, Switzerland
Camerini, U., University of Wisconsin, Madison, U.S.A.
Caraveo, P.A., Istituto Fisica Cosmica, Milano, Italy
Cassé, M., CEN Saclay, Gif-sur-Yvette, France
Cassiday, G., University of Utah, Salt Lake City, U.S.A.
Cavaliere, A., University of Roma, Italy
Cavallo, G., Istituto TE.S.R.E., Bologna, Italy
Cecchini, S., Istituto TE.S.R.E., Bologna, Italy
Cesarsky, C.J., CEN Saclay, Gif-sur-Yvette, France
Chaliasos, E., University of Athens, Greece
Chiuderi, C., University of Firenze, Italy
Cini, G., Istituto Cosmo Geofisica, Torino, Italy
Cowsik, R., Tata Institute of Fundamental Research, Bombay, India
Dardo, M., Istituto Cosmo Geofisica, Torino, Italy
Di Cocco, G., Istituto TE.S.R.E., Bologna, Italy
Dilworth, C., University of Milano, Italy
Drury, L., MPI für Kernphysik, Heidelberg, W. Germany
Dusi, W., Istituto TE.S.R.E., Bologna, Italy
Epstein, R.I., NORDITA, Copenhagen, Denmark
Ferrari, A., Istituto Cosmo Geofisica, Torino, Italy
Fomalont, F.B., NRAO, Charlottesville, U.S.A.
Forman, M.A., University of New York, Stony Brook, U.S.A.
Fowler, P.H., University of Bristol, England
Fry, W.F., University of Wisconsin, Madison, U.S.A.
Frontera, F., Istituto TE.S.R.E., Bologna, Italy
Gaisser, T.K., University of Delaware, Newark, U.S.A.
Galli, M., University of Bologna, Italy
Gavazzi, G., Istituto Fisica Cosmica, Milano, Italy
Giler, M., University of Lodz, Poland

Ginzburg, V.L., Lebedev Physical Institute, Moscow, U.S.S.R.
Giovannelli, F., Istituto Astrofisica Spaziale, Frascati, Italy
Haerendel, G., MPI für Extraterrestrische Physik, Garching, W. Germany
Haslam, G., MPI für Radioastronomie, Bonn, W. Germany
Hayakawa, S., University of Nagoya, Japan
Hill, P., MPI für Radioastronomie, Bonn, W. Germany
Horstman, H., University of Bologna, Italy
Hummel, E., University of Groningen, The Netherlands
Ikeuchi, S., Hokkaido University, Sapporo, Japan
Jodogne, J.C., Institut Royal Météorologique, Bruxelles, Belgium
Karakula, S., University of Lodz, Poland
Kellermann, K.I. NRAO, Green Bank, U.S.A.
Kembhavi, A.K., University of Cambridge, England
Kiraly, P., Central Research Institute of Physics, Budapest, Hungary
Koch, L., CEN Saclay, Gif-sur-Yvette, France
Kocharov, G.E., Physico-Technical Institute, Leningrad, U.S.S.R.
Kondo, I., University of Tokyo, Japan
Kota, J., Central Research Institute of Physics, Budapest, Hungary
Lachièze-Rey, M., CEN Saclay, Gif-sur-Yvette, France
Lee, M.A., University of New Hampshire, Durham, U.S.A.
Linsley, J., University of New Mexico, Albuquerque, U.S.A.
Mandolesi, N., Istituto TE.S.R.E., Bologna, Italy
Maraschi, L., University of Milano, Italy
Massaro, E., University of Roma, Italy
Medina, J., Grupos Cientificos CONIE INTA., Madrid, Spain
Meyer, P., University of Chicago, U.S.A.
Michalec, A., Universytet Jagiellonski, Krakow, Poland
Möbius, E., MPI für Extraterrestrische Physik, Garching, W. Germany
Montmerle, T., CEN Saclay, Gif-sur-Yvette, France
Moretti, E., Istituto TE.S.R.E., Bologna, Italy
Nash, W.F., University of Nottingham, England
Nishimura, J., University of Tokyo, Japan
Occhialini, G., University of Milano, Italy
Ormes, J.F., NASA/GSFC, Greenbelt, U.S.A.
Osborne, J.L., University of Durham, England
Pacini, F., Osservatorio Astrofisico, Firenze, Italy
Paizis, C., University of Milano, Italy
Palumbo, G., Istituto TE.S.R.E., Bologna, Italy
Panagia, N., Istituto Radioastronomia, Bologna, Italy
Paul, J., CEN Saclay, Gif-sur-Yvette, France
Pauliny-Toth, I.I.K., MPI für Radioastronomie, Bonn, W. Germany
Pérez-Peraza, J., Universidad Nacional Autonoma de Mexico, Mexico
Phillips, S., University of Durham, England
Pinkau, K., MPI für Extraterrestrische Physik, Garching, W. Germany
Pizzichini, G., Istituto TE.S.R.E., Bologna, Italy
Preuss, E., MPI für Radioastronomie, Bonn, W. Germany
Protheroe, R.J., NASA/GSFC, Greenbelt, U.S.A.
Puppi, G., University of Venezia, Italy
Ramaty, R., NASA/GSFC, Greenbelt, U.S.A.
Rao, Y.V., Emmanuel College, Lexington, U.S.A.
Rasmussen, I.L., Danish Space Research Institute, Lyngby, Denmark

LIST OF PARTICIPANTS

Rees, M.J., University of Cambridge, England
Reeves, H., CEN Saclay, Gif-sur-Yvette, France
Rjazhskaja, O., Lebedev Physical Institute, Moscow, U.S.S.R.
Roasio, R., University of Milano, Italy
Rochester, C.K., Imperial College, London, England
Salvati, M., Istituto Astrofisica Spaziale, Frascati, Italy
Sancisi, R., University of Groningen, The Netherlands
Sarkar, S., Tata Institute of Fundamental Research, Bombay, India
Scarsi, L., University of Palermo, Italy
Sequeiros, J., Grupos Cientificos CONIE INTA., Madrid, Spain
Sette, D., University of Roma, Italy
Setti, G., Istituto Radioastronomia, Bologna, Italy
Shapiro, M.M., Naval Research Laboratory, Washington, U.S.A.
Simon, G., Universität Gesamthochschule, Siegen, W. Germany
Sironi, G., Istituto Fisica Cosmica, Milano, Italy
Spada, G., Istituto TE.S.R.E., Bologna, Italy
Srinivasan, G., Raman Research Institute, Bangalore, India
Stoffel, H., MPI für Radioastronomie, Bonn, W. Germany
Stoneham, R.J., University of Cambridge, England
Thanmbyahpillai, T., Imperial College, London, England
Thielheim, K.O., University of Kiel, W. Germany
Treves, A., University of Milano, Italy
Trümper, J., MPI für Extraterrestrische Physik, Garching, W. Germany
Trussoni, E., University of Torino, Italy
Völk, H.J., MPI für Kernphysik, Heidelberg, W. Germany
Waddington, C.J., University of Minnesota, Minneapolis, U.S.A.
Wdowczyk, J., University of Lodz, Poland
Webber, W.R., University of New Hampshire, Durham, U.S.A.
Wefel, J., University of Chicago, U.S.A.
Wilkins, D., MPI für Astrophysik, Bonn, W. Germany
Windhortst, R., Sterrewacht Leiden, The Netherlands
Wolfendale, A.W., University of Durham, England
Yodh, G.B., National Science Foundation, Washington, U.S.A.
Zaninetti, L., University of Torino, U.S.A.
Zieba, S., Universytet Jagiellonski, Krakow, Poland

THE ORIGIN OF COSMIC RAYS (INTRODUCTORY REMARKS)

V.L. Ginzburg,
P.N. Lebedev Physical Institute,
Academy of Science of the USSR, Moscow.

The field of study, which by tradition is still called 'the problem of cosmic ray origin', is now at a watershed, at a turning point. For this reason the time chosen for the present Symposium is especially suited.

The above thesis needs to be confirmed and elaborated. To this end it will not be out of place first to dwell briefly upon the history of cosmic ray studies.[x)]

<u>1912</u>: the discovery of cosmic rays. As a matter of fact, the "dark" current in ionization chambers was studied even earlier but it seems correct to associate the actual discovery of cosmic rays with the flights of V. Hess. If a precise date of the discovery of cosmic rays is needed, August 7th, 1912, when Hess undertook his most successful flight, is suited best of all.
<u>1927</u> (approximately): the extraterrestrial origin of cosmic rays became accepted (it was supposed earlier that the observed ionization might result from the presence of radioactive elements in higher atmospheric layers). First indications of the fact that cosmic rays are not a hard γ-radiation appeared. Specifically, the geomagnetic effect and the existence of high-energy charged particles in the atmosphere were revealed which indicated the corpuscular nature of cosmic rays.
<u>1936</u> (approximately): the existence of the geomagnetic effect and, therefore, the fact that primary cosmic rays are high-energy charged particles was no longer in doubt.
<u>1939-41</u>: primary cosmic rays were proved to be for the most part protons. Such a conclusion was drawn from the results of measurement of the East-West asymmetry and from the study of primary particles using balloons.
<u>1948</u>: nuclei of some elements were discovered amongst the primary cosmic rays. By 1950 these data obtained with balloons (mainly with the aid of emulsions) had been confirmed and enlarged.

So, it was only in 1950 that the composition of primary cosmic rays became clear in the first approximation (for the electron flux only an upper limit was determined, which was of the order of one per cent of the total flux). A number of papers also appeared anticipating the further development of cosmic ray astrophysics (1934 Baade and Zwicky - a hypothesis on cosmic ray acceleration in supernova flares; 1949, Fermi - a statistical acceleration, the role of the interstellar magnetic field). Nevertheless, cosmic rays remained an object of secondary importance in astrophysics and for astrophysics. This is quite clear since there existed data on cosmic rays near the Earth only, and a high degree of cosmic ray isotropy did not permit the acquisition of observational data on their sources.

1950-53: emergence of 'cosmic ray astrophysics'. The scope of cosmic ray astrophysics I understand to be just the range of questions to which the present symposium is devoted. True, a more general name, "high-energy astrophysics" is used when one has in mind also gamma-astronomy and cosmic high-energy neutrino astronomy. It would be reasonable to understand the problem of cosmic ray origin only as the one concerning the origin of cosmic rays observed near the Earth.

What happened in 1950-53 consisted, in fact, in the establishment of the role of the synchrotron mechanism of electromagnetic wave radiation in space and, therefore, in establishing a connection between cosmic rays (or, more precisely, their electron component) and a certain considerable part of cosmic radioemission, and in some cases also, optical radiation. As a result the study of cosmic rays embraced the whole of our Galaxy and the Metagalaxy. Cosmic rays proved to be a Universal phenomenon and an exclusively important source of astronomical information. At the same time an energetic and a dynamical role of cosmic rays in supernova remnants, in the interstellar space and in radiogalaxies is very significant.[y)]

By the fifties the model of the cosmic ray origin which seemed then most probable, namely, the galactic halo model had already been discussed in detail (3-5). However, to prove this model it was necessary to make sure, firstly, that the cosmic ray energy density outside the Galaxy $W(CR, Mg)$ was much less than $W(CR, G)$ where $W(CR, G) \simeq 10^{-12}$ erg/cm^3 is the energy density of cosmic rays in the Galaxy (by 'Galaxy' we mean the volume $V \simeq 10^{68}$ cm^3, which corresponds to a quasi-spherical or to a somewhat flattened "cosmic ray halo" with characteristic dimension $R \simeq 3.10^{22}$ cm). The condition $W(CR, Mg) \ll W(CR, G)$ implies, evidently, that not a single metagalactic model of the origin of the major part of cosmic rays observed near the Earth is valid. Secondly - the existence of the cosmic ray halo would follow from the proof of the existence of a radio halo (when moving away from the galactic plane, the relativistic electrons responsible for the radiohalo lose their energy and the magnetic field strength may also decrease considerably; from this it is clear that the dimensions of the halo of cosmic rays, for the most part protons, may substantially exceed the dimensions of a somewhat bright radio halo).

To disprove convincingly the metagalactic model and to prove the existence of the halo turned out, unfortunately, to be a very difficult task and took almost a quarter of a century. But I am sure that now the work is done mainly owing to gamma-astronomical observations in the direction of the galactic anti-centre (which supports Galactic but not Metagalactic models) (ref. 6) and to astronomical studies of edge-on galaxies NGC 4631, NGC 891 and others, which have halos, and also to the observation of a radio halo in the Galaxy itself. One may hope that all the corresponding data obtained during the last few years will be presented at this Symposium. In any case, I cannot dwell upon this material now (see refs. 7-10 and the literature cited there).

I cannot guarantee, however, that everybody will agree with the statement on the existence of a convincing proof of the validity of the halo model (in the above-mentioned sense); to discuss this problem is one of our goals. But if one agrees with what has been said, an important phase in the study of the cosmic ray origin is already concluded. The opinion expressed in the beginning as to what concerns the turning point in the development of cosmic ray astrophysics should be understood in just this sense.

As has been said, this branch of astronomy is about 30 years old (one can say differently, subject to one's attitude: 'only 30 years' or 'already 30 years'). Everybody, evidently, realizes how much restricted we are in our possibility to glance into the future. However, it is natural and even necessary to think of what will become of high-energy astrophysics, say, by the beginning of the XXI'st century, or even by 2012 - a centenary of cosmic rays.

It is, of course, hardly possible to make a reliable prognosis and I for one do not claim this. Therefore, I shall restrict myself to enumeration, in the order of discussion, of some key problems and branches of further investigations.

1. Equipment on satellites and high-altitude balloons will make it possible to specify considerably the data on the chemical and isotopic composition of cosmic rays at different energies. In the first place one may hope for the determination of the chemical composition at energies up to 10^{12}-10^{13} eV/nucleon and the amount of radioactive nuclei, ^{10}Be, at $\frac{E}{mc^2}$ up to 10, i.e. already in the relativistic region. We are going to determine the electron, positron and antiproton spectra with high accuracy and over a wide energy range.

2. In order that one might obtain information on cosmic rays in the sources and on cosmic ray propagation (including generation, various secondary processes, etc.) from the data on primary cosmic rays near the Earth, it is necessary to perform calculations on the basis of diffusion models (8,11,12). The role of the galactic wind (convection) should be clarified, various plasma effects should be taken into account, etc.

3. The analysis of acceleration mechanisms is connected with what has been said above. Namely, we mean an analysis of acceleration in the explosion of a supernova itself, acceleration near pulsars and in turbulent supernova remnants, acceleration by shock waves (with account taken of diffusion) within young remnants and outside them. One of the main goals of investigations is now to clarify the role of acceleration by shock waves in young supernova remnants themselves (including their boundaries) and in interstellar space at distances up to 100 pc from the supernovae. By the way, I am inclined to think that acceleration proceeds mainly within young supernova remnants and acceleration by shock waves from supernova in interstellar space, and that acceleration connected with novae and other stars does not play an essential role (at $E \gtrsim 10^{10}$ eV). However, these questions are open and it is not easy to give reliable answers to them. A possible method is the registration of radio and gamma-rays from supernovae and the surrounding regions.

4. Radioastronomical observations of edge-on galaxies will help to specify the character of electron motion in the halo. There are also certain possibilities here for the Galaxy as a result of detailed radio mapping of the sky at different frequency bands. The study of synchrotron radiation for radio galaxies, active galactic nuclei and quasars is a separate question.

5. It is possible and even very likely that in the eighties gamma astronomy will undergo the same development as did X-ray astronomy in the seventies. In any case the perspectives of galactic and extragalactic gamma astronomy are most promising.

6. As to the cosmic rays of superhigh energy, $E \gtrsim 10^{17}$ eV, there exists a great vagueness. One of the possibilities is that particles with $E < 10^{19}$ eV are produced mainly in the Galaxy while particles with $E > 10^{19}$ eV come mostly from the Local Supercluster. To discover the origin of particles with $E > 10^{17}$ eV in different energy ranges one will have to undertake labour-consuming measurements of mean anisotropy and the directions of the paths of individual particles (anisotropy measurements are interesting and important also at lower energies). The problem of establishing the chemical composition of particles with $E \gtrsim 10^7$ is, of course, also urgent.

Besides, we should note that in the eighties, and probably even later, the maximum attainable energy with accelerators will correspond, in the laboratory system, to $E = \frac{2E_c^2}{mc^2} \simeq 2 \cdot 10^{15}$ eV (colliding beams of protons with an energy $E_c = 10^{12}$ eV in each beam). Therefore, at $E > 2 \cdot 10^{15}$ eV cosmic rays will evidently long remain the only source of particles. It is, of course, difficult to work in this region (suffice it to mention that the intensity of particles with $E > 10^{16}$ eV makes up no more than 10^2 particle/km² ster. h.), but there still exist certain possibilities such as by studying extensive air showers at mountain altitudes.

7. Realization of the DUMAND project will permit registration of neutrinos with energy $E \sim 10^{12}$ eV with a rather high angular resolution of about $1°$. This technique offers some exclusive opportunities, among which there is a detection of neutrinos from quasars and active galactic nuclei (10,13). Since it is only neutrinos (besides gravitational waves) that are able to penetrate deep into matter, it is high-energy neutrino astronomy that seems to offer a solution of the fundamental question of the origin of cores of quasars and active galactic nuclei (the dilemma is if the core is a black hole or a magnetoid-spinar (13)). Neutrinos from a supernova flare in the Galaxy could also be registered in the more modest under-ground installations which exist already.

Neutrino astronomy (and, in particular, high-energy neutrino astronomy), along with gravitational wave astronomy, is the last known reserve (in the sense of using essentially new channels of astronomical information). There can be no doubts as to the necessity and inevitability of the development of neutrino astronomy.

The above enumeration is, of course, rather conditional and incomplete (suffice it to mention also X-ray astronomy, the study of high-energy particles and photons from the Sun and from the magnetospheres of the Earth, Jupiter and other planets, etc.). What has been said is already enough, however, to realize the scale of work to be done. When in all the branches mentioned sufficient data are accumulated (a period of 20-30 years is probably enough and at the same time not too much) high-energy astrophysics will play a still more outstanding role in astronomy than it does today. Besides there is no doubt that some unexpected things are to be encountered, and this is one of the attractive features of Science. One can only envy those men who will see the astronomy of the XXIst century in all its richness.

NOTES

x) See, for example, the monograph (1) which includes as appendices a number of original papers, and also the collection of original papers (2).

y) The state of cosmic ray astrophysics as well as the history of the problem by 1958 are clear, for example, from ref. 3 and the Proceedings of the Paris Symposium (4). The situation in 1964 may become clear, I hope, from the book (5), which also contains a rather large bibliography.

REFERENCES

1. Hillas, A.M., 1972, 'Cosmic Rays', Pergamon Press.
2. Selected papers on cosmic ray origin theories. Ed. Rosen, S., Dover Publications, New York, 1969.
3. Ginzburg, V.L., 1956, Nuovo Cim. Suppl. 3, 38,; 1958, 'Progress in Elem. Particle and Cosmic Ray Physics', Amsterdam, 4, 339.
4. 'Paris Symposium on Radio Astronomy', Stanford Univ. Press, Stanford, 1959.
5. Ginzburg, V.L., and Syrovatskii, S.I., 'The Origin of Cosmic rays', Pergamon Press, 1964.
6. Dodds, D., Strong, A.W., and Wolfendale, A.W., 1975, 'Galactic γ-Rays and Cosmic Ray Origin', Mon. Not. R. astr. Soc. 171, 569-577.
7. Phil. Trans. Roy. Soc. London A277, 317, 1975.
8. Ginzburg, V.L., and Ptuskin, V.S., 1976; Rev. Mod. Phys. 48, 161, 675; 1976 Sov. Phys. Uspekhi, 18, 931.
9. Ginzburg, V.L., 1978, Sov. Phys. Uspekhi, 21, 155.
10. 16th International Cosmic Ray Conference. Conference Papers v. 1-14, Kyoto, Japan, 1979.
11. Ginzburg, V.L., Khazan, Ya. M. and Ptuskin, V.S., 1980, Astrophys. and Space Sci. 68, 295.
12. Carvalho, J.C., and Haar, D. ter., 1979, Astrophys. and Space Sci., 61, 3., Wallace, J.M., 1980, Astrophys. and Space Sci. 68, 27, Owens, A.J., and Jokipii, J.R., 1977, Astrophys. Journ. 215, 677.
13. Berezinsky, V.S., and Ginzburg, V.L., Mon. Not. R.A.S., 1980 (in the press).

The list of references is not representative and to a considerable extent must in more detail, than was possible in the text, present the speaker's opinion. I give it because I suppose a much more extensive literature will be cited in other reports.

REVIEW OF COSMIC RAYS

Peter Meyer
Enrico Fermi Institute and Department of Physics
The University of Chicago

I. INTRODUCTION

Photon astronomy is divided into areas covering different energy regimes, beginning with radio astronomy at the lowest, and ending with gamma-ray astronomy at the highest observable energies. The cosmic radiation on the other hand encompasses in a single area nuclear particles and electrons of astrophysical origin over a very wide energy range, extending from superthermal energies that just exceed those of the solar or stellar winds ($\approx 10^6$ eV) to particles that carry the largest quanta of energy observed in any astrophysical radiation, up to 10^{21} eV. Dealing with that wide a range in energy, one should not be surprised that the origin of this radiation and its behavior is likely to involve a variety of different physical phenomena. Also, the spectrum of cosmic rays has two dimensions, one in energy, and one in the species of the particles.

Today we witness rapid and exciting developments in this field. But, in spite of the many advances after several decades of research, we still remain uncertain about the most fundamental issue, the question of the origin of the cosmic radiation.

Three properties of the cosmic radiation are measurable to any desired degree of accuracy--at least in principle. The composition, the energy spectrum for each species, and the directional distribution for any species at any energy. Can one expect that improvement in the measurement of these quantities will lead to a major step toward the identification of the sources of the cosmic radiation? I believe the answer is positive, based on the experience of the past two decades and the findings of recent experimental and theoretical work. The emerging detailed analysis of the unique sample of matter represented by the high energy cosmic rays begins to pinpoint the sites and mechanisms that lead to their origin.

I take it to be my task in this review to address those questions in a broad sense. I shall attempt to summarize the results that form the basis for the work that lies ahead, pointing to the role that particle astrophysics is expected to continue to play if one wishes to understand the phenomena that govern astrophysical objects. After all, the very fact that nature accelerates nuclei and electrons to the enormous energies that we observe is by itself most remarkable.

Before discussing the present status and the outlook for the future, I shall briefly look back to the past to remind ourselves of the most important milestones that, at their time, were pioneering advances.

II. THE PAST

Cosmic ray research became a subject of astrophysics when it was recognized that the radiation is of extraterrestrial origin. Beginning at that time, intense efforts were made to elucidate the composition and the energy distribution of the radiation in increasingly finer detail, and by the end of the 1970s enormous progress had been made.

Looking back over the last 30 years we may describe the milestones as follows: By the time it became fully evident that the cosmic rays were of extraterrestrial and extrasolar origin, one also learned that they were strongly influenced by solar system phenomena. Indeed, it was the propagation process in the solar system and its first description by a simple diffusion model which led to the recognition that similar processes must take place on much larger scales in interstellar space. The end of the 1940s brought the discovery of a complex nuclear composition; the '50s gained the understanding of the solar influence through the first quantitative description of the solar wind, its properties, and its role in cosmic ray propagation.

The rapid progress in cosmic ray research that began around 1960 was in part triggered by two important technological developments: (1) the availability of earth satellites and deep space probes, and (2) the great advances in solid state electronics that permitted the design of sophisticated instrumentation with low weight and low power requirements for use on spacecraft and on high altitude balloons. I believe it is fair to say that researchers in the field of cosmic rays were the first to fully exploit and to advance these modern facilities that today play such crucial roles in many areas of astrophysics. Experiments of the 1960s brought, in rapid sequence, the elucidation of increasingly finer details of the elemental composition of the cosmic rays, the discovery of the very rare elements beyond the iron group which we today call the ultra-heavy (UH) nuclei, and the discovery and determination of the energy spectrum of the electron and positron components. Most significant among the many results was the recognition that the nuclear species are produced in nucleosynthesis processes, that their abundance distribution has considerable similarities with the distribution of solar system material, but that the energetic particles also exhibit significant deviations from solar abundances. As a reminder of this situation, Figure 1 shows the present status of the relative elemental abundances for the elements between hydrogen and nickel as measured at energies between about 100 and 300 MeV/nucleon together with two compilations of the "solar system" abundances. It is clear that this is the distribution for a sample of matter produced in thermonuclear processes in the interiors of stars. But, particularly due to the high abundance of spallation products (i.e., Li, Be, B and $21<Z<26$) and the possible charge dependence of the acceleration processes, the elemental distribution cannot discriminate in a unique manner among the different nucleosynthesis processes that have been proposed. Rather, such information is expected to come from measurements of the abundance distribution of the UH nuclei ($Z > 28$) and from observing the isotopic abundances of those nuclei that predominantly originate in the cosmic ray sources. Both these topics, active areas of present cosmic ray research, are discussed below.

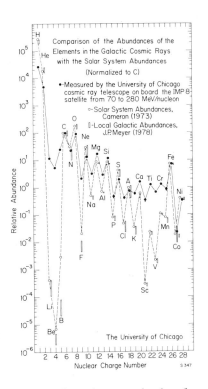

Fig 1: *The elemental abundance of the cosmic rays compared with 2 compilations of the solar system abundances.*

The accurate, and fully resolved measurement of the cosmic ray charge composition as it is shown in Figure 1 represents one of the milestones of past cosmic ray research. The deviations from solar system abundances are very interesting for several reasons. For example, the abundance of nuclei, originating predominantly in cosmic ray sources, appears to be influenced by the ionization potential of the particular atom, indicating preference for acceleration of nuclei with low first ionization potential. Even more importantly, the study of the spallation products has provided quite accurate knowledge of the average amount of interstellar matter traversed by the particles between source and observer. Comparison of the abundance of different secondary elements yields the distribution of the escape pathlengths with some accuracy.

Just a few years ago, when composition measurements were extended to energies of around 100 GeV/nucleon, it was found that the escape pathlength rapidly decreases with increasing energy. Figure 2 shows the measurement of the escape mean free path as a function of energy as it was obtained in one of the several composition measurements at high energy. The discovery of this rapidly decreasing escape mean free path with increasing energy has interesting repercussions with regard to the

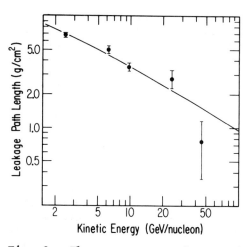 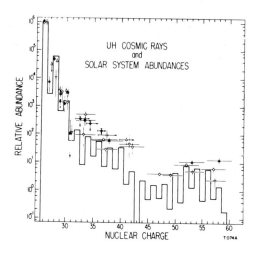

Fig. 2: The escape mean free path as a function of energy. The line is a powerlaw fit in total energy.

Fig. 3: Cosmic ray UH abundance between Fe and Nd compared with the element abundances in the solar system.

storage and containment time of particles in the galaxy, and hence the configuration of galactic magnetic fields which are the agent responsible for particle containment. Clearly, more accurate information and observations at still higher energy are needed to quantitatively apply this result to propagation models. This is therefore one area of intense research interest at the present time.

The discovery of cosmic ray nuclei beyond the iron group, and up to the actinides was first reported in 1965. The initial findings were based on the study of particle tracks in crystals of meteorite material. A few years later these nuclei could be identified in the contemporary flux of cosmic rays, and, through a series of experiments, their approximate abundance distribution was determined. Although these experiments were not able to resolve individual nuclear charges, they displayed an enhancement of elements around Platinum, a feature that is expected if the r-process of nucleosynthesis plays a dominant role in the production of these nuclei. These discoveries represented a major step forward in cosmic ray research and their details and consequences are not yet fully exploited. An example of abundance measurements up to $Z = 60$ compared with the solar system abundance is shown in Figure 3.

Another important, though negative result came from the search for antinuclei ($Z > 1$) in the cosmic radiation. Several experiments, using magnetic spectrometers, were able to put upper limits on the flux of antinuclei of 10^{-4} of the nuclei. The elements under study were He, C, and O. The discovery of even one antinucleus would have far reaching astrophysical consequences.

Among the fascinating questions of cosmic ray astrophysics that has been with us for many years is the origin and nature of the extremely high energy primary particles. Instruments able to directly measure energy spectra of individual nuclear species and of electrons now reach up to total energies of about 10^{12} eV. Cosmic ray observations beyond those energies have remained the realm of groundbased instrumentation, through studies of mu-mesons and, beyond 10^{14} eV, extensive airshowers. As a non-expert in this highly developed field, I can only sketch the past achievements for the three most important pieces of evidence (1) composition of the primary radiation, (2) its energy spectrum, and (3) anisotropies.

All evidence on the composition of very high energy primaries is indirect, mostly from observation and analysis of the structure of extensive airshowers (see the reviews by Watson, 1975 and Sreekantan, 1979). It appears that most observations are compatible with an approximate mixture of about 60% protons and 40% heavier nuclei in the primary particles, but the range of uncertainty is wide. A few observations have led to the claim that the abundance of heavier nuclei is enhanced with respect to protons at energies above 10^{13} eV. Very little is known of the composition above 10^{17} eV, and it remains to be seen whether this situation can be improved by new instruments in the years to come.

The accuracy for the shape of the energy spectrum has been steadily improved over the years, confirming a steepening of the spectrum around 2×10^{15} eV. While this steepening of the spectrum seems to be well established, its causes are not clearly understood. In more recent years, on the basis of somewhat limited statistics it was observed that the energy spectrum flattens again around 10^{19} eV. The status of the work on the energy spectrum is shown in Figure 4. If the very high energy cosmic rays are of universal extragalactic origin, this flattening of the

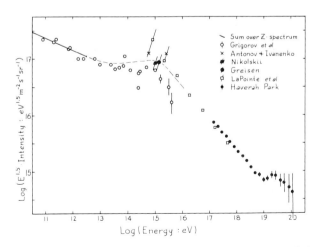

Fig. 4: *The energy spectrum of cosmic rays at high energies.*

spectrum is contrary to expectation, since particles that exceed 10^{19} eV should be attenuated due to inelastic collisions with photons from the universal blackbody radiation. One must be concerned whether the flattening of the spectrum around 10^{19} eV represents a real change in the primary spectrum or is introduced in the translation from measured shower size and structure to primary energy. Very little is known about the nature of nuclear interactions at those high energies. This question is still under debate among the experts. An alternate, though speculative explanation would place the sources outside the galaxy, but into relatively nearby objects. Neither of the observed changes in the slope of the spectrums at high energy lends itself to a straightforward astrophysical interpretation.

Observations of anisotropies pose a problem of a different nature. Only at energies above about 10^{14} eV are directional measurements entirely free of solar system influences. Early expectations that a galactic origin would lead to a readily observable anisotropy as the particle energy rises and its cyclotron radius becomes comparable to galactic dimensions were not borne out. In the recent past, anisotropies have been observed at energies beyond 10^{19} eV where the flattening of the spectrum occurs. The preferred arrival direction appears to be from the north galactic pole, which happens to be the direction towards the local supercluster. While these observations have led to numerous speculations, they are difficult to interpret.

The question of the origin of cosmic rays of very high energy has therefore remained in a state of flux (for details see Sreekantan,1979; Watson, 1975; and Wolfendale, 1979). While evidence has been mounting, corroborated by observations in gamma-ray astronomy, that low and intermediate energy cosmic rays are a galactic phenomenon, it remains difficult to accomodate the origin of the highest energy particles within the framework of a galactic model unless the prevalent views on particle containment in the galaxy and its vicinity are drastically revised. Understanding the origin of the highest energy cosmic rays therefore remains an important challenge for the future.

III. THE PRESENT

In active scientific fields, the present is a shortlived span of time, its achievements being quickly superceded and relegated into the past. This very much applies to cosmic ray research. Today several new avenues are being opened and new techniques are applied that lead to insights that were inaccessible even in the recent past. These provide the first glimpses into new facets of fundamental importance in understanding the origin of the radiation.

I shall divide this snapshot of the present status into two parts, first dealing with the origin question, and then with the containment in the interstellar medium, although these are not unrelated questions. The most important tool that leads to an understanding of the sources of the cosmic radiation is the precise determination of the composition of the particles, and the extrapolation of this composition to the sources. The following factors lead to the elemental and isotopic abundance distributions that one observes near earth and that were displayed in Figure 1:

a) the abundance distribution in the sources prior to, or during acceleration,

b) the charge dependence of the acceleration mechanisms, and

c) modifications that occur as secondary particles are produced and primary particles are lost in the collisions of source nuclei with the interstellar gas.

Item (c) depends on the amount of matter that the particles traverse. It is a most important input to study the nature of the interstellar medium and the particle containment, but it is an unwelcome complication in obtaining the source abundances. An indication for the role of item (b), the charge dependence of the acceleration, is found in the systematic increase of the cosmic ray abundance over the solar system abundance with increasing charge number Z. This increase is not only observed in the galactic cosmic rays, but also for energetic solar particles and is probably an effect of preferential acceleration. These two effects make it difficult to pinpoint the detailed nature of the nucleosynthesis processes that led to the composition of the source particles in spite of the fact that there now exist well resolved elemental abundance measurements for all elements from H to Ni of a precision that is better than for any other sample of extraterrestrial matter. Such insight, however, may be gained from investigations of the isotopic distribution, and from the element distribution of the UH elements. The UH elements, although extremely rare (see Figure 3), are not as contaminated with spallation products as the elements with $Z < 26$, since their abundances decrease almost monotonically with increasing Z. There is no equivalent to the Fe-peak. This area of research is now at the threshold of important advances from work carried out with experiments on two spacecraft, HEAO-3 and Ariel-6. Both these experiments are expected to provide the first resolution of individual nuclear charges in the UH range. Such charge resolution is needed to make unambiguous statements on the relative role of the r-process or other types of nucleosynthesis in the production of these elements. The earlier experiments, through the observation of an enhanced abundance of elements around platinum, and the actinides have indicated a dominant role of the r-process in producing the cosmic ray UH elements. But the Pt-peak is not far from a peak in Pb that would be expected from the s-process. A definitive answer can therefore be given only when Pt and Pb are individually resolved. Several other elemental abundances would greatly contribute to answer this problem. For example, the abundance ratio of Xe/Ba, which is 1 in the solar system, is expected to be 10 if the elements are exclusively produced in the r-process, and 0.1 if they are s-process products. The latest advances in this area will be discussed in more specialized papers at

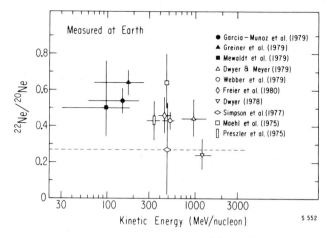

Fig. 5: *Measurements of the $^{22}Ne/^{20}Ne$ ratio as a function of energy.*

this conference.

Research to measure isotopic abundances of cosmic rays also progresses rapidly at this time. Several balloon experiments, and experiments on four spacecraft, IMP 7 and 8, ISEE-3, and HEAO-3 are directed toward measurements of isotopic abundances. The species for which we already have the first results on isotopic abundances include Ne, Mg, Si, and Fe, all of which predominantly originate in the cosmic ray sources. Additional results are available for the elements Be, B, and N, but I shall discuss these in connection with the questions of propagation (only the isotopes of hydrogen and helium had been studied in earlier years). These initial investigations of isotopic abundances have already

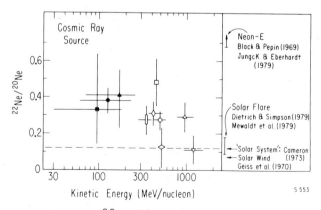

Fig. 6: *The $^{22}Ne/^{20}Ne$ ratio extrapolated to the cosmic ray sources.*

displayed one significant difference between the isotopic composition of solar system material and of the cosmic ray source nuclei. The results for the element Ne point to an overabundance of ^{22}Ne. In Figure 5 is shown the observed ^{22}Ne/^{20}Ne ratio measured at different energies by different observers. To obtain this ratio at the cosmic ray sources, one needs the help of a propagation calculation that takes into account the best information on the interaction cross-sections for spallation occurring in interstellar space. Figure 6 shows the result after this type of extrapolation, indicating that the ratio ^{22}Ne/^{20}Ne at the source of the cosmic rays is enhanced over the solar system abundances by about a factor of three. The isotopic ratio is also distinctly different from that found in the solar wind, and in solar flare particles. It is not sufficiently large to have any similarity with the Ne-E observed in some meteorites.

A display of the results for the ratio ^{26}Mg/^{24}Mg as measured at or near earth is shown in Figure 7, together with the expectation if the sources exhibit a solar system ratio. One experiment claims an enhancement of the neutron-rich isotope, and a second experiment marginally indicates this possibility. Obviously, further measurements are required that unambiguously determine even small deviations from solar system abundances for this important isotopic abundance ratio.

Few measurements of the isotopic abundance distribution of Si are available, and all are compatible with the solar system ratio of ^{30}Si/^{28}Si at the cosmic ray sources. However none of the experiments is sufficiently precise to rule out small but significant anomalies. These results are shown in Figure 8.

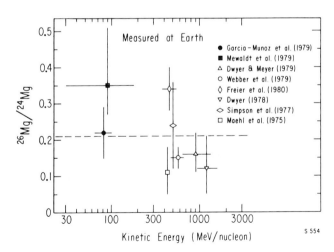

Fig. 7: Measurements of the ^{26}Mg/^{24}Mg ratio as a function of energy.

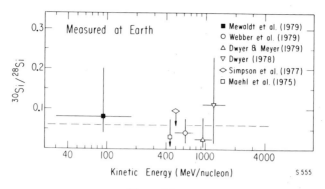

Fig. 8: Measurements of the $^{30}Si/^{28}Si$ ratio as a function of energy.

Major success has recently been achieved through the first convincing measurements of the isotopic abundance of the element Fe. Fe plays a special role in astrophysical nucleosynthesis processes as the end-product of exothermic nuclear burning. Its isotopic abundance distribution has been shown to critically depend on temperature, neutron concentration, and the density of the objects from which the cosmic rays have been ejected. The large enhancement of ^{56}Fe that exists in solar system matter, is now also established for the cosmic ray particles. Figure 9 is an example of a recent experimental result by the Cal Tech group. While containing only a small sample of particles it displays the advances in experimental technique by the excellent mass resolution that was achieved. It is important to note that several other recent experiments agree with the result that ^{56}Fe is by far the most abundant iron isotope in the cosmic rays. The quantitative question of the precise amount of admixture of ^{54}Fe and ^{58}Fe cannot yet be answered, except to say that neither isotope constitutes more than a few percent of ^{56}Fe.

It is quite clear from the results that are on hand that among the nucleosynthesis processes that have so far been theoretically investigated none can by itself explain the entire spectrum of cosmic ray abundances. Rather, as is the case for solar system material, a superposition of various processes is needed to describe the observations. For example, while the charge spectrum of the heaviest cosmic rays can be reasonably described by an r-process source, the nuclei with Z between 30 and 40 are found to be too abundant to fit any well understood model. The actinides, an important indicator for r-process material, show, however, a surprising underabundance in very preliminary results from the latest experiments.

Where has the present work on nuclear composition led us with respect to the problem of cosmic ray origin? Disregarding an extragalactic origin of the bulk of the cosmic radiation the two main potential sources that must be considered are (1) supernova explosions, (2) acceleration of particles from the interstellar medium. Clearly, these two sources may also simultaneously contribute in any mixture, since interstellar material, swept up in the shells of supernovae, might well be accelerated to

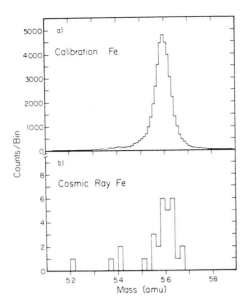

Fig. 9: The isotopic abundance distribution of Fe as measured by Cal Tech group.

cosmic ray energies. This is not the place to discuss these problems in detail. But we may summarize as follows. Measurements of the isotopic composition and of the UH composition point toward the fact that the sample of material that constitutes the primary cosmic rays has features in its composition that distinguish it from solar system material, but, perhaps more importantly also from interstellar material. The most convincing of these features are the over-abundance of ^{22}Ne, the apparent peak in the Pt region, and the high abundance of the actinides, if that is substantiated. Each of these features is difficult to explain if the cosmic rays were entirely of interstellar origin. It must be kept in mind, though, that they probe the interstellar medium of the past 10^6 to 10^7 years, rather than 10^9 years as is the case for solar system material. Hence, they must represent a sample of interstellar gas at a different stage of evolution. The work on isotopic and elemental abundances is discussed in the reviews by J.P. Meyer (1975), Waddington (1977), Müller (1977) and Balasubrahmanian (1979).

Cosmic ray propagation and containment are only indirectly linked to the question of origin. I shall review them here briefly, and begin with a discussion of the abundances of those nuclides that are spallation products, originating from collisions in interstellar space. The abundances of the stable secondary particles not only provide a measure of the average amount of material that the particles traverse during the time they are contained in the galaxy, but, through the requirement of consistency between different species, and measurements of the energy dependence of the abundances, they may yield the distribution of pathlengths (see Raisbeck 1979 for a recent discussion). The highly precise elemental abundances that new measurements provide, and the forthcoming work on isotopic abundances are greatly improving this knowledge. However, such efforts will be fully successful only if the knowledge of interaction cross-sections which enter any propagation calculation is improved to a similarly high precision. A particularly interesting sample of secondary nuclei is the radioactive nuclides with half-lives comparable to the cosmic ray containment time. These lend themselves as clocks for a direct measurement of the time of containment. Among the several isotopes that can in principle be used for this purpose, only the ^{10}Be ($\tau 1/2 = 1.6 \times 10^6$ years) abundance has been measured with sufficient accuracy. Several reliable measurements for the ^{10}Be/Be abundance now exist, which agree that the containment time of cosmic rays has an unexpectedly high lower limit of about 10^7 years at energies below 1 GeV/n. Combined with the determination of the escape length, this leads to an average density of the matter in which the cosmic rays spend their life of only about 0.3 g/cm^3. The evidence for a long cosmic ray containment is corroborated by recent measurements of the cosmic ray electron energy spectrum. The observed steepening of this spectrum around 30 GeV to a power law with spectral index of 3 or greater, interpreted as being due to synchrotron losses and Compton collision losses, also leads to a containment time of about 10^7 years.

The combined evidence of an energy dependent escape pathlength and of a minimum containment time of 10^7 years places stringent conditions on the characteristics of the interstellar medium in which the cosmic rays spend their life. The requirement for a low average matter density has led to models that include a dynamic galactic halo, and special low density regions as possible places where the cosmic rays dwell for a large fraction of their life. The proposed alternatives, and the spectrum of models that deal with cosmic ray confinement have recently been reviewed by Cesarsky (1980).

Finally, a word must be said about two rather special secondary components: positrons and antiprotons. Both of these components originate in inelastic collisions of, mainly, protons with the interstellar gas, and, if the proton spectrum is known, their spectra can be calculated. The positron spectrum has been measured up to a few 10's of GeV, and is compatible with a secondary source. The determination of the positron spectrum has led to some qualitative tests of solar modulation models. The recent observation of a strong electron-positron annihilation line, emitted from the region of the galactic center has raised the question of the possible role of the primary positrons and their origin. The presence of antiprotons in the cosmic rays was established experimentally only very recently. Their flux and energy spectra are not yet well determined. Preliminary results indicate, however, that the antiproton flux is compatible with expectations for a purely secondary component.

IV. THE FUTURE

The discoveries of the past years that were sketched in the preceding paragraphs form the basis for the future search toward understanding the origin of the cosmic radiation. They also show that today we are at the threshold of very important advances. Directions in which this research is likely to proceed in the next decade are well defined and the technology of experimentation has reached a stage where several new paths can be implemented. In addition, there is always the potential for unexpected discoveries. Such discoveries have often proven to be the most exciting. The future areas of endeavor that are expected to have great impact on the understanding of the origin of cosmic rays can be readily named:

1. The full exploration of the isotopic composition of cosmic ray nuclei, both of primary and secondary origin.

2. Precision measurements of the elemental composition, (a) in the UH regime, including the actinides, and (b) at lower atomic numbers, extending to yet unexplored high energies where interactions with interstellar material appear to become rare and where one may hope to narrow or even close the gap with airshower experiments.

3. The introduction of novel methods in the investigation of the spectrum, composition and arrival directions for cosmic rays with energies above 10^{14}eV and up to the highest energies.

We saw that the first exploratory experiments on the isotopic composition already contribute to the questions of nucleosynthesis and to the determination of timescales that are relevant to cosmic rays. Results that one may expect in a few years from isotopic studies of the elements Mg, Si, S, Fe and Ni will sensitively reflect the conditions under which these elements were synthesized and in addition provide one of the few available tests for the well developed theories of nucleosynthesis. Whether it will eventually turn out that the energetic particles originate predominantly in the interstellar medium, or are ejecta from supernova explosions, they represent a unique sample of material, sent on its way to the observer some 10^7 years ago, a time very short compared to the age of solar system material.

The first contributions to cosmic ray chronology using long lived radioactive isotopes have just been made with the abundance measurement of ^{10}Be. Several isotopes exist whose half-lives are comparable to the time of cosmic ray containment in the galaxy and which therefore are of interest for astrophysical studies. Examples of such, as yet unobserved, or marginally observed isotopes are: ^{26}Al, ^{35}Ar, ^{36}Cl, ^{41}Ca, ^{44}Ti, ^{49}V, ^{51}Cr, ^{53}Mn, ^{55}Fe.

The work on cosmic ray isotopes should not remain restricted to the very low energy regime where solar modulation effects and strongly energy-dependent cross-sections complicate the extrapolation to the sources.

Proposals exist to develop magnetic spectrometers that provide good mass resolution for isotopic abundances at energies well above 1 GeV/nucleon. The first pioneering experiments using these tools have already been carried out.

The topics of nucleosynthesis and cosmic ray chronology will greatly advance through detailed studies of the abundance of the UH elements. The next generation of experimental work is expected to yield this abundance distribution with individual element resolution and good statistical accuracy and thus contribute decisively to the nucleosynthesis problem, as was discussed above. Cosmic ray chronology gains a powerful new tool, once it becomes possible to measure, for example, the U/Th ratio or the abundances of the unstable nuclides ^{93}Np, ^{94}Pu and ^{96}Cm. Whether this generation of scientists will see a determination of UH isotopes is a question I would not dare to answer, but I would be surprised if inroads were not made into this promising and challenging regime, at least with the more abundant species of that group of elements.

Magnetic spectrometers are the tool that provides positron and antiproton spectra. The unique property of both these components comes from the fact that their production spectrum can be calculated from the known spectrum of their progenitors, high energy protons. Measurements of both components to high energies therefore yield independent information on particle propagation, and, in the case of positrons, on confinement. The discovery of a strong positron annihilation line from a region around the galactic center has again raised the question whether all positrons observed in the vicinity of earth are of secondary origin. This question may be answered with very precise measurements of the positron spectrum.

Moving up in energy, we expect the next decade of research to provide information of the elemental composition at energies exceeding 1 TeV/nucleon. This work will show whether the escape mean free path that was observed to decline between 10 and 100 GeV/n continues to decrease beyond 100 GeV/nucleon, thus leaving only source nuclei in the cosmic rays at those energies. Or, alternatively, shall one find a residual amount of matter that particles of all energies traverse? Clearly, such evidence permits quite detailed conclusions on the distribution of fields and matter around the cosmic ray sources as well as in the galaxy as a whole, since these are the agents that determine containment time and containment volume.

Nothing is known about the composition of the cosmic ray sources at and beyond 1 TeV/n. It is particularly interesting to ascertain whether this composition remains the same as observed at low energies, or, for example, becomes very abundant in Fe. Experiments that are now being developed will provide this information in the coming years, and may eventually extend sufficiently high in energy to provide an overlap with the lowest energies that can be observed in airshower experiments. Such overlap would provide a calibration that is needed if one wishes to put the conclusions on composition that are obtained indirectly from measurements of shower structure, on a solid base.

This brings me finally to a discussion of the future work on ultra high energy cosmic rays, a topic that will get a detailed treatment in a later paper. In this area, we can expect considerable advances when the new instrument of the Utah group, the Fly's Eye is in full operation. I recently learned that one year of operation of this instrument is expected to collect as much information as all the world's shower detectors together have gathered in the past. Not only can we hope to gain statistically significant information on the shape of the primary spectrum up to the highest energies, and on the distribution of arrival directions, but also on the primary composition. Through observation of atmospheric scintillations, the Fly's Eye investigates the shower structure in three dimensions, and hence extracts details of this structure that are needed to estimate the composition of the primary particles. Any information on the cosmic ray composition at energies much beyond 10^{14} eV will continue to come from indirect methods due to the paucity of particles.

V. CONCLUSION

Although a field with a long history, cosmic ray research today is in a phase where entirely new avenues are being opened. Experiments that were recently completed, that are underway, and that are in the planning stage let one expect that some of the crucial questions on the origin of this radiation will soon be answered. While much of the information will continue to come from particle observations, increasingly important contributions to the problem will be made through the work in gamma ray astronomy, X-ray astronomy and radio astronomy. Observations in these fields provide the means to gain insight into the phenomena that take place at the sites of potential sources.

This paper is the written version of a talk and is not a review of the field. Its contents are not balanced, are incomplete, and biased by the interests of the author. Important topics, like the anomalous low energy cosmic ray components, and the role of solar particle investigations in understanding acceleration mechanisms were omitted. No attempt was made to properly reference the large body of work on which this paper is based, except for refering the reader to a selection of review papers where references to the literature can be found. I wish to thank the many colleagues who provided me with preprints and the latest information on their work. I apologize to all of them for using, but not specifically quoting their work. My special thanks go to Dr. John Wefel for critically reading the manuscript and for many suggestions. Several of the figures were prepared by him. This work was supported in part by NASA under grant NGL-14-001-005.

REFERENCES*

Balasubrahmanian, V.K.: 1979, Proceedings of 16th ICR, Vol. 14, p. 121.

Cesarsky, C.: 1980, Annual Review of Astronomy and Astrophysics (in publication).

Meyer, J.P.: 1975, Proceedings of 14th ICR, Vol. 11, p. 3698.

Müller, D.: 1977, Proceedings of 15th ICR, Vol. 10, p. 474.

Raisbeck, G.M.: 1979, Proceedings of 16th ICR, Vol. 14, p. 146.

Sreekantan, B.V.: 1979, Proceedings of 16th ICR, Vol. 14, p. 345.

Waddington, C.J.: 1977, Proceedings of 15th ICR, Vol. 10, p. 168.

Watson, A.A.: 1975, Proceedings of 14th ICR, Vol. 11, p. 4019.

Wolfendale, A.W.: 1979, Pramana 12, p. 631.

*ICR stands for International Conference on Cosmic Rays.

ISOTOPES IN GALACTIC COSMIC RAYS

Hubert Reeves,
Section d'Astrophysique, Saclay
Institut d'Astrophysique, Paris

The data on isotopic ratios of elements in Galactic Cosmic Rays (GCR) is steadily improving and has recently reached the point where some information can be extracted, which has bearing on the problem of the origin of the cosmic rays. By and large, these data have generally confirmed the similarity between solar-type matter and GCR source, when spallation effects and selective acceleration are taken into account. The silicon and iron isotopic ratios, for

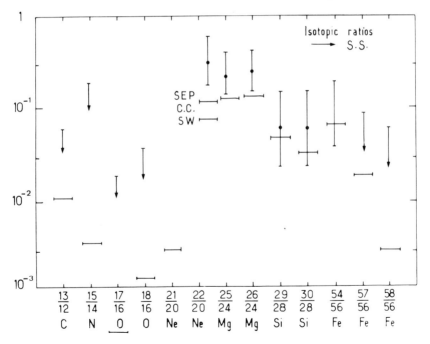

Fig. 1. Isotopic ratios in GCR (or their upper limits) compared to the same ratios in the solar system (horizontal bars). SEP is solar energetic particles, C.C is carbonaceous chondrites and SW is solar wind.

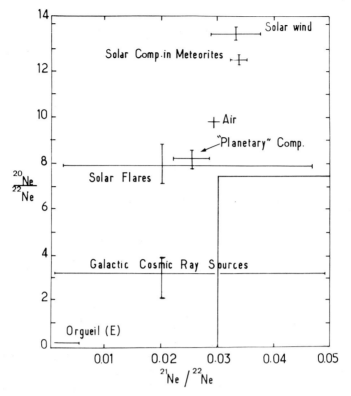

Fig. 2 Neon isotopic composition, Ne^{20}/Ne^{22} and Ne^{21}/Ne^{22}.

instance, are consistent with meteoritic ratios (Mewaldt et al. 1980a, 1980b). For iron, this is particularly important since two or perhaps even three different nucleosynthesis mechanisms are required to account for the species ^{54}Fe, ^{56}Fe, ^{58}Fe. For carbon, nitrogen and oxygen the problem is that the spallation corrections are large compared to the solar system ratios but nevertheless the present upper limits are not in contradiction with the solar system ratios (fig. 1).

The one isotopic ratio which definitely appears to differ is the neon ratio (Garcia Munoz et al. 1979a, Mewaldt et al. 1980a). Ironically, this is also the only element whose isotopic ratio is not known with certainty in the solar system. The situation is shown in figure 2. The solar wind ratio of $^{20}Ne/^{22}Ne$ is $\simeq$ 12-13 (Geiss and Bochsler 1979) while the value in the gas trapped in chondrites is about 8 (Black and Pepin 1969) not very different from the earth atmospheric value and the recently measured solar flare values (Mewaldt 1980). The GCR source value ($\simeq 3$) certainly differs from both of these values, whichever is representative of the solar value. (The low $^{20}Ne/^{22}Ne < 10^{-2}$ found in Orgueil (Eberhardt et al. 1979) probably originates from decay of ^{22}Na and may not have any bearing on the present discussion (fig. 2)).

There is also a fair probability that the magnesium ratios (Garcia-Munoz et al. 1979b) (both $^{25}Mg/^{24}Mg$ and $^{26}Mg/^{24}Mg$) are larger than the meteoritic ratios, although better data are needed to ascertain this question.

What do those ratios (Ne and perhaps Mg) teach us about the origin of the GCR? I can think of four possible scenarios by which these "anomalies" could have come about. I shall discuss them in turn, giving "pros" and "cons".

I - SELECTIVE ACCELERATION OF ISOTOPES

There is the possibility that we are not dealing with real compositional effects but with biases in the acceleration, just as is the case for elements in solar and galactic cosmic rays (Cassé and Goret 1978). In other words, is nature fooling us again? There are already some important examples of selective acceleration of isotopes. The $^3He/^4He$ ratio shows extreme variations (up to 10^4.) in solar flares and more modest variations (up to about 4) in solar wind. The $^{20}Ne/^{22}Ne$ shows selective acceleration effects in the solar wind (if the solar value is ~ 8) or in the solar flares (if the solar value is ~ 13). It is worth noticing that the Galileo mission to Jupiter is planned to measure this ratio there. The value obtained is likely to be extendable to the sun (Jupiter appears to have a solar composition in many other respects). The data will decide which, of the solar wind or the solar flares, is experiencing isotopic selective acceleration.

Are there other isotopic anomalies in the solar flares? There is a slight indication that magnesium may be such a case (Mewaldt 1980). If this is confirmed, we shall have to consider seriously this first scenario.

II - GALACTIC ENRICHMENT EFFECT

The GCR particles have ages of $\sim 10^7$ years (Garcia-Munoz et al. 1977, Wiedenbeck and Greiner 1980). If they represent a sample of ordinary galactic gas matter, they could differ from the solar system material (dating back to 4.5×10^9 years ago) simply because of gradual nucleosynthetic enrichment. More specifically this scenario requires that the abundance of ^{22}Ne should have increased by a factor from three to five during this period. It is fair to say that such large increments are not observed for other elements in stellar material of corresponding ages, even for secondary products like ^{13}C or N. This scenario would imply that ^{22}Ne is a product of small star nucleosynthesis (which are only becoming "ripe" in the last few billion years). The natural nucleosynthetic process would be:

a) CNO cycle which transform ^{12}C and ^{16}O in ^{14}N

b) The onset of helium burning with
$${}^{14}N(\alpha,\gamma) \ {}^{18}F(\beta^+,\nu) \ {}^{18}O(\alpha,\gamma) \ {}^{22}Ne \begin{array}{c}\longrightarrow (\alpha,\gamma){}^{26}Mg \\ \longrightarrow (\alpha,n){}^{25}Mg\end{array}$$

Thus, ^{22}Ne would be a secondary product and would be expected to increase at a slower rate than ^{20}Ne (which is a primary product of carbon burning). This process is advantageous in the sense that it would also explain the magnesium isotopic anomalies if confirmed. But the absence of N increase between the birth of the sun and now, may be a difficulty for this scenario.

III - SUPERNOVA INJECTION OF RECENTLY GENERATED MATERIAL

There is very little support for the idea that supernovae are contributing freshly brewed material to the GCR. The evidence for transuranic nuclei is vanishing with the new data of HEAO-C (Waddington 1980) and Ariel-5 (Fowler 1980). The long-sought r-process peaks are still to be established. Quite generally, from Fe to Pb the GCR source are hardly distinguishable from solar material (there is not even the selective acceleration bias). Thus the neon (and perhaps magnesium) anomaly would be the only remaining effect of the supernova! Invoking a supernova to account for this effect is like using a sledge hammer to kill a fly ... And one would be left with the problem of explaining why we do not get other anomalies.

The main reason for my lack of sympathy for this model is a question of strategy. This model is not vulnerable, in the sense that it can hardly be shown to be wrong. Too much freedom is left in the choice of the free parameters (zone mixing, etc ...). This is an unfortunate situation but one we have to live with. Personally, I would try any other solution before I fall back on this one, although it may still be the correct one.

IV - OB ASSOCIATIONS AND RELATED OBJECTS

To develop this scenario we take advantage of some recent developments in UV, X-ray and gamma-ray astronomy.

Because cosmic rays are charged particles, they are isotropized by galactic magnetic fields and loose the memory of their birthplace. But photons keep their original direction. For this reason gamma-ray observations are potentially highly informative in telling us where the cosmic rays originate (provided γ ray photons are produced by cosmic rays interactions).

The recent data of COS-B (Wills et al. 1980; Hermsen 1980) on photons with energies $\geqslant$ 100 MeV have revealed the presence of "gamma ray sources" which could well fulfill this hope. The one important piece of information is that those sources are all well within the galactic plane. The scale height is about 100 pc. This corresponds to the scale height of very young objects in the Galaxy: molecular clouds, OB associations, HII regions, etc. Hence the hint that the origin of the cosmic rays may well be related to these places of stellar births. Detailed analyses of the celestial coordinates of these sources gives in fact some correlation with the position of the OB associations with supernova remnants (Montmerle 1979).

Before this point is established however, one would have to convincingly establish that these gamma-ray enhancements do not simply reflect the concentration of matter in clouds. In other words, can they be quantitatively accounted for without invoking a local increase in the proton flux itself (the gamma ray source function is proportional to the product of the flux times the matter density)? Wolfendale (this Conference) has defended this view and has contested the presence of enhanced cosmic ray emissivities. It seems that for at least one source (ρ Ophiuchi) an enhanced emissivity is required. More work is needed to clear this very important issue ...

Blaauw (1964) has studied in detail the structure of OB associations. Their dispersion times is $\simeq 15 \times 10^6$ years, longer than the whole lifetime of stars more massive than 16 $M_\odot$. Stellar statistics indicate that in a typical OB association, from ten to twency stars will have time to undergo their evolution and die before dispersion. Upon dying, these stars generate a supernova remnant which disturbs the whole association (Reeves 1978, 1979). Such events appear to have played a role in the origin of the solar system and to be responsible for some of the meteoritic anomalies recently observed (Wasserbug 1978, Clayton 1976).

UV data from the Copernicus satellite have shown that O and B stars have very strong supersonic winds (2000 km/sec, $\dot{M} = 10^{-7}$ to 10^{-5} $M_\odot yr^{-1}$) giving rise to P Cygni profiles. Occasionally, among these O stars, a few Wolf-Rayet stars are present with rather unusual chemical composition: very low H/He ratio ($\leqslant 0.1$), enhanced N or C. Because of the high frequency of binaries in this population, it is believed that WR stars have lost their original atmosphere through Roche lobe overflow and are showing shells with freshly made nucleosynthetic products (of H or He burning) (see e.g. Van beveren and Packet 1979).

Thus, the following scenario, developed in detail later on this Conference by Cassé, Paul, Montmerle and Meyer. Acceleration takes place as a result of Fermi mechanism combined with Alfven scattering (Blandford and Ostriker 1978) either at the boundary of the stellar wind cavity (Cassé and Paul 1980) or in relation with shock-waves induced by the supernovae of the OB associations (Montmerle 1979).

The O stars have solar abundances. Energetically they meet the requirements of GCR. But how do we account for the ^{22}Ne anomaly?

We take advantage of the fact that ^{22}Ne is a normal product of

He burning after the CNO phase (same as described previously), to speculate that WR atmospheres could be enriched in ^{22}Ne with respect to ^{20}Ne. It is further known that these stars have extremely strong stellar winds ($\sim 10^{-4} M_\odot \text{yr}^{-1}$). Thus the sum of O stars and WR stars could inject a solar type sample of material with additional ^{22}Ne (and perhaps ^{25}Mg, ^{26}Mg). Observations of neon overabundances (if possible) in WR would help to promote this scenario.

No matter whether this specific ^{22}Ne enriching mechanism is the correct one or not, it seems fair to say that OB associations are likely to play an increasingly important role in the origin of cosmic rays. We are looking forward to the observations of gamma-ray lines (the 4.4. MeV line of ^{12}C in particular) in these regions to give us more information on this question.

REFERENCES

Blaauw, A. 1964, Ann. Rev. Astron. Astrophys. 2: 219.

Black, D.C., and Pepin, R.O., 1969, Earth Planet Sci. Lett. 6, 395, "Trapped neon in Meteorites II".

Blandford, R.D., and Ostriker, J.P., 1978, Ap.J. (Letters) 221, L29.

Cassé M., and Goret, P. 1978, Ap. J. 221, 703.

Cassé M., and Paul, J.A., 1980, Ap. J. 237, 236 "Local gamma rays and cosmic ray acceleration by supersonic stellar winds".

Clayton, R.N., Onuma, N., and Mayeda, T.K. 1976, "A classification of meteorites, based on oxygen isotopes" Earth Planet. Sci. Lett. 30, 10 Eberhardt, P.

Jungck, M.H.A., Meier, F.O., and Neiderer, J. 1979 "Presolar grains in Orgueil: evidence from Neon-E". Ap.J. 234 L 169.

Fowler, P. 1980, Bologna Conference on the Origin of Cosmic Rays.

Garcia-Munoz, M., Mason, G.H., and Simpson, J.A. 1977, Ap. J., 217, 859.

Garcia-Munoz, M., Simpson, J.A., and Wefel, J.P. 1979a, "The isotopes of neon in the galactic cosmic rays", Ap. J. 232, L95.

Garcia-Munoz, M., Simpson, J.A., and Wefel, J.P. 1979b, "The isotopic composition of neon and magnesium in the low energy cosmic rays", Kyoto 16th International Cosmic Ray Conference OG 7-13.

Geiss, J. and Bochsler, P. 1979, "On the abundances of rare ions in the solar wind" Proc. 4th Solar Wind Conf. Burghausen Springer Verlag.

Hermsen, W. 1980, Ph.D. Thesis, University of Leiden.

Mewaldt, R.A. 1980a, "Space craft measurements of the elemental and isotopic composition of solar energetic particles", Proceedings of the Conference on the Ancient Sun: Fossil Record in the Earth, Moon and Meteorites.

Mewaldt, M.A., Spalding, J.D., Stone, E.C., and Vogt, R.E. 1980b, "The isotopic composition of galactic cosmic ray iron nuclei", preprint.

Mewaldt, M.A., Spalding, J.D., Stone, E.C., and Vogt, R.E. 1980a, "High resolution measurements of galactic cosmic-ray Ne, Mg and Si isotopes", Ap. J. 235, L95.

Montmerle, T. 1979, Ap.J. 231, 95.

Reeves, H., 1978, "Supernovae contamination of the early solar system in an OB stellar association or the "Big Bang" theory of the origin of the solar system" in "Tucson Conf. on Protostars and Planets" ed. Gehrels.

Reeves, H. 1979 "Cosmochronology after Allende" Ap. J. 231: 229.

Vanbeveren, D., and Packet, W. 1979 Astron. Astrophys., 80, 242.

Waddington, J., 1980, Bologna Conference on the Origin of Cosmic Rays.

Wasserburg, G.J. 1978, Proc. Conf. on Protostars and Planets, Tucson, Arizona.

Wiedenbeck, M.E. and Greiner, D.E. 1980, A cosmic Ray age based on the abundance of Be. (Preprint).

Wills, R.D., Bennett, K., Bignami, G.F., Buccheri, R., Caraveo, P., D'Amico, N., Hermsen, W., Kanbach, G., Lichti, G.G., Masnou, J.O., Mayer-Hasselwander, H.A., Paul, J.A., Sacco, B., Swanenburg, B.N. 1980, Proc. COSPAR Symp on Non Solar Gamma Rays, Bangalore (India), Adv. Sp. Expl. $\underline{7}$, 43.

THE CHARGE AND ISOTOPIC COMPOSITION OF $Z \geq 10$ NUCLEI IN THE COSMIC RAY SOURCE

W. R. Webber
Space Science Center
University of New Hampshire
Durham, N. H. 03824

Interstellar fragmentation provides the greatest contribution to the abundance of cosmic ray nuclei with $Z = 17-25$ observed at earth. The usual procedure to estimate the source abundance of these nuclei is to correct for this interstellar fragmentation using a propagation model and a set of fragmentation parameters for Fe nuclei and its products. Only a fraction of these fragmentation parameters have actually been measured and the accuracy of these as well as the semi empirical parameters used for the unmeasured cross sections is no better than $\pm$ 10-15%. As a result the actual source abundances of these nuclei can only be deduced to an accuracy of $\pm$ 2-3% of the Fe abundance.

Recently during a calibration run of our cosmic ray isotope telescope at the BEVALAC in which a 9.3 g/cm^2 CH_2 target was used to fragment ~ 1 GeV/nuc Fe nuclei, both the isotopic and charge composition of the fragments were measured. It was noticed that the charge composition of the fragments was similar to that observed for cosmic rays at earth- not unexpected since the thickness of the target was equivalent to a slab length $\sim 2.5 \text{ g/cm}^2$ of H. In effect the target provided us with an intergrated measurement of the effects of interstellar propagation. Several relatively small and well known corrections can be made to this data to compare it directly with the observed cosmic ray composition. This comparison is shown in Table I. It is found that only 3 charges, S, A and Ca have finite source abundances and the source abundance accuracy is $\sim \pm 0.5$% of Fe for charges with $Z = 17-25$.

TABLE I
Comparison of Fe fragmentation at the BEVALAC with Cosmic Rays in Interstellar Space

Charge	BEVALAC – 9.3 g/cm² of CH₂ $\bar{E}$ = 850 MeV/nuc Measured*	Corrected for Rad Decay⁰	Normalized & Adjusted^X		Cosmic Rays Balloon 600 – 1000 MeV/nuc⁺	Source % of Fe
25	14.1	11.6	13.4 ± 0.6		13.1±0.4	−0.3±0.7
24	12.4	15.3	17.6 ± 0.7		17.7±0.5	+0.1±0.8
23	9.9	9.8	11.3 ± 0.5		10.4±0.4	−0.8±0.6
22	11.1	13.7	16.3 ± 0.7		17.2±0.5	+0.8±0.8
21	10.5	5.5	6.8 ± 0.4		7.5±0.3	+0.6±0.3
20	10.3	13.0	15.9 ± 0.7	Ca + Ar	25.6±0.6	+8.3±0.8
19	6.6	6.7	8.9 ± 0.5	+1.8	11.3±0.4	+0.5±0.6
18	6.0	6.7	9.0 ± 0.5	+1.0	16.6±0.6	+4.6±0.6
17	4.9	3.1	4.8 ± 0.4	+3.2	9.0±0.4	+0.8±0.5
16	5.1	7.0	10.2 ± 0.7	+3.5	33.0±1.5	+15.1±0.8
15	3.6	2.1	3.0 ± 0.3		7.5±1.5	

*Total Fe nuclei = 100. Data is extrapolated to the top of the telescope.
⁰Using measured individual isotopic cross sections.
^XThe BEVALAC secondary production in 9.3 g/cm² of CH₂ is normalized to the cosmic ray production in interstellar matter. Normalization factor = 1.166. An adjustment is also made to convert the slab length production at the BEVALAC to the equivalent exponential path length distribution in interstellar space. This factor ranges from 0.99 for Mn to 1.24 for S.
⁺These values represent an average of values from Israel et al (1979), Young (1979) and Lezniak and Webber (1978) in the energy range 600-1000 MeV/nuc.

These deductions regarding

the source composition may be compared with the isotopic composition data for cosmic rays that we have measured on balloon flights with this same telescope. Earlier, preliminary isotope data from our 1977 balloon flight was presented (Webber, et.al., 1979 a,b). In this paper we present combined data from the 1976 and 1977 flights. BEVELAC calibration data on the mass distribution peak shapes is used in this latest analysis. The overall mass resolution σ for the balloon data is virtually identical to that obtained at the BEVELAC calibration and slowly increases from a σ ∼ 0.28 AMU at C to ∼ 0.38 AMU at Fe. Data on the isotopic composition of all nuclei with Z = 6-26 is available but we will examine only S, A, Ca and Fe here as well as Ne and Mg. The balloon data is shown in Table II. Comparison of these isotopic abundances with those predicted on the basis of conventional propagation models leads to the following source abundances; $^{32}S \sim S = 13.7 \pm 2.0\%$; $^{36}A \sim A = 3.2 \pm 1.0\%$; $^{40}Ca \sim Ca = 8.5 \pm 1.6\%$ of Fe. For Fe we find $^{56}Fe = 91.7\%$; $^{54}Fe = 4.1 \pm 3.8\%$, and $^{58}Fe \lesssim 4.2\%$ of all Fe. The source abundances obtained in this manner are in good agreement with those obtained from the charge comparison. These source abundances are also consistent with average solar cosmic ray abundances (eg S = 16.8%, A = 3.2% and Ca = 8.4% of Fe according to Cook et.al., 1979).

TABLE II
ISOTOPIC ABUNDANCES (SELECTED ELEMENTS 1976 + 1977 DATA)

Isotope	Energy Interval (MeV/nuc)	Events	Events at Top of Atm	Events (MeV/nuc)	% of Element
^{20}Ne	436 - 589	358	1287	8.41	61.9
^{21}Ne		50 ± 15	169	1.19	8.7 ± 3.3
^{22}Ne	410 - 552	140	526	3.98	29.3 ± 2.7
^{24}Mg	484 - 664	564	1981	11.10	72.2
^{25}Mg		81 ± 20	278	1.85	12.0 ± 3.6
^{26}Mg	458 - 628	116	412	2.42	15.8 ± 1.6
^{36}A	536 - 730	26	87	0.45	45.9 ± 13.5
^{38}A	517 - 701	23	76	0.42	42.9 ± 13.3
^{40}A		5 ± 3	18	0.11	11.2 ± 7.5
^{40}Ca	568 - 788	46	162	0.73	45.3 ± 6.6
^{42}Ca	550 - 762	29	102	0.48	29.8 ± 5.9
^{43}		≤8	≤25	≤0.12	≤7.5
^{44}Ca		15	53	0.28	17.1 ± 5.7
^{54}Fe	662 - 921	25 ± 12	95	0.37	6.5 ± 3.8
^{56}Fe	646 - 899	275	1281	5.06	89.2
^{58}Fe	630 - 876	≤12	≤59	≤0.24	≤4.2

Using these same procedures we find the source abundance of ^{22}Ne to be 22.2 ± 2.7% of all Ne, or 4 σ higher than the measured solar cosmic ray ^{22}Ne abundance of 11.6% (Mewaldt et.al., 1979). The ^{21}Ne abundance we measure is consistent with a complete fragmentation origin. For ^{25}Mg and ^{26}Mg the source abundances are determined to be 7.5 ± 3.9% and 14.6 ± 1.9% of all Mg, respectively. The solar values for these isotopes are 10.1% and 11.2% of all Mg. A possible enhancement of ^{26}Mg of 2σ relative to the solar value, and in conjunction with the ^{22}Ne enhancement is evident.

References

Cook, W.R., Stone, E.C. and Vogt, R.E., Ap. J. (in press).
Israel, M.H., Klarmann, J., Love, P.L., and Tueller, J., 1979, Proc. 16th Int. Cosmic Ray Conf., Kyoto, 1, 323.
Lezniak, J.A., and Webber, W.R., 1978, Ap. J., 223, 676.
Menwaldt, R.A., Stone, E.C., Vogt, R.E., 1979, Ap., J., 231, L97
Webber, W.R., Kish, J., and Simpson, G., 1979a, Proc. 16th Int. Cosmic Ray Conf., Kyoto, 1, 424.
Webber, W.R., Kish, J., and Simpson, G., 1979b, Proc. 16th Int. Cosmic Ray Conf., Kyoto, 1, 430.
Young, J.S., 1979, PhD Thesis, University of Minnesota.

ISOTOPES OF COSMIC RAY ELEMENTS FROM NEON TO NICKEL

C.J. Waddington, P.S. Freier, R.K. Fickle and N.R. Brewster,
School of Physics and Astronomy, University of Minnesota,
Minneapolis, Minnesota, U.S.A.

Introduction. We are reporting here on the results obtained from a balloon exposure of a cosmic ray detector flown in 1977. This detector, described elsewhere, Gilman and Waddington (1975), Young (1979), measures elemental charge from scintillation and Cherenkov signals and mass from Cherenkov and total energy determined from a measure of residual range in nuclear emulsion. The charge resolution obtained ranged from 0.19 to 0.21 charge units between neon and nickel. This resolution was sufficient to ensure that all but a few percent of the nuclei were correctly identified, even for those elements of low abundance that have neighbors with high abundances, such as Cl or Al. The mass resolution obtained for those nuclei that stopped in the emulsions ranged from 0.40 to 0.70 amu for A between 20 and 60 amu. This was not adequate to uniquely resolve neighboring mass peaks in many cases, but was adequate to draw a number of conclusions regarding many of the more abundant elements.

Neon to Aluminum. Our results on these elements will be published shortly (Freier et al., 1980) and we only summarize them here. Both Ne and Mg show evidence for neutron enrichment relative to the solar system abundances. Neon relative isotopic abundances calculated at the "source" are ^{20}Ne: ^{21}Ne: ^{22}Ne = 69.0±8.9: 10.3±2.7: 20.7±3.4 compared with solar (Ne-A) values of 88.9: 0.3: 10.8. Similarly for magnesium at the source, we calculate for ^{24}Mg: ^{25}Mg: ^{26}Mg = 65±5.6: 17.5±3.6: 17.5±2.9, compared with solar values of 78.7: 10.1: 11.2. Na and Al, on the other hand, are both in good agreement with all the source abundances being ^{23}Na and ^{27}Al. The significance of the neutron enrichments, particularly if confirmed for Mg, has still to be clarified, although it may well be a consequence of a particular form of explosive Ne burning, requiring an enhanced metallicity in the source.

Silicon to Potassium. In view of the above results, it is of obvious interest to examine the next range of elements for similar effects. In particular, Si, S and Ar should have appreciable abundances and hence a good fraction of those observed should be of primary origin. We find, however, that Si and S are consistent with solar abundances, while Ar has no significant source abundances, as shown in the Table. The odd-charged elements, P, Cl and K have essentially no primary component and the isotopic distribution observed is quite consistent with that expected from propagation. No appreciable abundance of the clock isotope ^{35}Cl is observed, consistent with the observed absence of the longer lived ^{26}Al.

Calcium to Nickel. Iron and nickel are almost all primary in origin as is ^{40}Ca, but the heavier Ca isotopes are almost all secondary as are all the other elements in this charge range. The results for Ca are given in the Table and show an indication of an excess of ^{44}Ca at the source, which, if verified, would be another indication of high metallicity in the source. Iron is an important and difficult element that merits a full discussion. Mewaldt et al. (1980) have placed an upper limit on the abundance of ^{58}Fe of $\leq 6\%$, consistent with either a high or low metallicity. Our own results, based on an analysis of the shape of the iron peak and a comparison with the shapes of other strong mass peaks, such as that at Si, lead us to conclude the abundance of ^{58}Fe is $\leq 9\%$. Finally, Ni shows an interesting 1 to 1 ratio for ^{58}Ni to ^{60}Ni, unlike the 2.5:1 ratio in solar system material. This implies an intermediate metallicity, inconsistent with either the high or low values suggested above; see Mewaldt et al. (1980). It also suggests that ^{57}Ni may not be produced as abundantly as expected, casting doubt on the significance of a very low abundance of Co in the cosmic rays.

Table. Percentage source and solar system abundances (SS)

	Source	SS		Source	SS
^{28}Si	86±7	92.2	^{40}Ca	72±14	97
^{29}Si	14±4	4.7	^{44}Ca	23±7	2.07
^{30}Si	0±3	3.1	^{46}Ca	5±3	.003
Si/Fe	1.26	1.20	Ca/Fe	.079	.087
^{32}S	88±19	95.0	^{54}Fe	0±3	5.8
^{33}S	2.5±2	0.8	^{55}Fe	0±5	0
^{34}S	7±3	4.2	^{56}Fe	72±9	91.7
^{36}S	2.3±2.3	.01	^{57}Fe	<19±6	2.2
S/Fe	0.168	0.169	^{58}Fe	$\leq 9±4$	0.3
^{36}Ar	100±20	84.2	$^{56,57}Ni$	3.4±3	0
Ar/Fe	.016	.048	$^{58,59}Ni$	48±15	67.9
			$^{60,61}Ni$	45±15	27.39
			$^{62,63}Ni$	3±3	3.66
			Ni/Fe	.057	.058

References

Gilman, C.M., and Waddington, C.J.: 1975, Proc. Munich Conf. 9, 3166.
Freier, P.S., Young, J.S., and Waddington, C.J.: 1980, Ap.J. Lett. (in press).
Mewaldt, R.A., Spalding, J.D., Stone, E.D., and Vogt, R.E.: 1980, Ap.J. Lett. 236, L121.
Young, J.S.: 1979, Ph.D. Thesis, University of Minnesota.

This work supported by NASA under Contract No. NGR 24-005 050.

WOLF RAYET STARS AND THE ORIGIN OF THE ^{22}Ne EXCESS IN COSMIC RAYS

M. Cassé, J.A. Paul and J.P. Meyer

Section d'Astrophysique
Centre d'Etudes Nucléaires de Saclay, France

First order Fermi acceleration at the boundary between supersonic stellar winds from OB and Wolf-Rayet stars and the surrounding interstellar medium could be influential in the bulk energization of the local cosmic radiation (Cassé and Paul, 1980). Since wind acceleration is not supposed to accelerate thermal particles, a continuous injection of low energy particles (E $\sim$ 1 to 10 MeV/n) is required. We keep open the possibility that these particles may be injected from the interior of the stellar cavity, i.e. by the mass-losing star itself via a flare-like surface activity for instance. Observations of flare activity on hot and massive stars are mandatory to settle this idea. In this context, we expect that the CR reservoir is the surface of young and active stars and that the difference between the CR source (CRS) composition (i.e. corrected for propagation effects in the interstellar medium, ISM) and the surface composition of young stars (reflecting for most of them the present local ISM) is principally due to selective effects at injection depending on the atomic properties of the elements. This idea is supported by 3 arguments (Cassé and Goret, 1978, Meyer et al., 1979) i) the general resemblance between solar CR elemental abundances and elemental CRS abundances (see e.g. Mewalt, 1980) ii) the correlation between the (CRS/local galactic) abundance ratio and the first ionization potential and iii) the fact that dust grains must have been thoroughly destroyed in the medium from which CR are extracted. In the interstellar gas in which dust grains are present, Ni, Fe, Mg and especially Ca and Al are highly depleted (Salpeter, 1977) whereas they are normally abundant in CR.

The selective acceleration effects including a more subtle mass effect (Meyer et al., 1979) do not significantly alter the isotopic proportions of any given heavy element at the CR source. The isotopic composition is, therefore, the most genuine print of the thermonuclear origin of CR. At the present time, with our limited observations it seems that the isotopic CR composition inferred at the CR source is not strongly abnormal for the principal elements between H and Ni (see e.g. Balasubrahmanyan, 1979) except for Ne and possibly Mg. The ^{22}Ne/^{20}Ne ratio estimated at the CR source is thought to be about 3 times larger than the solar system isotopic ratio (see e.g. Balasubrahmanyan, 1979, and references therein). We are inclined to relate this peculiarity to the fact that Wolf-Rayet stars of the WC type, whose surface abundances are expected to be enriched in helium burning products - and especially ^{22}Ne - could contribute significantly to the CR injection and acceleration (Cassé et al., in preparation).

The role of WR stars in the CR energization has been illustrated in Cassé et al., (this conference). ^{22}Ne is believed to be the product of ^{14}N-burning through the sequence ^{14}N$(\alpha,\gamma)^{18}$F$(e^+\nu)^{18}$O$(\alpha,\gamma)^{22}$Ne. This chain of reactions starts before the 3α reaction (He-burning) and ends in the core of massive stars at the end of He-burning by ^{22}Ne$(\alpha,n)^{25}$Mg. Since, according to stellar models, the helium core is never in contact with the external convective envelope in normal (H-rich) stars, or even pure Helium stars (see e.g. Stothers and Chin, 1977), the only way to get ^{22}Ne at the stellar surface is to remove the envelope and expose the convective core. It seems to be the case for massive helium stars (WC stars) resulting from Roche lobe overflow in close binary systems followed by stellar wind mass-loss (Vanbeveren and Packet, 1979). Assuming that every atom of CNO initially present in the volume occupied by the Helium convective core has been converted into ^{14}N and subsequently into ^{22}Ne, the ^{22}Ne excess would be at the surface of a typical WC star of the order of 130 (relative to solar system)*. Since the ^{22}Ne excess inferred at the CR source is about 3, the contribution of WC stars has to be at maximum 3/130, neglecting other possible sources of CR ^{22}Ne as e.g. explosive hydrogen burning (Cassé et al., 1979, Audouze et al., 1980). The dilution, in the proportion $\sim 1/40$, of the ^{22}Ne-rich component with the bulk of CR expected to be of normal composition would lower the He and C (and/or O) excesses of the extra-component to a level compatible with CRS abundances. Lower mass Helium rich stars like nuclei of planetary nebulae are presently under study. A more quantitative estimate based on the work of Couch and Arnett (1972) lead essentially to the same conclusion (Cassé et al., in preparation). Finally the slight excess of ^{25}Mg and ^{26}Mg at the CR source found by Mewalt et al. (1979) would find an explanation in the framework of this model.

---

* The Ne elemental excess will not appear in the spectra of WC stars because Ne lines are not observable in the visible and UV ranges.

REFERENCES

Audouze, J., Chièze, J.P. and Viangioni-Flam, E. 1980, Astr.Ap. (in press).
Balasubrahmanyan 1979, 16th Int.Cosmic Ray Conf., Kyoto, 14, 121.
Cassé, M. and Goret, P. 1978, Ap.J. 221, 703.
Cassé, M., Meyer, J.P. and Reeves, H. 1979, 16th Int.Cosmic Ray Conf., Kyoto, 12, 114.
Cassé, M. and Paul, J.A. 1980, Ap.J., 237, 236.
Couch, R.G. and Arnett, W.D. 1972, Ap.J., 178, 771.
Mewalt, R.A. 1980 (preprint).
Mewalt, R.A, Spalding, J.D., Stone, E.C. and Vogt, R.E. 1979, Proc. 16th Int. Cosmic Ray Conf., Kyoto. 12, 86.
Meyer, J.P., Cassé, M. and Reeves, H. 1979, Proc. 16th Int. Cosmic Ray Conf., Kyoto, 12, 108.
Salpeter, E.E. 1977, Ann.Rev.Astr.Ap., 16, 267.
Spitzer, L. and Jenkins, B. 1976, Ann.Rev.Astr.Ap., 13, 133.
Stothers, R. and Chin, C.W. 1977, Ap.J., 216, 61.
Vanbeveren, D. and Packet, W. 1979, Astr. Ap., 80, 242.

CR-39 PLASTIC TRACK DETECTOR EXPERIMENT FOR MEASUREMENT OF CHARGE COMPOSITION OF PRIMARY COSMIC RAYS

Y.V. Rao, A. Davis and M.P. Hagan
Physics Research Division, Emmanuel College,
Lexington MA 02173, USA and
R.C. Filz
Air Force Geophysics Laboratory, Hanscom AFB MA 01731, USA

A study of the relative abundances and energy spectra of heavy cosmic rays and isotopic composition in the region of Fe peak can yield significant information concerning their origin, acceleration and interstellar propagation. In recent years solid state nuclear track detectors have been employed extensively to study heavy primary cosmic rays. Plastic track detectors necessarily have large geometric factors for heavy primaries, and a continuous sensitivity for the duration of an extended exposure. A balloon-borne experiment consisting of 1 m^2 passive detector array has been designed in order to obtain charge and energy spectra of primary cosmic rays in the region of Fe peak. Included in the array is a new type of nuclear-track-recording plastic, a polymer made from the monomer allyl diglycol carbonate (commercially known as CR-39). The stack was built as a set of nine modules. Three types of stack assembly was adopted for these modules: one consisting of 'pure' CR-39 plastic track detector: the next one, a composite assembly of CR-39 with three layers of 600 micron thick nuclear emulsions: and the last one with CR-39 and Lexan Polycarbonate. The payload was flown successfully in June 1979 from Eielson Air Force Base, Alaska. The flight was aloft for 3 hours 30 min at an average ceiling of 3 gm/cm^2 of residual atmosphere. An attempt to stabilize and orient the payload utilizing a biaxial magnetometer combined with and electrical rotator was unsuccessful. The failure to orient the payload in a stable position would prevent us from determining the true direction of each cosmic ray particle and trace it backwards through the earth's magnetic field using a computer tracing program. Recovery of the payload was routine and all materials were in perfect condition.

CR-39 from one of the modules has been etched in a solution of 6.25N Sodium Hydroxide Solution at 50°C for 120 hours. The etching was carried out in a precisely controlled bath that is stable to $\pm 0.1°C$. Optical scanning of CR-39 was performed using an Olympus SZ-3 stereo microscope. The top fifteen sheets have been scanned for stopping and relativistic cosmic ray nuclei. In this scanning 200 stopping particles and several relativistic primaries were detected. The relativisitic nuclei passed completely through the entire stack of thickness 14 gm/cm^2 with no diminution in ionization. Optical measurements on a sample of 50 stopping particles were carried out using Koristka R4 and Leitz/Ortholux Microscopes. We confined our measurements to only those events with high etch rates, and measurements were performed on only three pairs of cones and the ender. No attempt was made to follow the track to the top of the stack in these preliminary measurements. Fig. 1 shows the normalized track etch rate as a function of residual range. Two features stand out very clearly in this figure, i) a smooth variation of track etch rate with range is indicated for individual particles, ii) in the high etch rate region there is no evidence of saturation of etch rate, which is in excellent agreement with the results of Fowler et al [1]. The charge estimates for these particles should become available in the near future after the completion of calibration of CR-39.

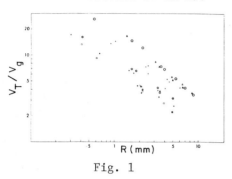

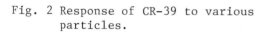

Fig. 1

Fig. 1 Normalized track etch rate as a function of residual range.

Fig. 2 Response of CR-39 to various particles.

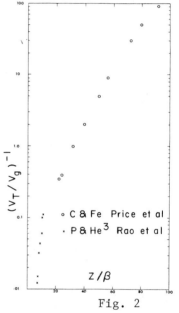

Fig. 2

Fig. 2 shows the response of CR-39 to various particles. Included in the figure are the data of Price et al [2] and our own data [3] from protons and He^3. It is clear from Fig. 2 that the normalized track etch rate approaches unity very slowly as Z/β decreases. We conclude that CR-39 will detect vertically-incident particles with Z/β as low as 8-10.

Emmanuel College authorship is sponsored by Contract F19628-79-C-0102 with AFGL. The authors thank E. Holeman for the programming assistance.

References:
1. Fowler,P.H.,V.M.Clapham,D.L.Henshaw,S.Amin,Proc.ICCR,Kyoto,11(1979)97.
2. Price,P.B.,E.K.Shirk,K.Kinoshita,G.Tarle,Proc.ICCR,Kyoto,11(1979)80.
3. Rao,Y.V.,A.Davis,M.P.Hagan,R.C.Filz,J.Blue,Bull.AM.Phys.Soc.25(1980)484.

SUPERNOVA AND COSMIC RAYS

John P. Wefel
Enrico Fermi Institute
University of Chicago
Chicago, Illinois, 60637 USA

I. Introduction:

The Supernova (SN) is one of the most important and most complex phenomena in astrophysics. Detailed observations of SN require advanced techniques of astronomy and high energy astrophysics, but the theoretical explanation of SN involves virtually every branch of physics. Supernovae, however, offer more than a challenging physics problem because SN are involved in the origin of most of the heavy elements, are the birthplaces of neutron stars, pulsars and probably black holes, control the structure of the interstellar medium, may be responsible for the birth of new stars (and possibly our own solar system), and, of greatest concern in this paper, are involved either directly or indirectly in the origin of the galactic cosmic rays.

Supernova were first observed as "guest stars", objects that appeared as new bright stars in the night sky only to disappear after a number of months. Recorded SN sightings go back at least as far as the ancient Chinese, and there have been several historical SN in our galaxy: Tycho (1572 AD), Kepler (1604 AD), and the Crab (1054 AD). Of course, there are many identified supernova remnants, among them Cassiopeia A which probably exploded $\sim$1700 AD but was not reported. Unfortunately, there have been no modern supernova in our galaxy, but $\gtrsim$ 100 SN have been observed in external galaxies. The optical observations of SN along with detailed studies of supernova remnants (SNR) remain the principal source of experimental information on the SN phenomenon.

It is impossible to review all aspects of supernovae within the space limitations of this paper. Rather a brief general overview will be presented, focussing on recent developments, followed by a discussion of the relationship between SN and galactic cosmic rays. I apologize at the outset to the many people whose papers will not be discussed, but I will attempt, where possible, to refer to papers or review articles through which the primary references may be obtained.

II. The Life of a Supernova -- A Brief Review

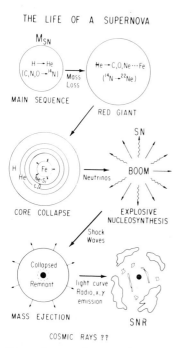

Fig. 1: Evolution of a typical Supernova

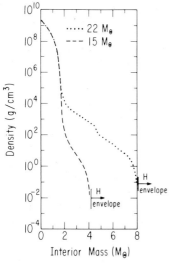

Fig. 2: Density profile for two stars following pre-supernova evolution

Supernovae are believed to be the death-throes of massive stars, occurring after the cessation of thermonuclear reactions. The life-cycle of a typical SN is illustrated on figure 1. The pre-supernova star evolves normally through the main sequence to the red-giant stage. Thermonuclear reactions then proceed at an ever increasing rate, converting the He to C and O and then burning these products, in turn, until the iron peak is formed. This leaves a star composed of the major alpha-particle nuclei, ^{12}C, ^{16}O, ^{20}Ne, ^{24}Mg, etc. and the iron peak in an approximate shell structure or "onion-skin" model as illustrated in the third panel of the figure. Gravitation then takes over and the core begins to collapse, increasing in temperature and density. Neutrinos are produced and carry away energy. The details of what happens next are still unclear, but somehow the star explodes, ejects the overlying layers of processed matter, and leaves a condensed remnant. The shock wave from the explosion moves through the mantle and heats the matter, initiating nuclear reactions (explosive nucleosynthesis) and providing the visual light display. The mantle itself expands and interacts with the local interstellar medium (ISM), eventually coming to rest as a supernova remnant. It is at these later stages that cosmic rays may be accelerated.

A. <u>Pre-Supernova Evolution</u>: This aspect of the evolution has been studied extensively with detailed evolutionary sequences both for complete stars and for helium cores of a variety of initial masses (see Arnett, 1977; Weaver, Zimmerman, and Woosley, 1978). Figure 2 shows the density profile for two stars. Note that both evolve similar cores, but the density gradient beyond the core is much steeper for the smaller star. Since each of the onion-skin shells is at roughly constant density, the $22M_\odot$ star contains more processed heavy elements by a factor of ~ 4. Integrating the amount of mass returned by each star of mass M over the mass distribution of stars in the galaxy, and over galactic history, Arnett (1978) has shown that the solar system abundances for the major isotopes are reproduced.

Stars may be divided into five mass ranges based upon their evolution and final state, as shown in Table 1. The lightest stars evolve so slowly that they remain on the main-sequence for the history of the galaxy. The next group contains stars that shed enough mass to become white dwarfs. Observations of young clusters suggest that an upper limit for this mass range is 6 ± 2 $M_\odot$, with the uncertainty reflecting our lack of understanding of mass loss processes. For an onion-skin configuration, a star must be massive enough to ignite the carbon core formed during helium burning. For stars below 8 ± 3 $M_\odot$, electron degeneracy pressure in the core can support the star from further collapse. Above ~ 8 $M_\odot$ carbon ignites non-degeneratively, and these stars evolve to the core collapse supernova discussed in this paper. Note that at the highest masses electron-positron pair formation in the oxygen core causes a change in the equation of state, leading to instability and explosion. The third line of the table shows a mass range which, fortunately, may not exist. If these stars do ignite carbon under degenerate conditions, a runaway results -- the carbon detonation supernova. This burns the entire core to iron which, if ejected, could lead to an "iron catastrophy."

TABLE 1: PRE-SUPERNOVA EVOLUTION

MASS RANGE ($M_\odot$)	EVOLUTION	END-POINT
$\lesssim 1$	main sequence only	-
~ 1 to 6 ± 2	mass loss	white dwarf
6 ± 2 to 8 ± 3	a) degenerate carbon core	a) carbon det. supernova
	b) pulsational mass loss	b) white dwarf
8 ± 3 to ~ 100	onion-skin structure	core collapse supernova
$\gtrsim 100$	e^+ e^- pair production	pair formation supernova

B. <u>Collapse, Explosion and Mass Ejection</u>: The fundamental problem facing the core collapse supernova model is the question of how to obtain an explosion with the ejection of the star's mantle. In recent years attention has focused on two classes of models: neutrino interaction models and hydrodynamic models. Neutrinos are the dominant energy loss mechanism for the core from carbon burning through the explosion. As the iron-nickel core collapses, temperature and density increase and neutronization occurs through $e^- + p \rightarrow \nu + n$ and $e^- +$ nucleus $\rightarrow \nu +$ nucleus'. This decreases the electron abundance, thus lowering the electron pressure and increasing the rate of collapse. At the high temperatures of the inner core, neutrino pair emission takes place. With the discovery of neutral currents in the weak interaction, the neutrinos take on added importance (see Freedman, Schramm and Tubbs, 1977). As collapse proceeds, the inner core, originally supported by electron pressure, evolves to a neutron gas. This implies a change in the effective adiabatic index, essentially a stiffening of the equation of state. The overlying matter falling into this stiffer core suffers a hydrodynamic 'bounce' with the formation of a shock wave.

How is the overlying material ejected? In the neutrino interaction model, the neutrinos streaming out of the core scatter from the mantle material transferring sufficient energy and momentum (it is hoped) to eject the mantle. However, detailed calculations show that the collapsing core of a massive star is not transparent to neutrinos, leading to a significant number of neutrino interactions before escape. This reduces the neutrino energy, and, since neutrino cross sections are proportional to E^2, reduces the effect of the neutrinos on the mantle. Further, the neutrinos are not emitted in a burst but are tied to the matter and come out over extended time periods. Calculations have obtained sufficient momentum transfer to slow or halt the infall of the overlying matter, but not an explosion. The hydrodynamic model offers an alternate approach, in which the reflected shock wave following the 'bounce' may drive off the mantle of the star. Recent spherical adiabatic collapse calculations by Van Riper (1978), assuming strong neutrino trapping and various equations of state, gave strong reflected shocks which steepened further in the star's density gradient (figure 2) providing ejection of the mantle. The results depend critically on the "elasticity" of the equation of state, giving a strong reflected shock only for adiabatic indices near 4/3. The typical energies in the explosion were a few times 10^{51} ergs, sufficient to power supernova light curves, and a dense remnant was formed.

C. <u>Supernova Light Curves</u>: The light curves of supernova are their visual "signature", and figure 3 shows 'typical' light curves (assuming any supernova is typical?) for type I and type II events, taken from observations collected by the Asiago Observatory (Barbon, Ciatti and Rosino, 1973). Both types show an initial peak in luminosity with the type II decreasing to an approximate plateau for several months, before a final decline. The type I, however, shows a long exponential tail which may persist for many months. The spectra of type I events show broad overlapping emission bands with little or no hydrogen emission. Type II's, on the other hand, clearly show the Balmer series in the first few weeks following maximum. Unfortunately, most supernova are discovered at or just past maximum light, so little is known about the pre-maximum epoch.

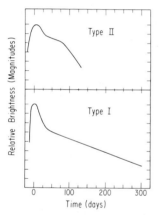

Fig. 3: 'Typical' SN Light Curves

Type II light curves have found an explanation, recently, in a shock wave model. Falk and Arnett (1977 and references therein) have followed the shock wave from core collapse into the star's mantle and envelope, tracking energy deposition, mass motions, radiative transfer, emissivity and opacity in a computer code combining numerical hydrodynamics and radiation transport. The peak in the light curve corresponds to the eruption of the shock wave through star's envelope with the light from the hot gas being absorbed and re-radiated by an circumstellar shell, thereby spreading the peak. The rest of the light curve is explained by the time evolution of the material which becomes progressively more transparent so that light from the mantle and eventually from the core itself can be observed. No energy source other than the shock wave is needed, provided there is a massive mantle/envelope with a density structure similar to the 22 $M_\odot$

star in figure 2. Comparison of these calculations with the detailed observations of supernova 1969ℓ in the galaxy NGC 1058 shows a remarkably good agreement.

The type I light curve has been explained recently by a mechanism involving radioactive decay. For a star whose density profile is similar to or even steeper than that shown for 15 $M_\odot$ on figure 2, the core collapse and explosion will process the inner 0.2-0.3 $M_\odot$ of mantle material to ^{56}Ni and eject it, provided the mantle/envelope does not stop the material. (This may require considerable mass loss during pre-supernova evolution.) Arnett (1979) has shown that the decay of the ^{56}Ni re-heats the matter after the peak so that the light curve decays initially as the ^{56}Ni mean life (8.8 days) until the γ-rays begin to escape, ~25 days past maximum. This is the transition to the exponential tail which is powered by the decay of ^{56}Co (half-life ~80 days). The ^{56}Co γ-rays escape, but the decay positrons provide the energy source. One important prediction of this model is that the γ-ray line emission from ^{56}Co should be observable over much of the exponential phase.

D. <u>Explosive Nucleosynthesis</u>: This nucleosynthesis takes place following the passage of the SN shock wave, which heats the mantle material. Explosive processing has been studied extensively in parameterized calculations, employing detailed nuclear reaction networks, to determine the conditions of temperature, density, neutron excess and expansion time scale required for the calculations to reproduce the solar system abundance distribution. Figure 4 shows a sketch of a pre-supernova configuration for a 22 $M_\odot$ star. The values of T_i indicated on figure 4 (in units of 10^9°K) are the temperatures derived from the parameterized studies for the carbon, oxygen and silicon shells to give the correct composition of the isotopes Ne-Si, Si-Ca, and the iron region, respectively (see summaries in Schramm and Arnett, 1973). Assuming that the different shells actually reach the temperatures indicated on figure 4, then the combination of pre-supernova evolution and explosive nucleosynthesis in massive stars does indeed form the majority of the stable isotopes through the iron peak.

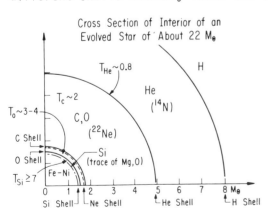

Fig. 4: Onion-skin configuration indicating burning shells, nucleosynthesis products and temperatures required for explosive processing.

Recent calculations, however, modify this simple picture. Weaver and Woosley (1979) find that the carbon region in their 25 $M_\odot$ star does not attain the required temperature but that explosive processing occurs in the neon rich region. Wefel <u>et al</u> (1980) find a large overproduction of the isotopes in the atomic mass range A = 69-77 from neutron reactions during explosive carbon burning, suggesting that explosive carbon burning cannot take place in most supernova events. Arnett and Wefel (1978) showed that the isotopes Ne-Si can be formed during <u>pre-supernova</u> evolution in high temperature, convective carbon burning shells, such as are shown on figure 4. Thus, the division between pre-supernova and explosive nucleosynthesis remains a gray area which contains many important problems (and probably a few surprises as well).

The role of massive star SN in producing the upper 2/3 of the periodic table is still unclear. The synthesis of these elements relies mainly on neutron capture reactions on, historically, a s(slow) or r(rapid) time scale relative to beta decay times. The rare proton rich species by-passed by neutron capture reaction are not included here (see Woosley and Howard, 1978). The bulk of the s-process isotopes are formed in lower mass stars (3 - 8 $M_\odot$) via thermal instabilities in high temperature shell burning. Each flash sweeps fresh material into the shell burning region where neutrons are liberated and subsequently captured by iron peak elements before the material is convected out of the shell (Truran and Iben, 1977). Although limited s-process environments do occur during the evolution of massive stars, the majority of the s-process material is <u>not</u> produced here.

The r-process requires intense neutron fluxes and short time-scales, as described by Norman and Schramm (1979). Such conditions might be found in the core of massive stars near the 'mass cut', the division between material which will be ejected and the matter that remains in the core, but the physics of this region is not well understood. Alternatively, an r-process event may occur in the helium burning shell if the shock wave heats the material to $T_{He} \sim 0.8$, as indicated on figure 4 (see also Thielemann, Arnould, and Hillebrandt, 1979), although such heating is still questionable. No completely acceptable site for an r-process event has yet been identified, but massive star supernova still remain the best prospect.

Massive star evolution provides both a supply of neutrons and heavy elements to capture the neutrons (seeds). Thus, neutron capture reactions are <u>expected</u> to be important. The seeds are the iron peak elements with which the star was originally endowed. The neutron excess resides in ^{22}Ne formed during core helium burning via $^{14}N(\alpha,\gamma)$ $^{18}F(\beta^+)$ $^{18}O(\alpha,\gamma)$ ^{22}Ne from the ^{14}N made from the initial carbon and oxygen in the star by CNO-cycle hydrogen burning. The evolution of these elements is indicated on figures 1 and 4. Investigations of neutron capture episodes are incomplete with only core helium burning and explosive carbon and explosive helium burning studied in some detail (see discussion in Wefel <u>et al</u>, 1980). It may turnout that the combined effects of neutron captures during pre-supernova and explosive evolution can reduce, substantially, the requirements placed upon the classical s- and r-processes.

E. <u>Supernova Remnants (SNR)</u>: Following the explosion, the ejected mantle/envelope material expands with velocities between 10^3 and 10^4 km/sec, impinging upon the local interstellar medium (ISM). The SN

material decelerates and eventually instabilities break the expanding
shell into pieces. A reverse shock wave forms and propagates into the
ejecta creating further turbulent motions. At this stage substantial
mixing probably takes place, and, if a magnetic field is present, it can
be drawn out into filiamentary structures by the instabilities. (For
further details see Chevalier, 1977).

Two of the best known SNR's are the Crab Nebula and Vela remnant,
both of which contain pulsars. The Crab shows pulsed emission from radio
frequencies to high energy gamma rays. On-the-other-hand, Cas A, represents a relatively young SNR ($\sim$300 years) which is not dominated by pulsar
emission. Optically, Cas A contains high velocity (5-10 thousand km/sec)
clumps of material along with much slower "quasi-stationary flocculi".
The spectra of the fast knots show that they are composed of heavy elements, O, S, Ar, Ca, with almost no hydrogen or helium. The flocculi,
however, show mainly H, He, and N with the latter two elements enhanced
relative to solar proportions. Cas A may be explained as a 10-30 $M_\odot$ star
that underwent significant mass loss (40-60%) prior to the explosion.
The fast moving knots are material processed through explosive oxygen
burning, and the flocculi sample material from near the hydrogen burning
shell (c.f. figure 4) where N and He are overabundant relative to H. Thus,
SNR provide an important tool for comparing the theory to observations.
Cas A is one well-studied case, but other SNR's need to be exploited.

The expanding SNR compresses the ambient ISM which eventually forms
into clumps or clouds. Inside the SNR shell the medium is hot and rarefied. Thus, after the SNR comes into pressure equilibrium, the ISM
around it consists of a large region of hot tenuous gas surrounded by
denser clouds. If the supernova rate is high enough, the entire ISM
evolves to a multi-component medium in which hot, tenuous gas (10^{-2} atoms/
cm^3, 10^{5-6}°K) occupies 70 to 80% of the volume with denser (10-100 atoms/
cm^3) clouds, each surrounded by a "warm" halo, composing the rest of the
ISM (McKee and Ostriker, 1977).

The interaction of supernova blast waves with cold clouds may be the
push necessary to begin the collapse of such clouds to form new stars and
planetary systems, possibly including our own solar system. This idea
has been stimulated by the discovery of material of anomalous isotopic
composition in meteorites. In particular, the demonstration that ^{26}Al
($T_{1/2} \sim 7 \times 10^5$ years) must have been active in the early solar nebula
places a supernova event within a few million years of the birth of the
solar system (see review by Schramm, 1978, and other articles in that
volume). Further, it may be that the dynamics and composition of regions
of active star formation, such as OB associations, are dominated by supernova from rapidly evolving massive stars. In such a scenario, the local
ISM could have a composition enriched in supernova produced heavy elements,
including r-process isotopes, from which may derive isotopic anomalies,
metal-rich stars, and possibly the galactic cosmic rays.

III. Cosmic Rays and Supernova:

The three main connections between cosmic rays and supernovae are the energy requirement, acceleration mechanism, and the detailed composition. The cosmic rays have an energy density in the galaxy (assuming uniformity) of ~ 1 ev/cm^3. However, cosmic rays leave the galaxy after a mean residence time of 10-20 million years (Garcia-Munoz, Mason, and Simpson, 1977), requiring replenishment to maintain a constant energy density. For a supernovae rate of ~ 0.1 per year and $\sim 10^{51}$ erg/SN, the cosmic rays can be maintained if $\sim 1\%$ of the SN energy is converted to energetic particles. This energy requirement was presented in detail by Ginzburg and Syrovatskii (1964) and with Ginzburg's rule (1=10 in astrophysics) has survived changes in supernova rates, cosmic ray confinement times and volumes, and SN energies.

Supernovae also offer a unique situation for charged particle acceleration, and, in fact, observations of synchrotron emission from both SNR's and pulsars is direct evidence that high energy electrons are accelerated. Observations of gamma rays with a spectrum characteristic of π° decay would provide evidence for proton acceleration as well. Models for galactic cosmic rays are constrained by the question of 'adiabatic losses'. In attempting to leave the expanding SNR, the cosmic rays interact and generate hydromagnetic waves which transfer energy from the particles to the medium. In the worst case, several orders of magnitude of particle energy may be lost before the SNR expands and cools sufficiently to damp the waves and allow the particles to escape. There is, however, some disagreement about the inevitability of this energy loss (see discussion by Schwartz and Skilling, 1978), but the problem is alleviated if particle acceleration does not begin until after several years of SNR expansion. Such a time delay is supported by measurements of the Co/Ni ratio in cosmic rays. The data imply that the electron capture isotope ^{57}Co ($T_{1/2}$ = 271 days) decayed prior to acceleration giving a minimum delay between nucleosynthesis and acceleration of 1-3 years.

Cosmic ray acceleration may take place either inside the SNR or beyond its borders. Scott and Chevalier (1975) proposed a second order Fermi acceleration mechanism operating in SNR such as Cas A. The fast moving knots generate turbulent magnetic fields which act as scattering centers. Matter evaporated from the knots or swept up by the expanding SNR is accelerated, beginning ~ 20 years following the explosion. Alternatively, particle acceleration by shock waves has been re-examined recently leading to an improved cosmic ray acceleration mechanism (see references in Blandford and Ostriker, 1978). In this "shock" model, particles at suprathermal energies are trapped by pitch angle scattering on either side of the shock and forced to make repeated transits across the boundary. The scattering centers are converging, giving particle acceleration by the efficient first order Fermi process. This "shock" model can operate anywhere, given a strong shock wave, and supernovae are probably the most prolific source of shock waves in the galaxy.

SN can provide both the energy input and acceleration for cosmic rays, but can they reproduce the cosmic ray composition? The elemental and isotopic composition of cosmic rays, both below and above the iron peak, is reviewed extensively in this volume (see chapters by Meyer, Reeves, and Fowler). Figure 5 shows a plot of the ratio of cosmic ray source abundances to solar system abundances (CRS/SS) as a function of the first ionization potential of the elements (see Casse and Goret, 1978). A similar plot versus the charge of the element would look like the dashed line, relative to the scale at the top of the figure.

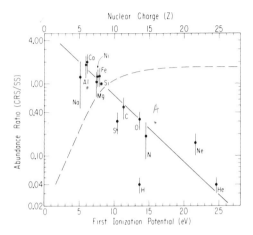

Fig. 5: Ratio of C.R. source to solar system abundances vs. first ionization potential. Dashed curve refers to scale at top and shows trend of a plot vs. Z.

The correlation between CRS/SS and first ionization potential suggests that a selective injection or preferential acceleration mechanism, based upon <u>atomic</u> properties of the elements, operates to modify the element distribution in the source region. This leads to the 'indirect' class of models in which the cosmic ray composition is derived by such preferential acceleration from material with approximately solar system composition. The models are indirect since acceleration can occur anywhere, and SN are involved only as the source of the accelerating shock waves.

'Direct' models begin with the correlation shown as a dashed line on figure 5 and attempt to explain the cosmic ray source composition from the SN itself. Hainebach, Norman, and Schramm (1976) worked out the details of one such 'direct' model in which the source composition resulted from pre-supernova evolution augmented by explosive nucleosynthesis and some mixing, prior to acceleration, with the ISM surrounding the star. They found that the abundances in a ~ 15 $M_\odot$ star were much closer to galactic cosmic ray source material than were more massive stars. A 15 $M_\odot$ star is about the average SN by number, implying either that SN may each accelerate the same mass of material as cosmic rays or that the smaller stars evolve to SNR configurations that are more apt to accelerate charged particles.

The 'indirect' and 'direct' models are representative of the two extremes of SN involvement in the origin of cosmic rays. Clearly, many models are possible between these extremes. For example, a 'direct' model utilizing a 22 $M_\odot$ star, which produces abundances close to those in the solar system, combined with preferential acceleration may very well provide the cosmic ray source abundances. This highlights the need for further characterization of the cosmic ray source, such as has been obtained recently through isotopic measurements and studies of transnickel cosmic rays.

The recent isotopic measurements important for SN models may be summarized as follows. At the cosmic ray source (relative to solar system abundances); (a) ^{22}Ne/^{20}Ne is enhanced by $\sim$3, (b) $^{25+26}$Mg/^{24}Mg may be enhanced by $\sim$1.5-2 (but this needs verification), (c) iron is mainly ^{56}Fe with ^{54}Fe and ^{58}Fe <10%, and (d) ^{40}Ca/Fe is normal within $\sim$30%. An enhancement of the neutron rich isotopes of Ne and possibly Mg does not follow easily from indirect models unless the material being accelerated is itself abnormal in composition. Direct models can accommodate these enhancements either from the helium burned zone (see figure 4) or from a star with increased metal content. The dominance of ^{56}Fe is comforting in either model and implies that the SN producing cosmic rays are no different from the rest in their behavior near the mass cut. The limits on ^{54}Fe are beginning to be restrictive, but ^{58}Fe could still be enhanced over its normal value (0.3%). The ^{40}Ca/Fe ratio is interesting here because Ca has a significantly lower ionization potential than iron. An enhancement of a factor of $\sim$2 would be expected, and the normal ratio observed must be explained by indirect models.

The UH (Z > 28) cosmic rays sample the s- and r-process components of the source material. Current data favor an r-process source for the Z > 70 region, but this interpretation has been questioned. For the elements just above iron, $30 \leq Z \leq 40$, the data show a mixture of s- and r-process components, indistinguishable with current data from the solar system mix. Indirect models predict a solar ratio of s- and r-process material, but direct models predict an r-process dominance, since the bulk of s-process elements are not formed in massive stars. The exception is the region just above iron where neutron capture episodes can produce some s-process material (Wefel, Schramm and Blake, 1977). Therefore, the presence of s-process material at Z > 70 becomes the key measurement, involving the separation of the r-process peak at $Z \sim 78$ from the s-process peak at $Z \sim 82$. Current results favor direct SN involvement, but the data now being collected by the satellites Ariel-6 and HEAO-3 should provide definitive results.

The enrichment of cosmic ray sources in ^{22}Ne, possibly ^{25,26}Mg, and the r-process dominance for Z > 70 lead to the speculation that the stars responsible for cosmic ray production may be enriched in 'metals', i.e. Z > 2 elements. An enrichment by $\sim$3 implies enhanced abundances, at about the same level, for the neutron rich isotopes of Ne, Mg, Si, S and for ^{54}Fe following the star's evolution. Such an enhanced metallicity cannot be present over the entire galaxy, but in local regions, such as regions of active star formation or OB associations, the ISM may be quite different. Rapidly evolving massive star SN can enrich the region such that later generations of stars are formed with higher metal contents. A similar scenario enhances the r-process isotopes, since these SN do not produce much s-process material. Acceleration of this local ISM as cosmic rays can take place via shock waves from second or third generation stars. The atomic properties of the elements may play a role in the injection/ acceleration. The presence of the refractory elements such as Al, Ca, Fe in condensed grain cores rather than in the gas phase may present a constraint unless the shock wave vaporizes these grains prior to acceleration of the material (see discussion by Dwek and Scalo, 1980).

This scenario, of course, may be purely imaginary, but certain aspects can be studied experimentally. In the cosmic rays, precise measurements of the isotopic composition of Mg, Si, S, and Fe are needed along with a measurement of the exact ratio of s- to r-process material as a function of atomic number. Observationally, such regions should be identified and studied with particular attention paid to the element/isotope abundances. These regions of active star formation and cosmic ray production should be observable in gamma emission from cosmic ray interactions and in γ-ray lines from radioactive isotopes. Current γ-ray experiments in space may be able to supply the data.

IV. Summary

Supernovae are one of the most important phenomena is astrophysics, and considerable progress has been made in recent years in understanding them. The pre-supernova evolution of massive stars continues to be studied in increasing detail, and the core collapse/explosion phase is now understood, at least to the point where reasonable physics in the hydrodynamic model does indeed provide an explosion with mass ejection. SN light curves, the dynamics of the expanding mantle, and the formation of SNR's have been explained and detailed comparisons of theory with experimental data are now possible. Acceleration mechanisms are available to produce the cosmic rays either directly from the SN matter or from a mix of SN debris and ISM. Recent cosmic ray isotopic data and element abundances for UH nuclei are providing increasingly stringent constraints on models for the synthesis of cosmic ray matter.

The present situation with respect to both theory and experiment is enormously exciting. Progress has been considerable, but much still remains to be done. Supernova certainly seem to be involved in the origin of cosmic rays, but exact models still elude our grasp. The decade of the 1980's promises to be a productive one for both supernova and cosmic rays.

V. Acknowledgements

Special thanks are due to David Schramm, without whom this paper would have been impossible, to J. A. Simpson, W. D. Arnett, J. B. Blake and M. G. Munoz for help and guidance, and to V. Zuell for typing a difficult manuscript. This work was supported by NASA grant NGL-14-001-006.

VI. REFERENCES

Arnett, W.D.: 1977, Ap. J. Suppl., 35, pp. 145-159.
⎯⎯⎯⎯⎯⎯⎯⎯⎯: 1978, Ap. J., 219, pp. 1008-1016.
⎯⎯⎯⎯⎯⎯⎯⎯⎯: 1979, Ap. J. Lett., 230, pp. L37-L40.
Arnett, W.D., and Wefel, J.P.: 1978, Ap. J. Lett., 224, pp. L139-L142.
Barbon, R., Ciatti, F., and Rosino, L.: 1974, "Supernovae and Supernova Remnants", ed. C.B. Cosmovici (Dordrecht: Reidel), pp. 99-118.
Blandford, R.D., and Ostriker, J.P.: 1978, Ap. J. Lett., 221, pp.L29-L32.
Casse, M., and Goret, P.: 1978, Ap. J., 221, pp. 703-712.
Chevalier, R.A.: 1977, Ann. Rev. Astron. Astrophys., 15, pp. 175-196.
Dwek, E., and Scalo, J.M.: 1980, Ap. J., in press
Falk, S.W., and Arnett, W.D.: 1977, Ap. J. Suppl., 33, pp. 515-562.
Freedman, D.Z., Schramm, D.N., and Tubbs, D.L.: 1977, Ann. Rev. Nucl. Sci., 27, pp. 167-207.
Garcia-Munoz, M., Mason, G.M., and Simpson, J.A.: 1977, Ap. J., 217, pp. 859-877.
Ginzburg, V.L., and Syrovatskii, S.I.: 1964, "The Origin of Cosmic Rays" (New York: Macmillan), pp. 194-200.
Hainebach, K.L. Norman, E.B., and Schramm, D.N.: 1976, Ap. J., 203, pp. 245-256.
McKee, C.F., and Ostriker, J.P.: 1977, Ap. J., 218, pp. 148-169.
Norman, E.B., and Schramm, D.N.: 1979, Ap. J., 228, pp. 881-892.
Schramm, D.N.: 1978, "Protostars & Planets", ed. T. Gehrels (Tuscon: Univ. of Arizona Press), pp. 384-398.
Schramm, D.N., and Arnett, W.D.: 1974, "Explosive Nucleosynthesis", eds. D.N. Schramm and W.D. Arnett (Austin: Univ. of Texas Press), pp. 45-229.
Schwartz, S.J., and Skilling, J.: 1978, Astron. Astrophys., 70, pp. 607-616.
Scott, J.S., and Chevalier, R.A.: 1975, Ap. J. Lett., 197, pp. L5-L8.
Thielemann, F.K., Arnould, M., and Hillebrandt, W.: 1979, Astron. Astrophys., 74, pp. 175-185.
Truran, J.W., and Iben, I.: 1977, Ap. J., 216, pp. 797-810.
Van Riper, K.A.: 1978, Ap. J., 221, pp. 304-319.
Weaver, T.A., Zimmerman, G.B., and Woosley, S.E.: 1978, Ap. J., 225, pp. 1021-1029.
Weaver, T.A., and Woosley, S.E.: 1979, Proc. 9th Texas Symp. on Rel. Astrophys., in press.
Wefel, J.P., Schramm, D.N., and Blake, J.B.: 1977, Astrophys. Sp. Sci, 49, pp. 47-81.
Wefel, J.P., Schramm, D.N., Blake, J.B., and Pridmore-Brown, D.: 1980, Ap. J., in press
Woosley, S.E., and Howard, W.M.: 1978, Ap. J. Suppl., 36, pp. 285-304.

WHAT CAN WE LEARN ABOUT COSMIC RAYS FROM THE UV, OPTICAL, RADIO AND X-RAY OBSERVATIONS OF SUPERNOVA 1979c IN M 100?

G.G.C. Palumbo and G. Cavallo
Istituto TESRE del CNR
Via De' Castagnoli 1, 40126 Bologna, Italy

Theories on the Origin of Cosmic Rays almost invariably invoke the Supernova (SN) phenomenon in its early phases as the cause for production and acceleration of high energy particles. So far only optical information about SNe has been available and from it there is no direct evidence of Cosmic Rays. It is not surprising then that models of Cosmic Ray production are still rich in free parameters. On April 19th 1979 a very bright ($\sim$12 mag) SN, labelled 1979c, was detected in the relatively nearby galaxy ($\sim$16 Mpc) M 100 ($\equiv$NGC 4321). This galaxy, incidentally has produced 4 SNe in 78 years. Event 1979c was followed quite intensively in the optical and UV (with IUE) regions of the spectrum as well as observed at radio and X-ray frequencies. A detailed account of these observations is in press (Panagia et al. 1980). Here we summarize only very briefly the results relevant to the present discussion.

While the SN photosphere, where the continuum and optical lines originate, was observed to expand at $v_{ph} \sim 10^4$ Km s^{-1} an external UV shell, where UV lines were formed, that about 1 week after maximum had a radius $R_{UV} \sim 2\, R_{ph}$, was observed to expand at $v_{UV} \sim (1 \div 4)\, 10^3$ Km s^{-1}. SN 1979c, identified as Type II had at maximum $M_{Bmax} = -19\overset{m}{.}4 \pm 0\overset{m}{.}8$. Seven days after maximum $R_{ph} \simeq 1.25\, 10^{15}$ cm. The most probable origin of the UV shell is from preexisting material ejected into space as a stellar wind by the progenitor star when it was in the red giant stage. The wind was estimated to be $\sim 10^{-4} M_\odot yr^{-1}$ to provide the observed $\sim 0.2 M_\odot$ of the UV shell. The shell therefore had a density of $\sim 5\, 10^6$ atoms cm^{-3}. From optical and UV measurements, we have $L_{TOT} \simeq 2.4\, 10^{43}$ erg s^{-1} and $E_{rad} \simeq 7\, 10^{49}$ erg. In the early phases ($\sim$10 days after max) radio measurements provided only upper limits. The same applies for X-ray observations. The best upper limit in the 0.5-4.5 KeV energy range comes from the Einstein Observatory HRI 239 days after max $L_X < 7.7\, 10^{39}$ erg s^{-1}. Almost exactly one year after maximum the SNR has been detected at the VLA radio telescope (K. Weiler, private communication) with flux $S_{6cm} = 5$ mjy corresponding to $L_R \sim 1.5\, 10^{36}$ erg s^{-1}.

As far as the production of Cosmic Rays is concerned we argue as follows:
1) There is evidence of an X-ray flash in the pre-detonation stage.

This comes from the presence of the UV shell. The X-ray flash, in fact, is the most likely cause of acceleration and ionization of the material surrounding the exploding star observed in the UV. The minimum energy required (if pure H) being $>5 \cdot 10^{45}$ erg, i.e., in the range of predictions for such an X-ray flash. One should remember that the X-ray flash is frequently associated with the shock wave which in Colgate model accelerates cosmic rays.

2) A substantial γ-ray ($\overline{E}_\gamma \sim 100$ MeV) flux ($\sim 10^{44}$ erg s^{-1}) has been estimated by Cavallo and Pacini (1980) from the reaction pp→ppπ° and $\pi^\circ \to 2\gamma$ in the specific case of SN 1979c. This flux was not observed because no γ-ray instrument available at the time was aimed at the SN. This is regrettable because detection of γ's from π° decay would have provided the best evidence of Cosmic Ray acceleration from SNe.

3) Because of the difference in velocity the photosphere eventually reached the UV shell. When this happened a reverse schock may have occurred (Mc Kee 1974). The X-ray luminosity in the 0.5-4.5 KeV band is $L_{isoth} \sim 6.5 \cdot 10^{37} n_1$. This again is a substantial emission of short duration which was not detected because no X-ray telescope was observing at the time.

4) The radio detection provides estimates of the magnetic field and of the total relativistic particle energy contained in the young remnant. With standard assumption of equipartition and ratio $K = W_p/W_e = 100$ (Ginzburg and Syrovatskii 1964) one finds $H \simeq 0.45$ Gauss and $W_{CR} \sim 10^{48}$ erg. For $K = 1$, $H = 0.12$ Gauss and $W_{CR} \simeq 7.5 \cdot 10^{46}$ erg. Apparently the radio observational frequency was very close to the syncrotron reabsorption turnover frequency with $\gamma_e \sim 190$. If no further injection occurs and one can somehow account for lifetime effects we can predict a t^{-5} decrease of the radio flux with time t. Also, when this remnant will have reached the size of Kepler's SNR both magnetic field and relativistic particle content will be of the same order of magnitude as in Kepler's.

In conclusion, Cosmic Rays are already present in the remnant after 1 year although ~ 2 orders of magnitude below the canonical requirement of $\sim 10^{50}$ erg per SNe. Moreover, one consideration is apparent from the above: when a nearby bright SN occurs X and γ-ray detectors should follow the first months of its evolution if one wants to learn more about Cosmic Rays from SNe.

References

Cavallo, G. and Pacini, F.: 1980, Astron. and Astrophys., in press.
Ginzburg, V.L. and Syrovatskii, S.I.: 1964, "The Origin of Cosmic Rays", Pergamon Press.
Mc Kee, C.F.: 1974, Ap. J. 188, 335.
Panagia, N. et al.: 1980, Month. Not. RAS, in press.

VERY HIGH ENERGY COSMIC RAYS

John Linsley
University of New Mexico, Albuquerque, NM, USA

Results from ground level and underground experiments on cosmic rays with energy 10^{12} to 10^{20} eV are reviewed. They show that the energy spectrum has two significant features, a 'knee' and an 'ankle'. The arrival directions of these cosmic rays at the solar system are anisotropic, features of the anisotropy appearing to be correlated with features of the spectrum. Detailed interpretation of this information awaits conclusive evidence regarding the composition of these cosmic rays. New results and prospective new results on the composition are described and discussed.

1. INTRODUCTION

Very high energy cosmic rays, those with energy per particle greater than 10^{12} eV, carry information that promises to be indispensable for deciding between theories of the origin of Galactic cosmic rays. Such cosmic rays may provide especially direct evidence on the magnetic field structure of the Galaxy, out to distances of some kiloparsecs from the solar system. Some of these cosmic rays, having energies greater than 10^{19} eV, appear to be extragalactic. There are difficulties in imagining an astrophysical setting in which acceleration to such great energies can occur at all. Moreover, the amount of energy required to fill the local supercluster with these particles at the observed level of intensity is quite considerable.

Up until now, the clearest result obtained from observations of these cosmic rays is the energy spectrum, meaning the distribution in energy per particle. Instead of being the featureless inverse power law that was at first anticipated, this spectrum (Figure 1) exhibits two structures, a 'knee' at $\sim 10^{15}$ eV and an 'ankle' at $\sim 10^{19}$ eV. Studies of the arrival directions of these cosmic rays show a definite pattern of energy dependent anisotropy, with evidence of correlation between this pattern and features of the spectrum.

These results already provide some guidance for our speculations about cosmic ray origin, so I will begin by presenting them, noting some of the experimental problems. At the same time I will go over conclusions that have already been reached. General reviews have been given by Sreekantan (1972) and Hillas (1975). The subject of anisotropy has been reviewed by Wolfendale (1977) and more recently by Kiraly et al. (1979b). A useful summary on the highest energy particles has been given by Watson (1980a).

In a sense, however, the existing results, although they have taken decades to obtain, have only brought us to where we can make informed plans for a new generation of experiments. We know now what kind of 'signals' are present, and about how strong they are, so we can tell how much it will cost in dollars and effort to make these signals stand out above the background. The principal requirement, for experiments belonging to the new generation, is a capability of determining spectra and anisotropies of resolved primary components, rather than a 'spectrum' and an 'anisotropy' for primaries that are unspecified except for having about the same total energy per particle.

The composition of cosmic rays selected to have equal energy is strongly biased in favor of heavy elements compared to cosmic rays having equal magnetic rigidity. The equal-energy mass spectrum is in fact approximately rectangular in the low energy region, as is shown by Table 1. Thus it is not implausible to imagine a change taking place, at higher energies, leading either to nearly pure H or to nearly pure Fe.

Table 1. Equal-energy mass spectrum from low energy data.

Mass number	1	4	12-16	20-40	52-58
Percentage*	43.1	20.6	13.1	10.6	12.6

*Assuming power-law spectra with differential exponent 2.6 and source-region charge composition (from Rasmussen 1974).

The resolving power I have in mind, for new-generation experiments, is the power to distinguish between the groups listed in Table 1, or between showers initiated by protons and those initiated by γ-rays. In the final section of this report I will discuss methods of investigating the composition of very high energy cosmic rays and describe some of the results that have been obtained.

2. ENERGY SPECTRUM

Figure 1 shows the dependence on energy E of the intensity of all particles with kinetic energy greater than E, multiplied for convenience

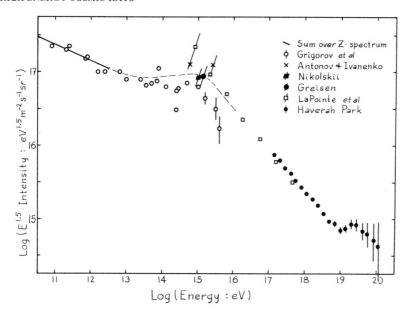

Fig. 1 Integral energy spectrum of cosmic rays.

by $E^{1.5}$. The solid line, 'sum over Z-spectrum', was derived from measurements in which the primary particles (nuclei) could be sorted according to charge (from Hillas 1979, Fig. 2). The open circles were obtained using an ionization calorimeter carried on a satellite (Grigorov et al. 1971). The remaining points were derived from measurements of extensive air showers. The large filled circle and the trapezoid are calorimetric results by Greisen (1956) and Nikolskii (1962), respectively. The energy was determined by adding the energy deposited in the atmosphere and the earth by the three major components: electrons (and photons), muons, and hadrons, using data obtained at sea level and mountain altitudes by a variety of techniques, with small allowances for neutrinos and excitation of nuclei. The crosses (Antonov and Ivanenko 1975) and squares (La Pointe et al. 1968) were derived similarly. The energy deposited in the atmosphere was determined empirically by measuring showers at various atmospheric depths, starting at 540 g cm^{-2} in case of the earlier experiment and 200 g cm^{-2} in case of the later one. The relatively small fraction of energy ($\simeq$10%) not accounted for by their track length integrals was evaluated by means of a hadronic cascade model. The filled circles are results obtained at Haverah Park (Cunningham et al. 1980, converted to integral form by Watson). In this case, also, the determination of energy was essentially calorimetric. The cascade model used to derive energy from the 'ground parameter' ρ_{600} was constrained to agree with such a number of independent measurements that it functioned essentially as an interpolation device.

The error bars for the Haverah Park points indicate Poisson-statistical standard deviations. The small scatter of the lower energy points shows the high relative accuracy that is typical of long-term air shower experiments. It is greater, of course, than the accuracy of the energy calibration in absolute units such as eV. Results from the only two comparable northern hemisphere experiments, those carried out at Volcano Ranch and Yakutsk, confirm the existence of an ankle. The Volcano Ranch and Yakutsk calibrations are also essentially calorimetric, but the methods differ from each other and from the Haverah Park method in important details. The Yakutsk method gives energies about 10% higher than the Haverah Park method, while the Volcano Ranch method has given energies about 20% lower. (In Figure 1 the error bars on the point due to Greisen correspond to ±20%.)

The detailed shape of the energy spectrum in the neighborhood of the knee has been investigated by several groups. Corresponding changes of slope are found in the number spectra of electrons, muons and atmospheric Cerenkov photons (Hillas 1975 and references therein). A typical result is the electron number spectrum measured at Chacaltaya by Bradt et al. (1965), shown in Figure 2.

Such changes in slope cannot be explained by assuming that above 10^{15}eV there is a change in the character of high energy interactions as they occur in air showers. A downward break at about the right energy is expected for open-galaxy models due to rigidity dependence of the diffusion coefficient (Ginzburg and Syrovatskii 1964). It is difficult, however, to account for the sharpness of the observed break on such a model (Bell et al. 1974). It has also been proposed that the knee corresponds to the limiting rigidity of a dominant source (Peters and Westergaard 1976). In either case, the effect on primaries with the low energy composition of Table 1 would be to produce secondary breaks at higher energies than the proton-cutoff break. The resulting spectra would disagree qualitatively with the observed one (Bell et al. 1974, Hillas 1979). Thus, if the knee results from magnetic processes a change in composition must already have taken place at somewhat lower energies. Alternatively, the knee, assuming that it is produced in the source region, may correspond either to a threshold for breakup of preferentially accelerated heavy nuclei, or to the threshold for energy loss by collisions with photons (Zatsepin et al. 1963, Hillas 1979). It has also been suggested that the knee is formed of cosmic rays from a different source or class of sources than those which produce the bulk of cosmic rays (Karakula et al. 1974).

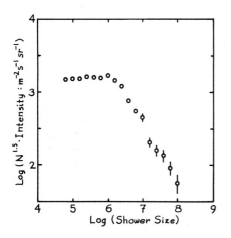

Fig. 2 Size spectrum of nearly vertical showers at Chacaltaya (depth 540 g cm^{-2}). The ordinate is integral intensity multiplied for convenience by (size)$^{1 \cdot 5}$.

VERY HIGH ENERGY COSMIC RAYS

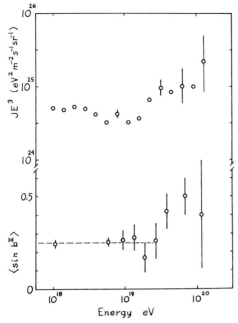

Fig. 3 Correlation between mean galactic latitude and energy. The dashed line indicates the expected mean for a random arrival direction distribution.

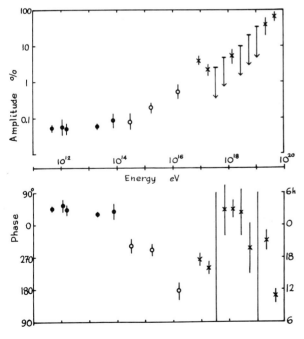

Fig. 4 Cosmic ray anisotropy

The other feature of the spectrum, the ankle just below 10^{19} eV, has frequently been associated with a crossover from Galactic to extragalactic cosmic rays. Data from Haverah Park (see Figure 3, from Lloyd-Evans et al. 1979) indicate that the additional flux arrives from high galactic latitudes, suggesting an association with active galaxies belonging to the Virgo cluster (Stecker 1968, Krasilnikov 1979).

3. ARRIVAL DIRECTIONS

Figure 4 shows results of Fourier-analyzing the variation of cosmic ray intensity with right ascension. Above is the amplitude, below is the phase, of the first harmonic. The 3 lowest-energy points, from left to right, are given by underground muon telescopes at Holborn (Davies et al. 1979), Poatina (Fenton and Fenton 1976), and Heber Mine (Bergeson et al. 1979). For these points, 'energy' means energy per nucleon, so the response is due mainly to primary protons. The next two points are given by air shower measurements at Mt. Norikura (Sakakibara et al. 1979) and Musala Peak (Benko et al. 1979). The following 3 (open circles) are from a survey of air shower data from many sources, by Linsley and Watson (1977), while the remaining 10 (crosses) are given by air showers recorded at Haverah Park (Lapikens et al. 1979a, Watson 1980b).

These results are shown as evidence that sidereal anisotropy is present, at levels that exceed noise, over most of the range from 10^{11} to 10^{20}

eV. The work of interpreting results of this kind is proceeding apace. At the lower energies the effects observed are nearly energy independent. Even at 10^{14} eV the Larmor radius of a proton approaching the solar system is a small fraction of a parsec, so one does not expect to be able to localize sources. Instead, the present goal is to determine the magnetic field configuration outside, but not far outside, the heliosphere (Wolfendale 1977, Kiraly et al. 1979a). It is hard to imagine how this could be done in any other way.

At an energy about equal to (somewhat less than) that corresponding to the knee in the spectrum, the magnitude of the anisotropy, as measured by the first harmonic amplitude in right ascension, begins to increase about as $\sqrt{E}$, while the direction of maximum intensity shifts to earlier times. These changes occur in an energy region where, according to most models, Galactic sources are still dominant. They can be used, therefore, as a basis for choosing between models of the source distribution and propagation mode for the bulk of Galactic cosmic rays.

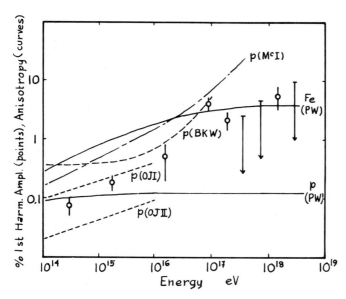

Fig. 5 Predictions of cosmic ray anisotropy. The points, taken from Figure 4, give the amplitude of the first harmonic of counting rate in right ascension. The curves give the anisotropy, $\partial = (I_{max} - I_{min})/(I_{max} + I_{min})$, according to Peters and Westergaard (PW) 1976, McIvor (McI) 1977, Bell et al. (BKW) 1974, and Owens & Jokipii (OJI,II) 1977. All of the predictions are for proton primaries except one of those by Peters and Westergaard.

Figure 5, taken from Lloyd-Evans et al., shows part of the data given in Figure 4, together with several predictions identified in the caption. It should be noted that the anisotropy equals the quantity measured times $(\cos\lambda \cos\delta)$, where λ is the latitude of the observation and δ is the declination of the upstream direction. Since δ is unknown, the measurements give lower limits. The two lowest curves clearly disagree with the evidence; the others are consistent with it.

Returning to Figure 4, the data above 10^{17} eV offer no easy interpretation. Looked at in detail, the evidence from Haverah Park is that above 10^{17} eV the simple trends shown by lower energy data break up into complex patterns (Lapikens et al. 1979a and references therein). The complexity has several aspects: 1) distributions in right ascension may contain significant harmonics higher than the first, 2) there may be significant changes in amplitude and phase from one 10° declination band to the next, and 3) there may be considerable differences in amplitude and phase between the first harmonics (summed over declination) for adjacent factor-of-2 energy bins. Such features are also shown, at lower significance levels, by earlier data from the Volcano Ranch experiment (Linsley 1975).

The most extreme example I can point out is shown in Figure 6. The data are from Haverah Park. The top curve shows a rather broad region of enhanced intensity near 18h for declinations 40-50°. At the same energy, neighboring declination bands are enhanced similarly but to a lesser degree. The middle curve is for an energy twice as great and a declination band displaced 20° toward the equator. The enhancement is highly significant in the band shown but is absent in neighboring bands. Assuming that the differences in right ascension and declination are due to magnetic deflections, one might expect by a naive extrapolation to find a very sharp enhancement at ~16h in the bottom curve (energy greater by another factor 2, declination 10° less), but one finds nothing of the sort. So this feature is not evidence for a localized power-spectrum source. It appears to be a modulation effect.

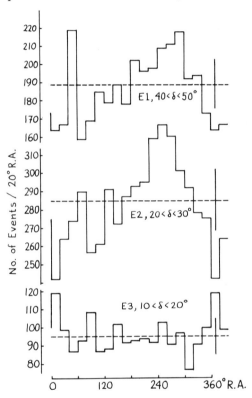

Fig. 6 Distributions in right ascension of shower directions for adjacent energy bins (Lapikens et al. 1979a, Pollock 1978). The mean energies of bins E1, E2, E3 are 0.9, 1.8, and $3.5 \cdot 10^{17}$ eV, respectively. Each distribution is for a different 10° declination band, as explained in the text.

Complexity of this general nature would be expected above 10^{17} eV even for a simple, regular Galactic field, as has been shown by the calculations of Karakula et al. (1972). It is expected all the more in light of evidence for numerous large-scale irregularities within a few kpc of the solar system (Kirshner 1980, Cassé and Paul 1980 and references therein). If one estimates that cosmic ray arrival directions will be disturbed by field irregularities over a range of 3R to 30R, where R is the Larmor radius, then the observations suggest that a substantial fraction of the primaries are protons. (Assuming a field of $3\mu G$, R is equal to $(.37E/10^{15}Z)$ parsec, where E is energy in eV and Z is charge number. For $Z\sim 26$ one expects that the streaming regime would persist up to $\sim 10^{18}$ eV, contrary to observation.)

A tendency of primaries with energy $>10^{19}$ eV to arrive from higher Galactic latitudes has already been mentioned. Figure 7 shows the arrival directions of the most energetic particles that have been detected by the 4 giant air shower arrays. In the southern hemisphere 2 well-known clusters are evident. In the northern hemisphere one can discern a much larger, rather diffuse group centered between the Galactic north pole and the anticenter. The zone $-30° < b^{II} < 30°$ is nearly vacant except near the spiral-out direction and near the anticenter.

It seems nearly impossible to sustain, against the evidence shown in Figure 7, a theory that cosmic rays are confined to galaxies, including their haloes. Such a theory would require that essentially no particles of this energy be protons or alpha particles, contrary to my previous argument and to strong evidence from shower profile fluctuations (Lapikens et al. 1979b and references therein). Even if one assumes for the purpose of discussion that all of the Figure 7 primaries are Fe-nuclei, the arrival directions do not support Galactic origin. 1) The tendency of the 3 clusters of points to be associated with the principal axes of Galactic symmetry is no greater than would be expected by chance. 2) The possible association between one cluster and the spiral-out direction is denied by the absence of any cluster near spiral-in. 3) Where symmetry would be expected, the cluster in the northern Galactic hemisphere is much larger than its southern counterpart.

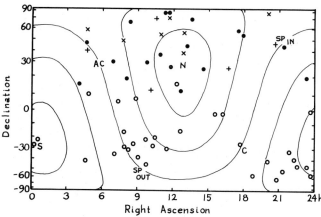

Fig. 7 Arrival directions of cosmic rays with energy $>5\cdot 10^{19}$ eV, from Krasilnikov (1979) with minor additions and amendments. The filled circles represent events recorded at Haverah Park; the open circles, those recorded at Sydney; the +'s and ×'s, those recorded at Volcano Ranch and Yakutsk, respectively. The contour lines show the Galactic latitude (b^{II}) at 30° intervals. The letters indicate the Galactic poles, the center and anticenter, and the directions 'spiral-in' and 'spiral-out'. The nonlinear declination scale compensates for differences in exposure.

4. COMPOSITION

Methods of determining the composition of air shower primaries depend on measuring secondary characteristics in addition to those used to give the energy of the showers. One such secondary characteristic is X_{max}, the depth of maximum development, another is the proportion of low energy muons at ground level. X_{max} is given directly by the time distribution of atmospheric Cerenkov photons at moderate core distances (Fomin and Khristiansen 1971, Orford and Turver 1976, Thornton and Clay 1979) and by the atmospheric scintillation technique (Bergeson et al. 1977). It is given less directly by time distributions of particles (Lapikens et al. 1979b), and indirectly by the shape of the radial distribution of particle flux at ground level (Linsley 1977, England et al. 1979). The median depth of production of low energy muons is expected to be another useful parameter. It can be derived rather directly from arrival time profiles (Blake et al. 1979, Aguirre et al. 1979) and indirectly from the shape of the muon radial distribution at ground level. For primary energies $<10^{15}$eV, measurements of air shower hadrons (Goodman et al. 1979) and of the high energy muons in air showers (Acharya et al. 1979) afford valuable information.

Possible constituents of the primary radiation at these energies range from γ-rays and neutrinos, through atomic nuclei of all reasonably abundant species, to dust grains containing 10^{10} nucleons or more. (Very high energy electrons are excluded because of synchrotron losses.)

Arguments for the presence, with detectable intensity, of very high energy γ-rays and neutrinos have been reviewed by Stecker (1973). Showers initiated by γ-rays are expected to be strongly deficient in muons. At energies of 10^{15}-10^{16}eV, where systematic searches have been carried out, the equal-energy abundance of γ-rays is so small ($\leq 2 \cdot 10^{-4}$) that they cannot be resolved with certainty from fluctuated nucleus-initiated showers (Firkowski et al. 1962, Toyoda et al. 1965). The average muon content of higher energy showers is consistent with the assumption that nearly all of them are nucleus-initiated, but an admixture of γ-ray showers up to ∿1% of the total cannot be ruled out at present.

The neutrino hypothesis (Berezinsky and Zatsepin 1969), that some or all of the largest air showers are produced by neutrinos, depends on the possibility that through continued increase with energy the cross section for inelastic neutrino-hadron collisions might become equal to the hadron-hadron cross section at cosmic ray energies ∿$3 \cdot 10^{21}$eV. Evidence against this hypothesis is afforded by the zenith angle distribution of large showers, which does not show an increasing proportion of very inclined showers ($\theta > 60°$) above 10^{18}eV. (Particles with a mean free path for shower initiation that is > the vertical depth of the atmosphere will be detected preferentially at large zenith angles for reasons of geometry.) This evidence can be strengthened by applying available measures of shower age (pulse risetime, lateral distribution) to neutrino candidates. (Neutrino showers will be observed preferentially as young showers, near maximum development, regardless of inclination.) However, these tests are statistical; it is impossible to say that any one of the observed large showers was not produced by a neutrino. Hence the 'resolving power' for separating primary neutrinos from nuclei becomes poor at the highest energies where the total number of events is small.

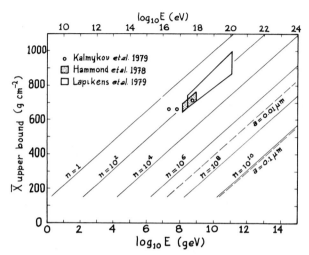

Fig. 8 Evidence that the primary particles above $\sim 10^{16}$ eV are atomic nuclei, not dust grains (Linsley 1980).

The dust grain hypothesis (Alfvén 1954, Hayakawa 1972), that the largest air showers are produced by relativistic dust grains, is disproven by evidence (Figure 8) that these showers require almost the entire thickness of the atmosphere for growth to maximum size. By the superposition principle, the depth of maximum development is determined by the energy per nucleon. For dust grains having the energy of these showers, the energy/nucleon would be $<10^{13}$ eV. Showers produced by such grains would reach maximum size high in the atmosphere and at sea level would consist almost entirely of muons.

The positive identification, at very high energies, of any of the 'exotic' constituents, γ-rays, neutrinos or dust grains, would be a discovery with important astrophysical consequences. Unaffected by magnetic fields, γ-rays or neutrinos might reveal the location and strength of powerful extragalactic cosmic ray sources. A discovery of neutrinos would automatically involve determining a fundamentally important interaction cross section. Dust grains, having a much larger ratio of mass to charge than nuclei, can reach the solar system from much greater distances, possibly from distant galaxies (Elenskiy and Suvorov 1977). Thus they might provide the most favorable means of detecting primordial antimatter, if it should exist. It is possible, using present data, to find an upper limit of the intensity of each of the exotic constituents as a function of energy, although this has not yet been done.

The mass spectrum of very high energy nuclei has a direct bearing on theories of cosmic ray acceleration, and in addition has great importance for interpreting features of the energy spectrum and the anisotropy. Methods of determining the mass of air shower primaries depend on one or another of the following three principles: 1) heavier nuclei have greater volumes, hence greater geometric cross sections, 2) heavier nuclei consist of a greater number of nucleons, hence the showers they produce are more regular in structure because of averaging, 3) the Lorentz factor of the primary particles, which for a given total energy is inversely proportional to the number of nucleons, determines the number of generations in the showers. This in turn determines the depth of maximum longitudinal development (optimum depth) and the proportion at ground level of low energy muons to electrons. The methods all make use of the superposition principle, which states that an average shower produced by a nucleus with energy E and mass number A is indistinguishable, except in early stages of development, from a superposition of A average proton-initiated showers each with energy E/A.

The result just shown in Figure 8 was obtained by the third method. The relation between optimum depth and primary mass is based on superposition and the so-called 'elongation rate (ER) theorem' (Linsley 1977). Another recent result having the same theoretical basis is one published by Thornton and Clay (1979). They claim that their measurements of optimum depth, derived from the pulse width of atmospheric Cerenkov signals, imply a change in primary composition between 10^{15} and $3 \cdot 10^{16}$ eV. Their data have been challenged on technical grounds by Orford and Turver (1979). However, it is pointed out by Linsley and Watson (1980) (see Figure 9) that the same conclusion reached by Thornton and Clay can be drawn from independent results by Antonov et al. (1979) using a technique to which the objections of Orford and Turver do not apply. The conclusion is that in this energy interval (located just past the 'knee' in the spectrum) the equal-energy primary composition changes from predominantly heavy to predominantly light. A more gradual change, from mixed composition (Table 1) to predominantly heavy (mostly Fe) in the energy interval 10^{12} to 10^{15} eV is implied by results of Grigorov et al. (1971) and Goodman et al. (1979).

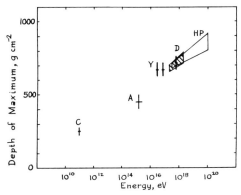

Fig. 9 Depth of maximum development of air showers, as a function of energy/particle. Sources are (Y) Kalmykov et al. (1979) and (D) Protheroe and Turver (1977). The point labeled (C) is calculated, using the known low energy composition and known properties of hadronic interactions. Point (A) is derived from results by Antonov and collaborators, reported in several publications since 1975. Point (HP) is derived from time profile measurements at Haverah Park. The derivation of points (A),(C) and (HP) will be described elsewhere (Linsley and Watson 1980).

The method illustrated by Figures 8 and 9 has the advantage that the relation between the optimum thickness and the average primary mass is model-dependent to only a small degree, and that the model-dependence is manageable by use

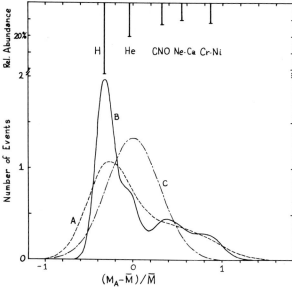

Fig. 10 Primary mass resolution attainable by measuring the muon content of large air showers ($N \sim 10^8$) at a moderately high elevation ($X \sim 850$ g cm^{-2}). M_A is the ratio of low energy muon flux to total flux at core distance 200m. $\overline{M}$ equals av(M_A). The line spacing (upper bar diagram) is derived from measurements of $\overline{M}$ vs shower size. The primary composition of Table 1 is assumed. Curve B includes 2 effects: bias in favour of lower-mass primaries, assuming that showers are selected using unshielded detectors, and line broadening due to fluctuations in shower development, estimated from simulations based on a range of hadronic cascade models. Curve A assumes an additional 30% reception fluctuations typical of early attempts to exploit this method (Linsley and Scarsi 1962, Toyoda et al. 1965). Curve C, representing the results of those attempts, is calculated assuming pure primary composition and 30% reception fluctuations.

of the ER theorem. However the resolving power is poor. The power to resolve, say, proton showers from those produced by alpha particles can be described in the same general terms used to describe methods for resolving neighboring isotopes at much lower energies; in terms, namely, of line width in relation to line separation. Figure 10 shows the theoretical resolving power afforded by one of the most promising methods. In order to achieve even this degree of resolution one will have to use shielded detectors with a very large combined area ($\sim 10^3$ m^2) distributed over an area of several km^2 at an altitude well above sea level.

ACKNOWLEDGEMENT

My thanks to Alan Watson, who was my guest during the writing of this, for help and good company. The work was supported by the National Science Foundation.

REFERENCES

Acharya, B.S., Rao, M.V.S., Sivaprasad, K. and Rao, Srikantha: 1979, Proc. 16th Int. Conf. Cosmic Rays, Kyoto, 8, pp. 312-317.

Aguirre, C., Anda, R., Trepp, A., Kakimoto, F., Mizumoto, Y., Suga, K., Inoue, N., Kawai, M., Izu, N., Kaneko, T., Yoshii, H., Nishi, K., Yamada, Y., Tajima, N., Nakatani, H., Gotoh, E., MacKeown, P.K., Murakami, K. and Toyoda, Y.: 1979, Proc. 16th Int. Conf. Cosmic Rays, Kyoto, 8, pp. 112-117.

Alfvén, H.: 1954, Tellus, 6, pp. 232-253.

Antonov, R.A. and Ivanenko, I.P.: 1975, Proc. 14th Int. Conf. Cosmic Rays, Munich, 8, pp. 2708-2713.

Antonov, R.A., Ivanenko, I.P. and Kuzmin, V.A.: 1979, Proc. 16th Int. Conf. Cosmic Rays, Kyoto, 9, pp. 263-268.

Bell, M.C., Kota, J. and Wolfendale, A.W.: 1974, J. Phys. A, 7, pp. 420-436.

Benko, G., Kecskemety, K., Kota, J., Somogyi, A.J. and Varga, A.: 1979, Proc. 16th Int. Conf. Cosmic Rays, Kyoto, 4, pp. 205-209.

Berezinsky, V.S. and Zatsepin, G.T.: 1969, Phys. Letters, 28B, pp. 423-424.

Bergeson, H.E., Cassiday, G.L., Chiu, T.-W., Cooper, D.A., Elbert, J.W., Loh, E.C., Steck, D., West, W.J., Linsley, J. and Mason, G.W.: 1977, Phys. Rev. Letters, 39, pp. 847-849.

Bergeson, H.E., Cutler, D.J., Davis, J.F. and Groom, D.E.: 1979, Proc. 16th Int. Conf. Cosmic Rays, Kyoto, 4, pp. 188-193.

Blake, P.R., Connor, P.J., Nash, W.F., Mann, D.M. and O'Connell, B.: 1979, Proc. 16th Int. Conf. Cosmic Rays, Kyoto, 8, pp. 82-87.

Bradt, H., Clark, G., La Pointe, M., Domingo, V., Escobar, I., Kamata, K., Murakami, K., Suga, K. and Toyoda, Y.: 1965, Proc. 9th Int. Conf. Cosmic Rays, London, 2, pp. 715-716.

Cassé, M. and Paul, J.A.: 1980, Astrophys. J., 237, pp. 236-243.

Cunningham, G., Lloyd-Evans, J., Pollock, A.M.T., Reid, R.J.O. and Watson, A.A.: 1980, Astrophys. J. (Letters), 236, pp. L71-L75.

Davies, S.T., Elliot, H. and Thambiapillai, T.: 1979, Proc. 16th Int. Conf. Cosmic Rays, Kyoto, 4, pp. 210-214.

Elenskiy, Ya.S. and Suvorov, A.L.: 1977, Astrofizika, 13, pp. 731-735 (trans. Astrophysics, 13, pp. 432-434).

England, C.D., Lapikens, J., Norwood, H.M., Reid, R.J.O. and Watson, A.A.: 1979, Proc. 16th Int. Conf. Cosmic Rays, Kyoto, 8, pp. 88-93.

Fenton, A.G. and Fenton, K.B.: 1976, Proc. Int. Cosmic Ray Symposium on High Energy Cosmic Ray Modulation, Tokyo, pp. 313-315.

Firkowski, R., Gawin, J., Zawadzki, A. and Maze, R.: 1962, J. Phys. Soc. Japan, 3, pp. 123-

Fomin, Yu.A. and Khristiansen, G.B.: 1971, Yadernaya Phys., 14, pp. 654-

Goodman, J.A., Ellsworth, R.W., Ito, A.S., MacFall, J.R., Siohan, F., Streitmatter, R.E., Tonwar, S.C., Vishwanath, P.R. and Yodh, G.B.: 1979, Phys. Rev. Letters, 42, pp. 854-857.

Ginzburg, V.L. and Syrovatskii, S.I.: 1964, "The Origin of Cosmic Rays" (Oxford: Pergamon).

Greisen, K.: 1956, in "Progress in Cosmic Ray Physics", ed. J.G. Wilson (New York: Interscience), 3, pp. 1-141.

Grigorov, N.L., Gubin, Yu.V., Rapoport, I.D., Savenkov, I.A., Yakovlev, B.M., Akimov, V.V. and Nestorov, V.E.: 1971, Proc. 12th Int. Conf. Cosmic Rays, Hobart, 5, pp. 1746-1751.

Hammond, R.T., Orford, K.J., Protheroe, R.J., Shearer, J.A.L., Turver, K.E., Waddoup, W.D. and Wellby, D.W.: 1978, Nuovo Cim. 1C, pp. 315-

Hayakawa, S.: 1972, Astrophys. Space Sci., 19, pp. 173-179.

Hillas, A.M.: 1975, Physics Reports, 20C, pp. 59-

Hillas, A.M.: 1979, Proc. 16th Int. Conf. Cosmic Rays, Kyoto, 8, pp. 7-12.

Kalmykov, N.N., Nechin, Yu.A., Prosin, V.V., Fomin, Yu.A., Khristiansen, G.B. and Berezhko, I.A.: 1979, Proc. 16th Int. Conf. Cosmic Rays, Kyoto, 9, pp. 73-78.

Karakula, S., Osborne, J.L., Roberts, E. and Tkaczyk, W.: 1972, J. Phys. A, 5, pp. 904-915.

Karakula, S., Osborne, J.L. and Wdowczyk, J., 1974, J. Phys. A, 7, pp. 437-443.

Kiraly, P., Kota, J., Osborne, J.L., Stapley, N.R. and Wolfendale, A.W.: 1979a, Proc. 16th Int. Conf. Cosmic Rays, Kyoto, 4, pp. 221-225.

Kiraly, P., Kota, J., Osborne, J.L., Stapley, N.R. and Wolfendale, A.W.: 1979b, Rivista del Nuovo Cim., 2, pp. 1-46.

Kirshner, R.: 1980, Nature, 284, pp. 597-598.

Krasilnikov, D.D.: 1979, Proc. 16th Int. Conf. Cosmic Rays, Kyoto, 8, pp. 26-31.

Lapikens, J., Lloyd-Evans, J., Pollock, A.M.T., Reid, R.J.O. and Watson, A.A.: 1979a, Proc. 16th Int. Conf. Cosmic Rays, Kyoto, 4, pp. 221-225.

Lapikens, J., Walker, R. and Watson, A.A.: 1979b, Proc. 16th Int. Conf. Cosmic Rays, Kyoto, 8, pp. 95-100.

La Pointe, M., Kamata, K. Gaebler, J., Escobar, I., Domingo, V., Suga, K., Murakami, K., Toyoda, Y., and Shibata, S.: 1968, Can. J. Phys., 46, pp. S68-S71.

Lloyd-Evans, J., Pollock, A.M.T. and Watson, A.A.: 1979, Proc. 16th Int. Conf. Cosmic Rays, Kyoto, 13, pp. 130-135.

Linsley, J. and Scarsi, L.: 1962, Phys. Rev. Letters, 9, pp. 123-125.

Linsley, J.: 1975, Proc. 14th Int. Conf. Cosmic Rays, Munich, 2, pp. 598-603.

Linsley, J. and Watson, A.A.: 1977, Proc. 15th Int. Conf. Cosmic Rays, Plovdiv, 12, pp. 203-208.

Linsley, J.: 1977, Proc. 15th Int. Conf. Cosmic Rays, Plovdiv, 12, pp. 89-96.

Linsley, J.: 1980, Astrophys. J. (Letters), 235, pp. L167-169.

Linsley, J. and Watson, A.A.: 1980, Phys. Rev. Letters, to be published.

McIvor, J.: 1977, M.N.R.A.S., 179, pp. 13-

Nikolskii, S.I.: 1962, Proc. 5th Interamerican Seminar on Cosmic Rays, 2, paper XLVIII, pp. 1-4.

Orford, K.J. and Turver, K.E.: 1980, Phys. Rev. Letters, 44, pp. 959-961.

Owens, A.J. and Jokipii, J.R.: 1977, Astrophys. J., 215, pp. 677-684.

Peters, B. and Westergaard, N.J.: 1977, Astrophys. Space Sci., 48, pp. 21-46.

Pollock, A.M.T., 1978, Ph.D. Thesis, University of Leeds.

Protheroe, R.J. and Turver, K.E.: 1977, Proc. 15th Int. Conf. Cosmic Rays, Plovdiv, 8, pp. 275-280.

Rasmussen, I.L.: 1974, in "Origin of Cosmic Rays", eds. J.L. Osborne and A.W. Wolfendale (Dordrecht: D. Reidel), pp. 97-133.

Sakakibara, S., Ueno, H., Fujimoto, K., Kondo, I. and Nagashima, K.: 1979, Proc. 16th Int. Conf. Cosmic Rays, Kyoto, 4, pp. 216-220.

Sreekantan, B.V.: 1972, Space Science Reviews, 14, pp. 103-174.

Stecker, F.: 1968, Phys. Rev. Letters, 14, pp. 1016-1018.

Stecker, F.: 1973, Astrophys. Space Sci., 20, pp. 47-57.

Thornton, G. and Clay, R.: 1979, Phys. Rev. Letters, 43, pp. 1622-1625.

Toyoda, Y., Suga, K., Murakami, K., Hasegawa, H., Shibata, S., Domingo, V., Escobar, I., Bradt, H., Clark, G. and La Pointe, M.: 1965, Proc. 9th Int. Conf. Cosmic Rays, London, 2, pp. 708-711.

Watson, A.A.: 1980a, Quart. J. Roy. Astron. Soc., 21, pp. 1-13.

Watson, A.A.: 1980b, private communication.

Wolfendale, A.W.: 1977, Proc. 15th Int. Conf. Cosmic Rays, Plovdiv, 10, pp. 235-251.

Zatsepin, G.T., Nikolskii, S.I. and Khristiansen, G.B.: 1963, Proc. 8th Int. Conf. Cosmic Rays, Jaipur, 4, pp. 100-123.

DIFFUSION OF HIGH ENERGY COSMIC RAYS FROM THE VIRGO CLUSTER

M. Giler,
University of Lodz, Poland
J. Wdowczyk,
Institute of Nuclear Research Lodz, Poland
A.W. Wolfendale,
University of Durham, U.K.

The origin of the highest energy cosmic rays is unclear but the observation that many are seen to arrive from high Galactic latitudes supports the idea of Extragalactic sources. The characteristic features to be explained by any origin model are: the flattening of the energy spectrum above 10^{19} eV and the apparent presence of a directional anisotropy which increases systematically with energy.

Here we propose a model in which the particles above about 10^{18} eV are protons which are generated in galaxies in clusters and which diffuse in a tangled intergalactic magnetic field. Diffusion from

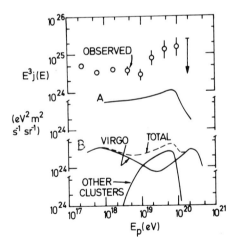

Figure 1. Primary cosmic ray spectrum. The observations are an average of data from Volcano Ranch, Yakutsk and Haverah Park (see Wdowczyk and Wolfendale, 1979 for references). The model predictions are: A Virgo cluster predominates, $D = 1.38 \times 10^{34} E_{19} \text{cm}^2 \text{s}^{-1}$
B Other clusters also contribute, $D = 3 \times 10^{35} E_{19}^{1.5} \text{cm}^2 \text{s}^{-1}$
(E_{19} is the proton energy in units of 10^{19} eV)

non-uniformly distributed sources in the presence of interactions with the 2.7K radiation then generate spectra of the right general shape. If the diffusion coefficient is $\sim 10^{34} cm^2 s^{-1}$ at $10^{19} eV$ then the VIRGO cluster predominates but if D is higher by one or two orders other clusters will also contribute.

Figure 1 shows the measured energy spectrum and that as predicted. The energy input required for the VIRGO cluster is $1.38 \times 10^{25} E_{19}^{-2.17} s^{-1} GeV^{-1}$ (i.e. $\simeq 6.5 \times 10^{44}$ erg s^{-1} above 1 GeV). Such a value is somewhat higher than appears to be the case for our own Galaxy when scaled up to the mass of the cluster; this problem is removed if the spectral slope for VIRGO is a little less or if unusual galaxies (unusual from the Cosmic Ray standpoint) exist in VIRGO (M87?). In any event it is necessary to assume that the Galaxy is not producing Cosmic Rays above 10^{18} eV at the present time.

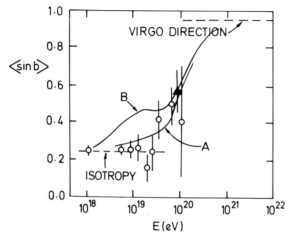

Figure 2. Anisotropy direction versus energy. The ordinate is the mean of the sine of the Galactic latitude of the arrival directions. O : Haverah Park expt. (Lloyd-Evans et al., 1979), Δ : Yakutsk (Krasilnikov, 1979). A and B as in Figure 1.

The models which fit the spectral shape also give a reasonable fit to the anisotropy results, although the latter may not be an accurate representation of the extragalactic anisotropy below about 10^{19} eV because of residual deflections by the Galactic field. Figure 2 shows the fit to the direction of the anisotropy; the fit to the amplitude is similarly reasonable.

References

Krasilnikov, D.D., 1979, Proc. 16th Int. Cosmic Ray. Conf. (Kyoto) 8, 26
Lloyd-Evans, J., Pollock, A.M.T., and Watson, A.A., 1979, Proc. 16th Int. Cosmic Ray Conf. (Kyoto) 13, 130.
Wdowczyk, J., and Wolfendale, A.W., 1979, Nature, 281, 5730, 356.

ON THE DETECTION OF HEAVY PRIMARIES ABOVE 10^{14} EV

T. K. Gaisser and Todor Stanev*
 Bartol Research Foundation of The Franklin Institute,
 University of Delaware, Newark, Delaware 19711   and
Phyllis Freier and C. Jake Waddington
 School of Physics and Astronomy, University of Minnesota,
 Minneapolis, Minnesota 55455

Knowledge of the chemical composition is fundamental to understanding the origin, acceleration and propagation of cosmic rays. At energies much above 10^{14} eV, however, the detection of single primary cosmic rays is at present impossible because of their low flux, and the only source of information is from the cascades initiated by energetic primary particles in the atmosphere--the extensive air showers (EAS). A similar situation exists for the study of hadronic interactions above 10^{15} eV. A recent EAS experiment (Goodman et al., 1979) suggests the possibility that the spectrum becomes increasingly rich in heavy nuclei as the total energy per nucleus approaches 10^{15} eV. Above that energy the overall spectrum steepens and the question of composition is almost completely open.

New air shower experiments, such as the University of Durham fast-timing Cerenkov array (Orford and Turver, 1976), the Soviet air Cerenkov experiment of Khristiansen et al. (1979), and the Utah Fly's Eye (Cassiday et al., 1978), show great promise for obtaining rather direct information both about the properties of hadronic cross sections and about composition at 10^{17}-10^{18} eV (and even higher in the case of Fly's Eye). These techniques map longitudinal development, including the early stages, of individual showers. Even for these experiments, however, interpretation is not straightforward because the actual beginning of the shower cannot be seen. In the case of Fly's Eye, for example, it has been proposed that the atmospheric depth at which the shower reaches 1/4 maximum ($y_{1/4}$) be used as a measure of the depth of shower initiation. Simulations will thus continue to be essential to interpretation of EAS experiments. Particularly important in this context is a proper treatment of nuclear breakup and especially of pion production in nucleus-nucleus collisions.

Following Dixon, Turver and Waddington (1974) our strategy is to use real interactions of nuclei at moderate energies (10 GeV/nucleon) as the basis of an extrapolation to EAS energies. We assume that the fragmentation probabilities are independent of interaction energy and depend only on the nature of projectile and target nuclei. In a Monte Carlo approach this enables us to choose fragmentation histories from the sample of real events. Direct use of the data is inadequate for pion production because

* On leave from INRNE, Sofia, Bulgaria.

of its known energy dependence. Instead we reduce each nucleus-nucleus collision to an equivalent number of nucleon-nucleon collisions. This enables one to use the energy-dependence implied by one's favorite model of nucleon-nucleon collisions. The number of nucleon-nucleon collisions is related to the type of fragmentation by giving each released nucleon a certain probability of interacting to produce pions. The fraction C_1 of unbound nucleons that interact can be estimated from data by comparing the multiplicity of charged mesons per released nucleon with the multiplicity in pp collisions in the appropriate energy range. We find C_1 decreasing from 0.4 for Z = 8-10 to 0.1 for Z = 26.

As an illustrative example of the use of these results we have compared the average and the standard deviation in depth of maximum (y_{max}) and in $y_{1/4}$ for the superposition model and for the model described above. In both cases the same Monte Carlo program was used to generate subshowers from individual nucleon-nucleon collisions. In the realistic model fluctuations in $y_{1/4}$ are about twice as big as for superposition and are comparable to the separation between $<y_{1/4}>$ for adjacent mass groups, a circumstance which may complicate the analysis of the experiments mentioned at the beginning.

These considerations are also relevant to experiments with thin calorimeters in which the primary charge can be measured directly but in which not all the energy is deposited in the calorimeter. Fluctuations in early cascade development due to nuclear breakup will contribute to fluctuations in the relation between E(total) and E(visible) for primaries with Z > 1. We have in mind the Japanese-American emulsion chamber collaboration (Jones et al., 1980).

We are grateful to the U. S. Department of energy (TKG), to the U.S. National Science Foundation (PF and CJW), and to NSF and the Bulgarian State Committee for the Promotion of Science and Technical Progress (TS and TKG) for partial support of this research, a more complete account of which is in preparation. Relation to previous work (Tomaszewski and Wdowczyk 1975) is discussed in an Appendix (unpublished).

References

Cassiday, G. L., et al.: 1978, Cosmic Rays and Particle Physics-1978 (Proc. Bartol Conf., ed. T. K. Gaisser) AIP Conf. Proc. #49, p. 417.
Dixon, H. E., Turver, K. E., and Waddington, C. J.: 1974, Proc. Roy. Soc. A339, 157.
Goodman, J. et al.: 1979, Phys. Rev. Letters 42, 854.
Jones, W. V., et al.: 1980, Bull. Am. Phys. Soc. 25, 546. See also Dake, S. et al.: 1979, Proc. 16th Int. Cosmic Ray Conf. (Kyoto) 6, 330.
Khristiansen, G. B.: 1979, Proc. 16th Int. Cosmic Ray Conf. (Kyoto) 14, 360 and references therein.
Orford, K., and Turver, K. E.: 1976, Nature 264, 727; see also Hammond, R. T., et al., N.C. 1c, 315 (1978).
Tomaszewski, A. and Wdowczyk, J.: 1975, Proc. 14th Int. Cosmic Ray Conf. (Munich) 8, 2899.

INTERSTELLAR AND INTRACLUSTER TUNNELS AND ACCELERATION OF HIGH-ENERGY COSMIC RAYS

A. Ferrari, Max Planck Institut für Extraterrestrische Physik, Garching, Germany, and Istituto di Cosmogeofisica del CNR, Torino, Italy

A. Masani, Istituto di Fisica Generale, Torino, Italy

The present evidence about the origin of high-energy cosmic rays is that two ranges exist: one below 10^{18} eV related with galactic sources, and one up to 10^{20} eV, corresponding to extragalactic processes (Király et al. 1979). However the continuity of the spectrum indicates that the physical mechanisms must be correlated. In the global energetics the spectral range above 10^{12} eV is irrelevant and the bulk of CR energy is actually provided by supernovae, pulsars, X-ray binaries, etc. in the Galaxy. Nevertheless none of these objects seems capable of producing CR above 10^{15} eV/nucleon. We have investigated the possibility that the acceleration at higher energies is statistical, taking place over a hierarchy of scales. A model has been developed in terms of the quasi-linear theory of particle acceleration by MHD turbulence (Kulsrud and Ferrari 1971). Cosmic rays with $\lesssim 10^{15}$ eV/nucleon injected by single sources into interstellar space undergo momentum diffusion by stochastic interaction with long wavelength MHD perturbations; small wavelength modes provide pitch-angle scatterings. These MHD perturbations, Alfvèn and fast magnetosonic waves, are generated by the dynamic interaction of supernova remnants with the interstellar medium. From observations (Jokipii 1977), the range of possible wavelengths extends from the proton gyroradius to the size of the so-called "superbubbles", up to 100 pc, with a power-law spectrum. Correspondingly acceleration is efficient up to 10^{18} eV/nucleon; in fact for $B = 10^{-3} \div 10^{-5}$ G, $n = 0.01 \div 1$ cm^{-3} and $L = 0.1 \div 100$ pc, we find that the acceleration timescale

$$\tau_{acc} = (B/\delta B)^2 (c/v_A)^2 L/c (\varepsilon/eBL)^{4-\alpha}$$

($\varepsilon = eBL$ is the maximum allowed energy) is always shorter than the time scales of losses and turbulent structure lifetimes; $\alpha = 3.5 \div 4$ is the spectral index of turbulent modes. Contrary to the original Fermi mechanism, in this theory the time scale for acceleration up to any energy ε is fixed by the final phase, previous steps being negligible.

A similar scenario can be envisaged in the extragalactic range, simply assuming that galaxies, especially those with active nuclei, are the localized sources injecting particles with energies $\varepsilon \lesssim 10^{18}$ eV. Magnetic inhomogeneities can be produced by: (1) protruding SN cavities, (2) galactic winds, (3) galactic wakes due to Kelvin-Helmholtz instabilities for the relative motion in the intergalactic medium. It is not possible at present to measure a spectrum of MHD perturbations in IGM and their filling factor in clusters is also difficult to estimate; data will come from radio and X-ray observations. For instance we already know that in the Abell cluster 1367 a 300 kpc radio halo has been detected (Gavazzi 1978) neither connected with radio galaxies, nor with the cluster centre. It likely indicates the occurrence of particle acceleration in IGM.

For the acceleration to energies above 10^{18} eV, a rate of energy supply $\sim 10^{41}$ erg/s is required in a volume equal to the Virgo cluster. Referring for instance to the case of galactic wakes, the energy released in turbulence by a streaming galaxy can be estimated to be

$$\dot{E}_g = 10^{44} \eta \, (M_{11} \, v_{g,8}^3 \, D_{24}^{-1}) \text{ erg/s},$$

assuming a typical dispersion velocity $v_g = 10^{3.}$ km/s and a stopping distance $D = 1$ Mpc; η is an efficiency factor. With 10^3 galaxies in a cluster, the energetic supply is quite adequate (this is also true for not rich clusters as our Local Group). Eventually in a stationary situation the power-law spectrum should not differ from that measured in the ISM. Acceleration of stochastic origin of particles leaking from galaxies is then possible.

Scattering from MHD inhomogeneities has been similarly used by Wdoczyk and Wolfendale (1979) in a model of CR propagation in the Virgo cluster.

The acceleration time scale must be here confronted essentially with that for γp losses against the cosmic background radiation. For $B \cong 10^{-6}$ G, $n \cong 10^{-6}$ cm^{-3}, $L = 0.1 \div 1$ Mpc, acceleration is faster up to a few times 10^{20} eV/nucleon. This continuous acceleration allows also to explain the flattening of the energy spectrum above 10^{19} eV as a piling up of CR when balance is reached between stochastic acceleration and γp losses; at higher energies the spectrum should then drop rapidly. An alternative explanation of this flattening, also compatible with our model, has been forwarded by Wdoczyk and Wolfendale in terms of propagation effects in turbulent IGM.

Finally we find that in this scenario measurements of arrival direction anisotropies in the range $10^{16} \div 10^{17}$ eV can be fitted with the large scale structures of the ISM; correspondingly anisotropies above 10^{19} eV could provide an indirect tool for studying the structure of IGM in our Local Supercluster.

References

Gavazzi G.: 1978, Astron. Astroph. <u>69</u>, 355
Jokipii J.R.: 1977, Proc. 14th Int. Cosmi Ray Conf. (München), <u>6</u>, 2
Király et al.: 1979, Rivista Nuovo Cimento <u>2</u>, 1
Kulsrud R.M. and Ferrari A.: 1971, Astroph. Space Sci. <u>12</u>, 302
Wdoczyk J. and Wolfendale A.W.: 1979, Nature <u>281</u>, 356

FEATURES OF THE HIGH ENERGY ELECTRON SPECTRUM

J. Nishimura
Institute of Space and Aeronautical Science,
University of Tokyo, Tokyo, Japan.

Recent studies of the spectrum of high energy primary electrons using emulsion chambers have been made with an exposure factor some 10 to 100 times larger than those obtained by other experimental devices. (Taira et al., 1979, Nishimura et al., 1980). The total exposure is now almost 6 m^2-day-str., including the recent exposure at Palestine, Texas in May 1980, and it is now quite possible to extend observations of the electron spectrum into the TeV region with reliable accuracy in the next few years.

Because high energy electrons lose their energy rapidly through synchrotron radiation and inverse Compton processes, an observed high energy electron could not have traversed far from its origin. Regarding this point, several works were presented at the time of Kyoto Conference in 1979. (Giler et al., 1978, 1979, Nishimura et al., 1979, Webber et al., 1979).

One is the effect of a small number of discrete sources which could contribute to the high energy electrons, assuming that supernovae are the sources of cosmic-rays. Assuming that supernovae occur in the Galaxy at a rate of one per 30-100 years, one would expect only several supernovae to be the sources of observed electrons in the TeV energy range. Thus the electron intensity above 1 TeV will show large fluctuations from a smooth power law behaviour due to the small number of discrete sources which are capable of contributing to the observed flux. We would therefore expect to observe humps in the spectrum correlating to the individual sources. Results of some detailed calculations assuming the random occurrence of supernovae and taking the frequency mentioned above are shown in Ref. I.

Other aspects of the analysis of features of the high energy electron spectrum is the effect of the deficiency of short path lengths which was obtained by an analysis of the composition of heavy primaries in cosmic-rays. The results have demonstrated that the spectrum is depressed quite appreciably beyond 100 GeV. Such results seem to be difficult to reconcile with the data near 1 TeV observed by our group, and throw doubts on the likelihood of electrons and heavy primaries being produced in the same sources.

The path length distribution is known to be energy dependent. These works have been carried out by assuming that the path length distribution is the same at all energies. Here we mention that the effect is highly dependent on the model of propagation. As an example, in the case of double leaky box model (nested leaky box model), the deficiency of short path lengths is caused near the source, and the corresponding energy is higher than that of the observed electrons. This causes the depression to be smaller than that calculated by the previous authors. Furthermore, the leakage probability from the source region is higher for higher energy electrons, and this also increases the flux of electrons at high energies.

The results of calculation taking account of these factors are shown in Fig. 1. This demonstrates that the effect of the deficiency of the short path lengths is highly dependent on the propagation model, and could yield agreement with the observed data in the several hundred GeV region for a certain model of propagation of cosmic-rays.

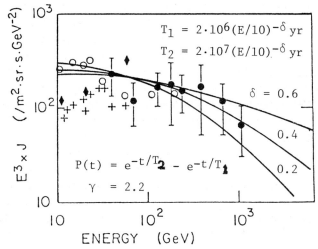

Fig. 1. Calculation with double leaky box model

References

Giler, M., Wdowczyk, J., and Wolfendale, A.W.: 1978, J.Phys., G. $\underline{4}$, 269 1979, Proc. Int. Cosmic Ray Conf., Kyoto, 507.
Nishimura, J., Fujii, M., and Taira, T.: 1979, Proc. Int. Cosmic Ray Conf. Kyoto, 488, (referred to as Ref. I).
Nishimura, J., et al.: 1980 Ap. J. (in press).
Taira, T., et al.,: 1979, Proc. Int. Cosmic Ray Conf., Kyoto, 478.
Webber, W.R., Goeman, R.A., and Tushak, S.M., 1979, Proc. Int. Cosmic Ray Conf. Kyoto, 495.

ULTRA HEAVY COSMIC RAYS

P. H. Fowler, M. R. W. Masheder, R. T. Moses, R. N. F. Walker
and A. Worley
Department of Physics, University of Bristol

INTRODUCTION

The existence in the cosmic rays of ultra-heavy (UH) nuclei with $Z \gtrsim 30$ was established by two separate experiments in 1966. Fleischer et al.[1] first demonstrated the fossil tracks of such nuclei in certain meteoritic crystals and shortly afterwards Fowler[2] established their existence in present-day cosmic rays with the detection of their tracks in photographic emulsion which had been exposed during a high altitude balloon flight. The fluxes of such nuclei are very low, only $\sim 10^{-4}$ of that of iron, and the most suitable method of detection to date has been the analysis of the tracks formed by these particles in very large (several m^2) arrays of plastic detector material, notably Lexan polycarbonate. Such exposures on balloons and on Skylab[3] have provided practically all present knowledge of the UH cosmic rays. Unfortunately, the charge resolution obtained was disappointing, even though scrupulous care was taken in the handling and etching of the material, and the charge scale itself of necessity had to be based on a considerable extrapolation from the iron peak and could not be used with great confidence. The situation now, however, is in the process of being transformed. We have two satellite experiments devoted to the study of UH cosmic rays and in operation at the moment. These are the Bristol University experiment on Ariel 6 launched on 3rd June 1979 and the joint group under Israel, Waddington and Stone on HEAO-C launched in September 1979. It is therefore appropriate, I believe, if I devote this review to the new preliminary results and a comparison of this material with the published data.

THE BRISTOL ARIEL VI EXPERIMENT

The Ariel VI project offered the opportunity of a two-year exposure in orbit of a comparatively small counter experiment. The satellite was launched by NASA on the 100th Scout launch vehicle on 3rd June 1979 from Wallops Is., Virginia into a near-circular 625 km orbit inclined at 55°. The salient features of the detector, which is novel in a number of ways, are shown in Fig. 1. A spherical

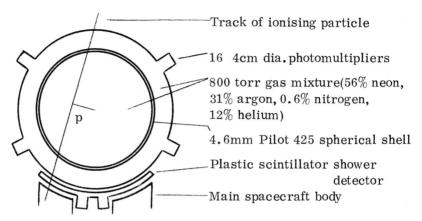

Fig. 1. Schematic cross-section of Ariel VI cosmic ray detector.

vessel of diameter 75cm contains a gas scintillation mixture and a thin spherical shell of Pilot 425 plastic, and forms a single optical cavity viewed by sixteen photomultipliers. The spherical symmetry of the detector has three significant consequences. It enables the detection of particles over the full 4π steradians, although naturally in close Earth orbit the aperture is restricted, to about 8.5 steradians in this case. Secondly, the track geometry is characterised by a single quantity, the impact parameter p. Finally, the acceptance of all angles of incidence brings the important complication that the photomultipliers must of necessity intercept a fraction of the incoming particles; it is essential to remove such events from the data.

The passage of a typical cosmic ray nucleus results in the emission of Cerenkov radiation from each transit of the shell of Pilot 425 and scintillation from the whole path of the particle in the gas, which fills the space both inside and outside the shell. Both of these processes are well understood and the response of the detector should be accurately proportional to the square of the particle charge, except, perhaps, for the very highest charges along whose tracks electron recombination with the positive ion column can increase the scintillation yield. This effect of high ion density along the track, readily observable under laboratory conditions with fission fragments, will be discussed more fully in a later publication. The determination of the particle charge relies upon the estimation of each of the two components from the photomultiplier outputs. This can be achieved because the two components are emitted over markedly different timescales.

The gas is a mixture of argon, nitrogen, neon and helium with partial pressures as indicated in Fig. 1. The helium, included to facilitate ground leak tests, does not contribute materially to the light output. The scintillation consists largely of the band spectrum of nitrogen in the near ultra-violet, the nitrogen being activated by excited argon states formed by the cosmic ray particle. Further argon excitations are contributed via excited neon states also formed by the cosmic rays. Both the scintillation and the Cerenkov emission fall largely in the absorption band of the Pilot 425 and are wavelength shifted to ~ 425nm and reradiated promptly and isotropically. This process is crucial in ensuring the near-homogeneous distribution of light within the cavity. The trapping within the shell by total internal reflection of the reradiated light is prevented by a lightly frosted finish on the outer surface. Any blue light has a lifetime in the optical cavity of ~ 20ns, which therefore dominates over the other timescales in the case of the Cerenkov emission. In contrast, the mean duration of the scintillation from the gas is ~ 300ns, which is largely due to the time for the excitation transfer from argon to nitrogen. The difference between these two timescales is significant and enables the proportions of each component in the composite Cerenkov/scintillation photomultiplier pulse to be determined. Measurements are made of the total light output and that part of the output received during a 400ns gate which opens 80ns after receipt of the fast leading edge of the Cerenkov pulse, and therefore contains little contribution from Cerenkov radiation.

In addition to the main sphere there is a plastic scintillator detector lying between the sphere and the main spacecraft body whose prime function is to flag those events with trajectories that pass through the spacecraft. This is necessary because an electron shower containing, say, N electrons can simulate the pulse due to a nucleus of charge $\sqrt{N}$, and can only develop when the products of a high energy nuclear interaction pass through several radiation lengths of matter. A study of the spacecraft had revealed several areas containing significant paths of high-Z materials, in particular from the Ni-Cd battery and the considerable numbers of tantalum capacitors arranged in rows.

ESTIMATION OF CHARGE FOR INDIVIDUAL COSMIC RAYS

It is well known that a single pair of Cerenkov and scintillation measurements do not suffice for the individual determinations of Z and β, due to the double-valued nature of the Cerenkov/scintillation ratio. This occurs as a result of the relativistic rise in ionization at high energies. For the range of velocities up to $\beta c \sim 0.99$, however, which contains the major fraction of all events, the ratio is a monotonically increasing function of β and a solution is possible. In this velocity range the determination of Z, β and p would therefore appear to require three independent measurements. The total and delayed signals introduced above represent the only two measurements available, yet a solution is possible in this detector since appropriate dimensions have been chosen for the inner and outer radii of the plastic shell so that β and p behave as far as possible as a single parameter. The major part of the uncertainty in the estimation of Z is not in the mismatching of the well-populated parts of the curves, but in the ambiguity associated with the relativistic rise. Thus the addition of a further optical cavity to make individual β and p determinations possible would not have addressed itself to the major part of the problem. In any case, this course of action would have complicated this essentially simple experiment to an unacceptable degree so that it could not have matched the flight opportunity that was offered.

RESULTS

We are reporting the results of our analysis of ~ 300 days of real time and recorded data. The events on the experimental high priority store (HPS) have not yet been adequately evaluated but, due to irregularities in the tape recorder playback and command systems, should eventually provide a small amount of additional data. The live time that we cover here corresponds to an exposure of ~ 400 m^2sr days for most events, an exposure of the same order as that of the Skylab plastic experiment. Our experiment has the advantage that it covers the full charge band Z > 32 with nearly uniform efficiency. The choice of satellite altitude and orbital inclination has the result of exposing the detector to various regions where high fluxes of trapped radiation are found. The region of the South Atlantic Anomaly, where as expected the high flux of background protons severely disturbs the cosmic ray charge estimates, is excluded from all analyses. Occasional effects have also been observed in the southern auroral belt.

Charge estimates for all events are made on the basis of comparison of the total and delayed signals with those of iron nuclei. The Z = 26 calibration is established for each day's data using about 10,000 iron peak events. No more frequent calibration is necessary, since the performance of all components of the experiment appears to be remarkably unchanging. The gas scintillation output, for example, has decayed by only 3% over the period covered here, and shows only a transient 7% fall during the $14°C$ temperature rise typically experienced by the detector sphere during an all-Sun period. This stability may be partly due to the low temperature ($-18°C$) at which the sphere is usually maintained.

The behaviour around and below the iron peak enables one to judge the resolution and interpret the high charge data. In any charge spectrum with good statistics the outstanding features are the skewness of the Fe and Ni peak and the dramatic fall in abundance of ~ 10^4:1 between Fe and Z = 32. The relativistic rise at high energy produces the exponential tail to high Z and causes the Ni peak to appear only as a shoulder. Zn, with its much lower abundance of ~ $6 \; 10^{-4}$ of Fe[4], is smothered almost completely. The coefficient of the exponential tail is consistent with that expected for a particle energy spectrum of the form $N(>E) \propto E^{-1.5}$ and the variation of the energy loss in the gas with particle energy. The values of ΔZ due to this effect are proportional to Z and so dominate at the higher charges. The tail would still be serious even if the Poisson fluctuations were much reduced, and one would only gain considerably in resolution if the high energy particles could be individually recognised with certainty.

Fig. 2 displays the charge spectrum for events with apparent charge $32 < Z_{app} < 50$ using the cut off restriction, cut off $\geq$ 3GV. An odd-even effect

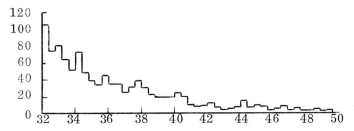

Fig. 2. Histogram of Z_{app} for $32 < Z < 50$ and cutoff $> 3GV$.

is apparent throughout. The overall fall with increasing Z_{app} is enhanced by the tail from Ni and Zn. An odd-even effect is expected on the basis of our resolving power as determined on Fe, and indicates that these even nuclei are more common than their odd-numbered neighbours. However, significant abundance values of $_{35}$Br and $_{37}$Rb must be expected as they each have two stable or long-lived isotopes. Particularly prominent and therefore relatively abundant are $_{38}$Sr and $_{44}$Ru. Analysis of these measurements is considered in the next section.

A problem for the lower charges in this band results from the effects of the priority threshold, which was centred between 34.0 and 36.5 for practically all of the data. Only for $Z_{app} > 37.5$ are events always given the highest priority whatever their geometry or velocity, and hence only for such particles is the recording efficiency unity throughout. For lower charges the recording efficiency is a strong function of both its charge and the instantaneous flux of nuclei with $Z \geq 20$.

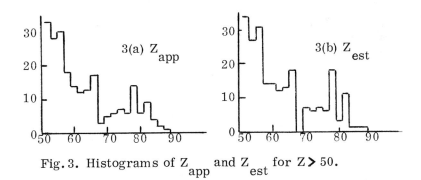

Fig. 3. Histograms of Z_{app} and Z_{est} for $Z > 50$.

The most interesting results, as expected, come in the highest charge band. Fig. 3a displays the histogram of apparent charge plotted in bins of width $\Delta Z_{app} = 2.0$ charges, a figure we consider appropriate to the errors of charge assignment which are largely due to the effects of high energy particles. In Fig. 3b the events have been rebinned to produce the histogram for our best estimate Z_{est} of the actual charge of the particles by removing the effects of the relativistic rise using the assumption that the energy spectrum of these nuclei is similar to that of Fe and Ni[5]. The ten nuclei in Fig. 3a with apparent charges $84 \leqslant Z_{app} \leqslant 88$ are consistent with the tail expected from $_{82}$Pb and $_{78}$Pt, and only two nuclei with $Z_{est} > 88$ remain, with charges of 98 and 114. The major part of the discussion of these results will be presented after the charge spectrum at the cosmic ray source has been calculated. It will be useful, however, to compare directly the measured spectrum of Fig. 3b with those of other experiments, and we consider the following features, present both in Figs. 3a and 3b, significant:-

  (a) very few events with $Z_{est} > 90$
  (b) the prominent peak at $Z_{est} \sim 78$
  (c) the low abundance of nuclei with $Z_{est} \sim 70$
  (d) a feature at $62 \leqslant Z_{est} \leqslant 66$, based on ~ 70 observed events
  (e) the expected abundance feature at $52 \leqslant Z_{est} \leqslant 56$.

TABLE 1

Charge	Present Experiment		Balloon Data (Top of Atmos.)		Skylab	
26	10^6		10^6		10^6	
50	6.0 ± 1.2		*7.5 ± 2.3			
52	6.8 ± 1.3		*6.2 ± 2.0			
54	5.4 ± 1.2		*5.5 ± 1.7			
56	6.2 ± 1.2		*5.4 ± 1.7			
58	2.8 ± 0.9	5.6 ± 1.2	*5.8 ± 1.7	10.6 ± 2.1		
60	2.8 ± 0.9		*4.8 ± 1.4			
62	2.4 ± 0.9		*3.1 ± 1.0			
64	2.6 ± 0.9	8.6 ± 1.3	*1.8 ± 0.7	6.1 ± 1.2	*3.0 ± 1.0	
66	3.6 ± 1.0		*1.2 ± 0.5		*2.7 ± 0.9	
68	−0.4 ± 0.8		0.6 ± 0.4		*0.8 ± 0.4	
70	1.4 ± 0.6	2.2 ± 0.8	0.8 ± 0.3	2.4 ± 0.6	*1.6 ± 0.5	3.3 ± 0.8
72	1.2 ± 0.6		1.0 ± 0.3		*0.9 ± 0.4	
74	1.4 ± 0.6		1.1 ± 0.3		0.9 ± 0.3	
76	1.2 ± 0.6	6.3 ± 1.1	1.5 ± 0.4	4.4 ± 0.8	1.6 ± 0.4	3.8 ± 0.6
78	3.7 ± 1.1		1.8 ± 0.5		1.3 ± 0.4	
80	0.6 ± 1.0		1.3 ± 0.4		0.9 ± 0.3	
82	2.2 ± 0.8	3.1 ± 0.9	0.9 ± 0.4	3.0 ± 0.6	0.6 ± 0.3	2.3 ± 0.5
84	0.3 ± 0.4		0.8 ± 0.3		0.8 ± 0.3	
86	0.2 ± 0.3		0.5 ± 0.3		0.2 ± 0.2	
88	0.2 ± 0.2	0.8 ± 0.5	0.3 ± 0.2	1.9 ± 0.5	0	0.8 ± 0.25
90	0.4 ± 0.3		1.1 ± 0.4		0.6 ± 0.3	

Based on 178 events $50 < Z < 65$ 84 events $50 < Z < 65$ 104 events $Z > 65$
and   82 events $Z > 65$ 96 events $Z > 65$

Table 1 compares our data with those of previous experiments. Asterisks indicate the figure was subject to significant and rapidly charge-dependent corrections. Numbers using wider charge bins are also given because of the correlated errors in our data.

The spectrum of Fig. 3b can be compared directly with that from Skylab and with balloon data as is shown in Table 1. Points (b),(c) and (d) above are manifest in both of these other sets of data, but the Skylab estimate for $Z \sim 63$ was considered by the authors to be uncertain, owing to the rapidly changing efficiencies near the threshold of the plastic, Lexan. The agreement thus supports the Lexan charge scale employed in the Skylab and balloon experiments.

THE CHARGE SPECTRUM AT THE SOURCES

A particularly simple form of the 'leaky-box' model can explain many abundance features for $Z \lesssim 26$ very adequately; it is appropriate to see whether such a model may apply to the UH cosmic rays as well. For this model we suppose that the sources of the cosmic rays are essentially uniformly distributed and that the cosmic rays themselves are contained in a volume that contains interstellar hydrogen. We ignore energy loss by ionisation and take a value for the escape mean free path $X_o = 5 \text{gcm}^{-2}$. In Table 2 we give the detected spectrum and the computed secondary and source spectra. Also shown is the abundance spectrum for the solar system from Cameron, 1973[6]. The tabulation shows that, with the exception of the highest charges with $Z \gtrsim 78$, a substantial fraction of observed UH nuclei are in fact secondaries from interactions in interstellar space. This unfortunately hinders the detection of source features, as the proportions of true primaries are rather small. This arises because the interaction lengths involved fall in the range $1 \lesssim \lambda_z \lesssim 2 \text{gcm}^{-2}$ compared with the adopted value for the escape mean free path, $X_o = 5 \text{gcm}^{-2}$. Although any lower value of X_o would be equally consistent with the UH data, higher values would result, for some charges, in negative source abundances.

DISCUSSION

A natural charge spectrum with which to compare the cosmic ray results, and thereby hopefully gain insight into the conditions at the sources, is the solar system composition of Cameron. Earlier authors[7,8] have made such a comparison for UH cosmic ray plastic track data and have concluded that the source appears to be enriched in the material which has undergone nucleosynthesis by the rapid neutron capture process. The apparent flux of actinides ($Z \gtrsim 90$), which can only be made by such an r-process, gave some support to this view. More recently, however, it has been suggested[9] that, at least for the elements $3 \lesssim Z \lesssim 26$, the cosmic ray source spectrum could be simply explained as a sample of normal interstellar material, modified only by the effects of differing ionization potentials for the various elements. The solar system composition is taken, in the absence of better data, as the required interstellar composition. In the light of our new data, we should consider whether this simple and attractive model could also apply to the UH nuclei, or whether enrichment in say r-process material is needed to explain the data.

TABLE 2

Charge	Present Experiment						Cameron Solar System Abundances
	N_{app} (a)	N_{app} (b)	N_{est}	N_{sec}	N_{source}		
26	2.4×10^6	5.0×10^6	5×10^6	5×10^6	10^6		10^6
32				290			170
34	315		590	220	88 ± 12		92
36	183		330	144	46 ± 8		68
38	156		315	97	55 ± 8		39
40	95		180	60	32 ± 7		37
42	43		66	48	5 ± 5		5.6
44	44		94	40	15 ± 5	} 26 ± 4	2.5
46	33		63	25	11 ± 4		2.1
48	21		37	22	4.5 ± 4	} 4.8 ± 3	2.2
50	16		30	29	0.3 ± 4		4.6
52		33	34	17	5.6 ± 3.1		8.5
54		28	27	17	3.2 ± 2.6	} 13.8 ± 3.9	7.3
56		30	31	16	5.0 ± 2.6		6.2
58		18	14	6	2.8 ± 2.0	} 5.0 ± 1.6	1.8
60		14	14	8	2.2 ± 1.7		1.03
62		12	12	10	0.7 ± 1.7		0.31
64		13	13	9	1.4 ± 1.7	} 6.2 ± 1.9	0.44
66		17	18	5	4.1 ± 1.8		0.52
68		3	-2	4	-1.6 ± 1.5		0.34
70		5	7		4.6 0.6 ± 1.6	} < 2.0	0.30
72		6	6		5.4 0.3 ± 1.6		0.30
74		7	7		3.2 1.5 ± 1.6		0.23
76		6	6		4.5 0.6 ± 1.6	} 8.6 ± 2.7	1.36
78		14	18		2.7 6.5 ± 2.4		2.2
80		6	3		1.9 0.4 ± 1.6		0.7
82		9	11		0.7 4.4 ± 2.0	} 5.7 ± 1.8	5.0
84		4	1		0.1 0.5 ± 1.2		0.1
86		2	1		– 0.5 ± 0.8		–
88		1	1		– 0.5 ± 0.5	} 2.0 ± 1.2	–
≥ 90		2	2		– 1.0 ± 0.7		0.10

The figures listed in columns (a) and (b) for N_{app} refer to the actual numbers of events used for the determination of the charge spectrum. Column (a) uses data for geomagnetic cutoff values greater than 3.0GV only. Column (b) uses all data. The errors shown are due to Poisson statistics alone and are quite strongly negatively correlated in adjacent charge bins. Thus wider bins are used at higher charges to display the features of the source spectrum.

The comparison between the source spectrum and the solar system abundances may be summarised as follows:

(a) the UH component as a whole has an abundance, when normalised to iron, close to that of the solar system. This has long been apparent and is again present in the results of this experiment. The total fluxes in the bands of charge $34 < Z < 50$ and $Z > 50$ each fall within a few percent of the solar system figures, in the latter case being based on 230 observed events. The low fluxes we observe in the actinide and $Z \sim 70$ regions are also what would be expected if the two spectra had a common origin. The 18 events around $Z \sim 70$ in the raw data appear to be spallation products.

(b) the main features of the solar system composition are the two peaks around $Z = 54$ and $Z = 80$ corresponding to neutron closed shells at production with $N = 82$ and 126, respectively. We would expect, on the basis of the model, to see these in the cosmic ray source and indeed, they are both clearly present. The ratio between the abundances in the two peaks, however, differs markedly from the solar system value, showing a preference for the higher charges. For the ratio $\frac{52 \leq Z \leq 56}{74 \leq Z \leq 84}$ we obtain 0.95 ± 0.3 compared with the solar system value of 2.5.

(c) the solar system abundances would lead us to expect a particular distribution within each peak; in the case of the lower peak, a fairly uniform distribution, and for the peak around $Z = 80$, a preference for lead ($Z = 82$) rather than platinum ($Z = 78$). Our results indicate that the upper peak is predominantly platinum. The Pt/Pb ratio of 1.5 ± 0.7, which has survived nearly unchanged from the raw data, is to be compared with the solar system value of ~ 0.45. In the case of the lower peak we can present no evidence that the distribution is other than similar to that of the solar system.

(d) the most striking abundance anomalies occur at charges $Z \sim 44$ and $Z \sim 64$. In each region, the cosmic rays are approximately five times overabundant when compared with the solar system.

The overabundances at $Z \sim 44$, 64 and 78 are not explainable with the simple model with solar system abundance values. They can, however, all be considered as supporting the evidence for an enrichment in r-process nuclei first suggested by the passive data. This interpretation of the anomalies at $Z \sim 44$ and 64 requires elaboration. It is noteworthy that both of these charge regions, which correspond to two mass regions around $A=105$ and $A=164$, are the sites of small solar system abundance anomalies. A number of authors have advanced explanations for these solar system features in terms of contributions from fission fragments. The feature around $A=105$ is interpreted by Ohnishi[10] as due to the lighter fission fragments from β-induced and spontaneous fission in the source during the final phase of r-process nucleosynthesis. The parent nuclei have a wide range of masses around $A=250$, but since the mass of the lighter fragment is very nearly independent of that of its parent the feature at $A \sim 105$ remains relatively sharp. For the heavier peak centred on $A=164$, which is both broader and more pronounced, Steinberg and Wilkins[11] invoke fission of superheavy elements with A ~ 300. This is a higher mass than would be associated with the closed shell and abundance peak with N=184, Z=94 during the r-process, because these authors consider that such nuclei would tend to decay to two fragments with mass ratio close to unity, rather than the familiar asymmetric decay which occurs at other masses. The observed feature is produced by the more massive fragments from fission of a relatively narrow band of parent masses.

In the cosmic ray source each of these peaks is overabundant and approaches 10^{-5} of the flux of iron nuclei, a figure comparable with the main peak at Z ~ 78. Thus, if explained in terms of fission, the source abundance of the transuranic species would need to be at least of this magnitude. Such a figure is consistent with models involving a cyclic r-process in which the number of neutrons per seed nucleus and the r-process duration are together sufficient not only to raise the masses of the seed nuclei to high values, at which point they undergo fission, but also to allow their daughter fragments to repeat the process. In such models (see, for example[12] and [13]) the peak centred at N=184 accumulates an abundance comparable with that of the peak at N=126, which produced the platinum feature. The accompanying amount of material with A ~ 300 synthesized in the same process is thought to be small, and thus any explanation of our abundance anomalies in terms of a significant contribution from nuclei with A ~ 300 would raise the problem of accounting for the presumed even higher abundance of nuclei with A ~ 280. The combined abundance of Xe, Ba and Ce enables an upper limit of ~ 1.5 10^{-5} of Fe to be placed upon the size of the abundance peak at N=184, A=280 in such circumstances, if all these nuclei are ascribed to fission fragments this figure is little greater than that suggested by the rare earth abundance feature. It is thus perhaps attractive to consider the N=184 abundance peak itself as the source of the main anomalies, the original mass of 280 being distributed between two peaks at A=105 and 164 and about ten neutrons. This, of course, would be contrary to the views of Steinberg and Wilkins (ibid) who consider that fission will be nearly symmetric for these nuclei.

The present best estimate of the flux of cosmic ray nuclei with Z ≥ 90 comes from the Skylab data summarised in Table 1. On this basis we would expect to have observed four such events in our present sample. Two highly charged particles were detected with Z_{app}=98 and 114 although at this stage, clearly, neither event should be taken wholly at face value. The event with Z=98 could be a perfectly normal actinide. Its light was well distributed in the main sphere and there was no accompanying signal from the plastic scintillator. The highest charge event, however, did activate the plastic scintillator detector and accordingly its interpretation as an electron shower cannot be excluded.

Finally, we must consider the source spectrum for charges in the range 32 ≤ Z ≤ 42. As already noted, their general abundance is similar to that of solar system material. In this charge band the balloon-borne counter experiments of Israel et al. (ibid) provide the best existing source of data. Table 3 shows abundance values from Israel, the present experiment and the solar system.

TABLE 3

Charge	Present Experiment Detector	Present Experiment Source	Israel et al. 1979	Cameron Solar System All	Cameron Solar System r-process	H.B.S.
26	10^6	10^6	10^6	10^6	10^6	
28	$5\ 10^4$	$5\ 10^4$	$5\ 10^4 \pm 4\%$	$5.8\ 10^4$	0	
29	-	-	-	650	0	
30	-	-	600 ± 100	1500	0	1500
31	-	-	103 ± 35	58	0	41
32	-	-	94 ± 33	138	0	106
33	-	-	39 ± 12	8	0	2.4
34	118 ± 8	88 ± 11	33 ± 12	80	34	36
35	-	-	44 ± 15	16	9	1.3
36	67 ± 6	46 ± 8	34 ± 12	65	11	20
37	-	-	23 ± 11	7.1	3.6	2.1
38	64 ± 6	55 ± 8	33 ± 13	32	0.9	27
39	-	-	5 ± 5	6.0	1.6	1.7
40	36 ± 5	32 ± 7	16 ± 9	34	8.4	4.7
41	-	-	-	1.7	1.0	0.0
42	13 ± 4	5 ± 5	-	4.8	1.3	0.4
	Based on 730 events $Z \geqslant 33$		54 events $Z \geqslant 31$			

Table 3 gives comparisons between the present experiment and the existing balloon data of Israel et al. for the lower charges. Also shown are Cameron's abundance values and separately those from the r-process alone. The final column shows a computed H.B.S. spectrum normalised at Zn.

There is general agreement between the two sets of cosmic ray data, but in our results Z=32, which is marked in Israel's data, does not appear strongly, although it may be partially masked by the relativistic rise tails of Ni and Zn. Our results show that all the even charges $34 \leqslant Z \leqslant 40$ are important in the source spectrum, with Z=34 the most abundant, and that compared with the solar system, Z=38 is somewhat enhanced in the cosmic rays. There is a rapid fall for Z > 40.

Models of r-process nucleosynthesis in which parameters are chosen to fit the high charge data do not normally produce these lighter elements with A < 78. An exposure to a far smaller integrated neutron flux is required to make nuclides of mass $58 \leqslant A \leqslant 76$. This avoids the otherwise rapid build-up to the closed shell at N=50 due to the prompt capture of neutrons without intervening β-decays, by the Fe seeds, which produces a marked peak in the region of A=80 but very little material with lower mass. The strength of the feature associated with N=50 is, however, model dependent. If, for example, more seed nuclei are continuously introduced the peak remains strong. If not, then the whole feature can wither as the nuclides undergo β-decay and capture further neutrons. Clearly, therefore, one or more additional processes must be involved in the production of nuclides in the range $58 \leqslant A \leqslant 76$ both for the cosmic ray source and for the solar system material.

From a study of abundance values from the balloon data of Israel et al.[4], Wefel, Schramm and Blake[14] concluded that a form of the s-process that occurs in massive stars, the helium burning s-process (HBS) may well be significant in this charge region. This process meets the requirement for a limited neutron exposure introduced above. The authors drew attention to the fact that these massive stars are the supernova progenitors. Thus, if supernovae are the source of the bulk of the galactic cosmic rays, material made in the HBS might be expected to be present in significant quantities in the cosmic rays. Furthermore, such stars are also the source of the abundant Fe that pervades the galaxy and solar system and is also such an abundant and important constituent of the cosmic ray source. No less than 5% of the total mass (and energy) of the cosmic rays is contributed by Fe nuclei, which are considerably overabundant when compared with hydrogen.

Comparison of the cosmic ray spectra in Table 3 with solar system abundance values shows a high level of agreement between the two sets of data. However, when compared with the r-process contribution alone the fit is poor, not only for Z=32 as expected, but also for Z=38. This species is prominent in our source spectrum but is almost by-passed by the r-process, since there is only one accessible nuclide, $^{88}_{38}$Sr. In addition, the prominence of the peak at Z=34 is not matched in the experimental data. The HBS process on its own is expected to be responsible for a considerable fraction of the solar system material for the range $32 \leqslant Z \leqslant 40$ and its normalization is usually fixed by demanding that it be responsible for all $^{58}_{26}$Fe. Since the abundance of cosmic ray $^{58}_{26}$Fe is not well-known, we can instead choose to normalise to, say, cosmic ray Zn. Such a spectrum is given in Table 3. Noteworthy features are the strong odd-even effect, rather similar abundances for the even nuclei with Z=34, 36 and 38, and the rapid fall-off in abundance for $Z \geqslant 40$. On its own the HBS yield has strong similarities to the cosmic ray source spectrum for $30 < Z < 40$, but the best fit is to the mix represented by the solar system, which contrasts to the situation for $Z > 50$.

CONCLUSION

The Ariel VI experiment provides for the first time data of reasonable statistical weight over the entire band of UH cosmic ray nuclei. Comparison of the measured and source abundances are made with solar system material. There are striking similarities and equally striking divergences. The production of nuclides of the solar system and of the cosmic rays both appear to require a number of processes of nucleosynthesis. However, it seems clear that the mix of processes involved has to be substantially different for the two types of material, and the cosmic ray source appears rich in r-process material that is believed to be synthesized in supernova explosions. As new data become available from the satellite experiments, so important advances in understanding can be expected in the near future.

REFERENCES

1. Fleischer, R.L., Price, P.B., Walker, R.M., Maurette, M., and Morgan, G.: 1967, J. Geophys. Res. 72, pp. 355-366.
2. Fowler, P.H., Adams, R.A., Cowen, V.G., and Kidd, J.M.: 1967, Proc. Roy. Soc.A. 301, pp. 39-45.
3. Shirk, E.K., and Price, P.B.: 1978, Astrophys. J. 220, pp. 719-733.
4. Israel, M.H., Klarmann, J., Love, P.L., and Tueller, J.: 1979, Proc. 16th Int. Cosmic Ray Conf., Kyoto 12, pp. 65-69.
5. Fowler, P.H., Alexandre, C., Clapham, V.M., Henshaw, D.L., O'Sullivan, D., and Thompson, A.: 1977, Proc. 15th Int. Cosmic Ray Conf., Plovdiv 11, pp. 165-173.
6. Cameron, A.G.W.: 1973, Space Sci. Rev. 15, pp. 121-146.
7. Price, P.B., and Shirk, E.K.: 1975, Proc. 14th Int. Cosmic Ray Conf., Munich 1, pp. 268-272.
8. Fowler, P.H.: 1977, Nucl. Instr. and Meth. 147, pp. 183-194.
9. Cassé, M., Goret, P., and Cesarsky, C.J.: 1975, Proc. 14th Int. Cosmic Ray Conf., Munich 2, pp. 646-650.
10. Ohnishi, T.: 1978, Astrophys. Space Sci. 58, pp. 149-165.
11. Steinberg, E.P., and Wilkins, B.D.: 1978, Astrophys. J. 223, pp. 1000-1014.
12. Schramm, D.N., and Fowler, W.A.: 1971, Nature 231, pp. 103-106.
13. Blake, J.B., and Schramm, D.N.: 1974, Ap. Space Sci. 30, pp. 275-290.
14. Wefel, J.P., Schramm, D.N., and Blake, J.B.: 1977, Astrophys. Space Sci. 49, pp. 47-81.

THE HEAVY NUCLEI EXPERIMENT ON HEAO-3

W.R. Binns[a], R. Fickle[b], T.L. Garrard[c], M.H. Israel[d],
J. Klarmann[d], E.C. Stone[c] and C.J. Waddington[b],
[a] McDonnell Douglas Research Laboratories, St. Louis, MO,
[b] University of Minnesota, Minneapolis, MN,
[c] California Institute of Technology, Pasadena, CA,
[d] Washington University, St. Louis, MO.

The third High Energy Astronomical Observatory, HEAO-3 was launched on the 20th Sept., 1979 into a 496 km, 43.6° orbit, and has since been successfully returning data from all three experiments on board. One of these experiments, that intended to study the heavy and ultra heavy nuclei in the cosmic radiation, is described here.

A schematic view of the instrument is shown in Fig. 1, from which it can be seen that the array is double-ended and consists of three main elements. Two pressure chambers, filled with an argon-methane mixture $\approx$ 850 torr ($1.08 \times 10^5 N/m^2$) each contain two x-y hodoscopes, made of wires spaced 1 cm apart, and three parallel plate ionization chambers. Between the pressure chambers is mounted a Cherenkov counter composed of two layers of Pilot 425 radiator looked at by eight photomultipliers. The array has a total geometry factor of 5.9 m^2sr for events that traverse at least two of the hodoscope pairs and two of the seven charge-measuring detectors. In practice, this means that approximately 5×10^4 iron nuclei are detected per day.

This instrument was designed to achieve a charge resolution of 0.3 charge units over the charge range of $17 \leq Z \leq 120$, which requires a measurement accurate to 0.6% at $Z = 100$. The spacing of the hodoscope wires, the amount of gas between each plane of the ionization chamber, and the thickness of the Cherenkov radiators were all chosen with this requirement in mind. Response maps of the ionization chambers and the Cherenkov detectors have been prepared using that in-flight data available. The response of the ion chambers has been found to be only weakly spatially dependent, showing a gradient which, for all positions more than 8 cms from the edge, is nowhere greater than 0.05% per cm. The Cherenkov chamber has a much larger spatial variation. Over some 80% of the area the gradients are less than 1% per cm, but near the corners they are larger. The accuracy with which these gradients are measured will steadily improve as further in-flight data are obtained. The Cherenkov response also shows a long term temporal dependence of ± 3%, which is closely, but not uniquely, related to the temperature variations. This variation can be measured with great accuracy from the in-flight iron nuclei $\lesssim$ 0.2%.

Examination of selected data shows that the instrument has the intrinsic resolution needed. To date we have resolved individual elements up to iron and even-charged elements up to strontium. It only remains to accumulate sufficient data to determine the corrections with the required accuracy. From one year of data we should have about 100 iron nuclei per cm^2 of detector surface suitable for calibration. That should be sufficient to achieve resolution of individual elements. It will then remain to verify the charge scale by identifying individual element peaks all the way up the charge range, since we cannot rely on theories of energy loss at such high Z values.

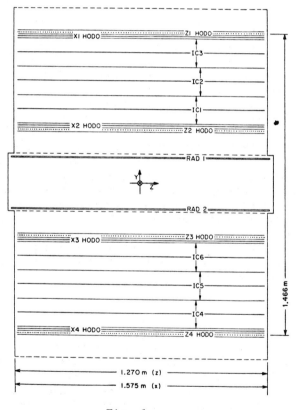

Fig. 1

Supported in part by NASA Contracts NAS8-27976, 7, 8 and grants NGR 05-002-160, 24-005-050, and 26-009-001.

PROPAGATION STUDIES RELATED TO THE ORIGIN OF COSMIC RAYS

R. Cowsik
Tata Institute of Fundamental Research, Bombay 400005, India.

Propagation of cosmic rays is discussed with the intent of deriving results relevent to the origin of cosmic rays. Starting from a brief description of the methods for demodulating the effects of the solar wind on the spectra of particles, we describe an accurate method for correcting for spallation effects on the cosmic-ray nuclei during their transport from the sources subsequent to their acceleration. We present the composition of cosmic rays at the sources and discuss its implications to their origin. We discuss briefly the effects of stochastic acceleration in the interstellar medium on the relative spectra of primaries and secondaries in cosmic rays and show that the observation of decreasing relative abundance of secondaries with increasing energy rules out such phenomena for galactic cosmic rays. The spectrum of cosmic-ray electrons is discussed in terms of contributions from a discrete set of sources situated at various distances from the solar system on the galactic plane. We show that unless there are at least 3.10^4 sources actively accelerating cosmis rays in the Galaxy the spectrum of electrons would have a premature cut-off at high energies. Finally we point out some important questions that need to be clearly resolved for making further progress in the field.

I. INTRODUCTION

We are at the threshold looking out at the nineteen eighties during which decade one feels hopeful that the problems posed by the discovery of cosmic rays early during this century would find a clear resolution. This confidence stems from the tremendous progress that took place during the seventies, not merely in the acquisition of data of excellent quality in the field of cosmic rays and many other related fields but also in the development of theoretical frame-work for the interpretation of the data. In this essay we shall review some of these developments and show how these studies have characterised the sources so well that their identification is imminent. A good part of this progress has been due to the age-old belief that cosmic ray sources should also be sources of high energy gamma rays and the recent discoveries using the SAS-II ans COS-B instruments have stimulated much new thinking.

This is not a review paper but a personalised account of ideas in the field ; unfortunately we have not been able to cover all the important work in the field nor have we been able to give credit to the originators of several important ideas which have been worked on and developed further over the recent years. Here we wish to substantiate the thesis that bulk of the cosmic rays below air-shower energies ($\sim 10^4$ GeV/n) originate in a very large number of discrete sources ($\sim 10^5$) strewn about the galactic disc. The cosmic rays which are accelerated in these sources interact with the matter surrounding the sources before leaking into the interstellar space where they suffer further interactions before escaping into the intergalactic space. These interactions generate secondaries like Li, Be, B and also gamma rays which serve as excellent diagnostic tools. Since our view of the true cosmic-ray spectra at low energies in the interstellar space is obscured by modulation effects due to solar wind and magnetic field our presentation here follows the reverse sequence describing first the methods for demodulating the solar effects and then methods for correcting the effects of spallation etc to get the actual spectrum and composition of the particles at the site of their acceleration. We then discuss the implication of the source spectra and source composition. Finally, we point out that there are two questions related to the cosmic ray origin which stand out and which have to answered clearly before further progress can be made in this field. These questions are ;

a) Is the residence time of cosmic rays in the interstellar space a decreasing function of energy ?

b) What really are the physical processes involved which choose certain ions, in preference to others from the available matter, for acceleration to cosmic ray energies ? In other words what is the injection mechanism that is operative in cosmic-ray accelerators ?.

II. DEMODULATION OF SOLAR WIND EFFECTS.

While the available measurements of the energy spectra of comic-ray particles are confined to those performed with detectors on satellites and balloons at heliocentric distances of the order 1 AU, these spectra elsewhere in the solar systems and especially in the near interstellar space are of much astrophysical interest. In order to predict the spectra at various heliocentric distances, starting from the observed spectra near the Earth, account must be taken of the effects of particle propagation through the radially diverging magnetised solar wind plasma. These effects are usually described in terms of a spherically symmetric steady state established by the balance of the inward diffusion of particles with an outward convection combined with the associated energy losses (Parker 1966). Very recently this simple picture is undergoing a drastic change with the discovery that the 11-year modulation of the neutron monitor counting rates (representing the galactic cosmic-ray intensity at $\langle E \rangle \sim 5$ GeV) are strongly correlated with the size of the polar coronal holes during the same period (Hundhausen et al. 1980). The correlation is so good that it establishes for the first time a connection between cosmic ray variations and a measurable parameter of the Sun, namely, the coronal holes which are known to be the sources of major streams of fast solar wind and establish a connection between interplanetary magnetic structure and the general magnetic field of the Sun. This correlation can be interpreted as non-spherically symmetric effects such as entry (Svalgaard and Wilcox 1976) and drift (Jokipii et al. 1976). This development is so recent that a full understanding of its consequences to the cosmic ray field will take some time and it may be worth while to consolidate the basic results obtained from solving the spherically symmetric transport equations of Parker (1966). These may indeed be adequate to correct for the relative modulation effects between isotopes of the same species at $E \gtrsim 300$ MeV/n. With these limitations the transport of the cosmic rays in the solar system is described by the equation

$$\frac{V}{r^2}\frac{\partial}{\partial r}(r^2 U_i) - \frac{K_i p^{\alpha_i}}{r^2}\frac{\partial}{\partial r}\left(r^2 \frac{\partial U_i}{\partial r}\right) - \frac{2V}{3r}\frac{\partial}{\partial p}(p U_i) = 0 \tag{1}$$

Here, V = velocity of the solar wind

$K_i p^{\alpha_i}$ = diffusion constant for the species i

p = momentum per nucleon or per electron

and U_i = number density of particles at heliocentric distance r and momentum p.

Let the spectral observations near the earth be parametrised in the form

$$U_i(0,p) \equiv N_i(p) = \frac{N_i p^2}{(p_i^{\alpha_i+3/2} + p^{\alpha_i+3/2})^{\nu}} \tag{2}$$

Then the solution at any r is given by (Cowsik and Lee 1977)

$$U_i(r,p) = \frac{N_i p^2}{(p_i^{\alpha_i+3/2} + p^{\alpha_i+3/2})^{\nu}} \cdot M\left(\nu_i, 2, \frac{w \times p^{3/2}}{p_i^{\alpha_i+3/2} + p^{\alpha_i+3/2}}\right) \tag{3}$$

Where M(a,b,z) = confluent hypergeometric function which equals one at z = 0,

x = r in units of the solar modulation boundary

and $w = (1 + \frac{2}{3}\alpha_i) r_0 V K_i^{-1}$

special cases of interest for this galactic cosmic-ray modulation are

$$\nu = 1 \quad U_i(r,p) = N_i(p) \frac{1}{\eta}(e^{\eta} - 1) \tag{4}$$

$$\nu = 3/2 \quad U_i(r,p) = N_i(p) e^{\eta/2}[I_0(\eta/2) + I_1(\eta/2)]$$

$$\nu = 2 \quad U_i(r,p) = N_i(p) e^{\eta}$$

where $\eta = \omega \times p^{3/2} (p_i^{d_i+3/2} + p^{d_i+3/2})^{-1}$ and I_0 and I_1 are modified Bessel functions.

Hopefully we will soon understand the effects of drifts adequately to have a full-fledged 3-dimensional solution including all effects. In the mean while the cosmic ray data obtained by various groups are of sufficiently high quality to warrant an uniform treatment of the modulation effects, that is generally agreed upon, to obtain the interstellar spectra.

III. CONSTRAINTS ON STOCHASTIC ACCELERATION IN THE INTERSTELLAR MEDIUM

The spectrum of particles in the interstellar space results from the combined effects of injection of particles by the sources and a variety of transformations these particles undergo before they escape from the Galaxy. Before discussing in detail the effects of nuclear spallation we wish to show that cosmic ray particles are not accelerated further in the interstellar medium stochastically by the socalled Fermi-mechanism (Fermi 1949). In this acceleration process cosmic rays grain energy through repeated collisions with magnetized clouds of gas that more about in the interstellar space. The interest in this process, in the context of galactic cosmic rays waned with the accumulation of radioastronomical evidence showing that the velocities of the interstellar clouds were too small and their relative spacing too large to accelerate cosmic rays significantly within the residence time of $\sim 10^7$ years. However the interest in the stochastic acceleration of cosmic rays in the interstellar space has been rejuvinated for two reasons : (1) Jokipii (1977) has considered the possibility that a spectrum of hydromagnetic waves may be present in the intestellar medium and these may provide for a rather rapid rate of acceleration ; (2) Axford et al. (1977) and Blandford and Ostriker (1978) have discussed the possibility that the shocks caused by the debris of supernova explosions extended to distances of ~ 100pc and that these shocks coupled strongly to the cosmic-ray gas leading to $\sim 10\%$ gain in the energy of the particles in each encounter. Now we proceed to show that if indeed such processes are effective the ratio of the fluxes of the secondary to that of the primary nuclei in cosmic rays will increase logarithmically with energy contrary to observations.

The energy spectrum of particles subject to Fermi acceleration is controlled by the well-known differential equation

$$\frac{\partial f}{\partial t} = -\frac{f}{\tau} - a \frac{\partial}{\partial E}(Ef) + KE \frac{\partial^2}{\partial E^2}(Ef) + I \qquad (5)$$

If particles are injected into the system at $t = 0$ with $E = E_i$ then setting $I = \delta(t) \cdot \delta(E-E_i)$ we get the Green's function of the problem to be (Cowsik 1979)

$$f(E,E_i,t) = \frac{1}{2\pi E} \left(\frac{E}{E_i}\right)^{\frac{K+a}{2K}} \exp\left[-(\ln E - \ln E_i)^2/4Kt\right] \cdot \exp\left[-(\delta + \frac{1}{\tau})t\right] \qquad (6)$$

with $\delta = (K+a)^2/4K$. Now the spectrum of cosmic rays, F, in the steady state is obtained by integrating the eq. 6 over time.

$$F_{1,2}(E, E_i) = \left[4K(\delta+\tfrac{1}{\tau})\right]^{-1/2} \frac{1}{E} \left(\frac{E}{E_i}\right)^{\frac{K+a}{2K} \pm \sqrt{(\delta+\tfrac{1}{\tau})/K}} \tag{7}$$

The + sign in the exponent obtains for $E < E_i$ and the - sign for $E > E_i$. It is convenient to rewrite eq. 7 as

$$F_{1,2}(E, E_i) = \frac{A}{E} \left(\frac{E}{E_i}\right)^{\eta \pm \gamma} \tag{8}$$

Notice here that acceleration occurs irrespective of the sign of a, the term emphasised by Fermi. The second order term causes the particles to spread rapidly in 'lnE' space and thus plays an essential role in the acceleration process.

If during this acceleration process the particles traverse matter they will generate secondary particles through nuclear interactions. An example of this process is the generation of elements like Li, Be and B in the interactions of the cosmic ray carbon nuclei. If ξ be the production rate of the secondaries S above the production threshold E_{th}, we may write in the steady state

$$0 = -\frac{S}{\tau} - a\frac{d}{dE}(ES) + KE\frac{d^2}{dE^2}(ES) + \xi F(E, E_i) \tag{9}$$

Noticing that eq. 8 solves the homogeneous part of this equation we get

$$S = \int F(E, E_p) \xi F(E_p, E_i) dE_p$$

$$= \int_{E_{th}}^{E} F_2(E, E_p) \xi F_2(E_p, E_i) dE_p + \int_{E}^{\infty} F_1(E, E_p) \xi F_2(E_p, E_i) dE_p$$

$$= \frac{A^2 \xi}{E} \left\{ \ln \frac{E}{E_{th}} E^{\eta-\gamma} + \frac{1}{2\gamma} E^{\eta-\gamma} \right\} \tag{10}$$

Thus the ratio of secondaries to primaries has the behaviour

$$\frac{S}{F} = A\xi \left\{ \frac{1}{2\gamma} + \ln E/E_{th} \right\} \tag{11}$$

In Fig. 1, this theoretical expectation is compared with the observations taken from a recent compilation of Ormes and Frier (1978)

Whereas the stochastic processes develop logarithmically increasing L/M ratios the observations indicate a fall off at high energies thus ruling out the effectiveness of the stochastic acceleration processes in the interstellar medium. These results has been generalized to processes where τ, a and K are arbitrary functions of energy, with essentially the same conclusions (Cowsik 1980).

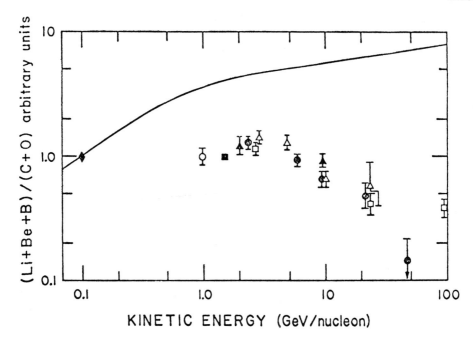

Fig. 1 - The predictions of the Fermi-theory are compared with the observations of L/M ratio in cosmic rays. The strong discrepancy rules out effective stochastic acceleration in the interstellar medium.

IV. NUCLEAR SPALLATION EFFECTS DURING PROPAGATION

The basis transport equation for cosmic ray nuclei is a second order partial differential equation (Ginzburg and Synovatskii 1964). This is in general too complicated to handle and Cowsik et al. (1966, 1967) introduced the concept of vacuum path length distribution which gave the probability that a cosmic ray particle survived for a time t after injection when all interactions were switched off. Since then invariably all calculations of the abundances of cosmic rays including effects of spallation and losses of energies due to ionization have been performed using this technique. However, this technique has mainly two difficulties : first, in this approach unless the pathlength distribution is an exponential or a convolution of exponential distributions the primary, secondary and thertiary nuclei can not be proven to have the same path length distribution. Second, the exact calculation of all the nuclear abundances in cosmic rays becomes extremely cumbersome. Since all nuclei more massive than the particular one of interest can contribute to its flux through nuclear spallation or radioactive decay one has to propagate all the more massive nuclei through the path-length integral first. In this procedure the propagation of calculational errors and the effects of uncertainty in the cross-sections is more severe. For these reasons consider the equation where the loss of particles from the volume of interest is parametrised by a leakage lifetime τ_i (Cowsik and Wilson 1973, 1975, Wilson 1979)

$$\frac{dN_i}{dt} = -\frac{N_i}{\tau_i} - N_i \beta c n_H b_i + \sum_{j>i} \beta c n_H \sigma_{ij} N_j + S_i \qquad (12)$$

Where N_i = density of nuclei of the ith kind
$1/\tau_i$ = escape probability per unit time
$\beta c n_H b_i$ = nuclear break up and decay probability
$\beta c n_H \sigma_{ij}$ = spallogenic and radioactive production rate of nuclei of the ith kind by nuclei of the jth kind
S_i = average source strength per unit volume.

Consider a diagonal matrix D_{ij} with elements $(\tau_i^{-1} + \beta c n_H b_i)$ and a triangular matrix T_{ij} with elements $\beta c n_H \sigma_{ij}$ for $j > i$ and zero for $j < i$. Defining $M_{ij} = T_{ij} - D_{ij}$, we can write

$$\frac{dN_i}{dt} = S_i + M_{ij} N_j \qquad (13)$$

The relative constancy of cosmic rays over geologic times (see Honda 1979 for a review) allows us to assume a steady state ie. $dN_i/dt = 0$ and write for the abundances of cosmic rays at the sources simply as

$$S_i = - M_{ij} N_j \qquad (14)$$

If we wish to calculate the expected composition N_j for any assumed source composition S_i, we can use

$$N_j = - (M^{-1})_{ji} S_i \qquad (15)$$

If cosmic rays are sequentially contained in ν regions each characterised by a transformation matrix M_{ij}^{α} we have

$$S_i = \left\{ \prod_{\alpha=1}^{\nu} M^{\alpha} \right\}_{ij} N_j \qquad (16)$$

This corresponds to the vacuum pathlength distribution

$$f(t) = C \sum_{\alpha=1}^{\nu} \frac{(-1)^{n_\alpha} t^{n_\alpha - 1} e^{-a_\alpha t}}{\prod_{\substack{\beta=1 \\ \beta \neq \alpha}}^{\nu} (a_\beta - a_\alpha)^{n_\beta} (n_\alpha - 1)!} \qquad (17)$$

Where n_α are the number of region with identical escape probability $a_\alpha = \tau_\alpha^{-1}$. A particular case of this was explicitly discussed earlier by Cowsik and Wilson (1973, 1975). This was called the nested-deaky box for cosmic rays.

Now, in the above equations τ_i are unknown. One usually assumes that τ_i are independant of the particle species and then one varies τ (and in some cases n_H also) till the abundances of certain elements such as Li, Be and B at the sources becomes a minimum or indeed zero.

The most important result that has come out during the last decade is that the relative composition of the cosmic rays at the sources do not show any sizeable variation with energy in the range 1-100 GeV, though τ itself seems to decrease beyond a rigidity of ~ 10 GV/c or ~ 5 GeV/n for the heavy nuclei (see Orth et al. 1978 for a recent discussion). One of the very important questions that faces us now is the way in which $\tau(R)$ is to be interpreted. Before discussing this question in the next section we will now proceed to discuss the implication of the source abundances of cosmic rays shown in Table 1.

Table 1 - Abundances of cosmic-ray nuclei at the sources derived using a pathlength distribution which has an exponential form but truncated below 1.3 gcm^{-2} (Silberberg 1980, Garcia-Munoz and Simpson 1979).

Element	C. Rays	Galactic	C.R./Gal.
He	3070 ± 200	$(2.08 \pm 0.46) \times 10^4$	0.148 ± 0.034
C	100 ± 1	100 ± 23	1.0 ± 0.23
N	6 ± 1	17.7 ± 7.7	0.34 ± 0.16
O	128 ± 3	177 ± 38	0.72 ± 0.16
F	0.7 ± 0.5	7.15	
Ne	16 ± 2	20.8 (1.7)	0.77 (1.7)
Na	2.7 ± 1	0.43 ± 0.07	6.3 ± 2.6
Mg	29.5 ± 2	8.08 ± 0.23	3.65 ± 0.27
Al	4.4 ± 1	0.65 ± 0.03	6.8 ± 1.57
Si	28.0 ± 2	7.69 ± 0.23	3.64 ± 0.28
P	0.6 ± 0.4	0.074 ± 0.015	
S	3.8 ± 0.6	3.46 ± 1	1.10 ± 0.36
Cl	0.3 ± 0.2	0.036 (1.6)	
A	0.8 ± 0.3	0.69 (1.7)	1.16 (1.8)
K	0.7 ± 0.3	0.028 ± 0.009	
Ca	3.7 ± 0.5	0.48 ± 0.05	7.71 ± 1.42
Fe	30.5 ± 0.2	6.77 ± 0.46	4.51 ± 0.43
Ni	1.7 ± 0.3	0.37 ± 0.046	4.59 ± 1.0

This procedure yields basically two parameters regarding the cosmic ray residence in the Galaxy. A path length distribution with an exponential fall-off with $\Lambda = 5$ g cm^{-2} with the path lengths below ~ 1.3 g cm^{-2} cut-off, seems to fit the data best. The cut-off at short pathlengths is necessitated by having to fit simultaneously the secondaries of medium nuclei like C, N, O and those of the heavy nuclei like Fe. The nested leaky box model where cosmic-ray nuclei spend part of their time in a containment volume just arround the sources generates naturaly the low pathlength cut-off as due to the convolution of two exponentials. On the other hand the interpretation that typical sources are far apart faces the difficulty that in this case the source function for the secondaries is the product of the interstellar gas density and the steady state density of the primaries. Thus the secondaries can arrive at the earth even after passing through only a very short path length. Under these conditions one would not be able to fit the data over the whole range from Li to Fe.

The second parameter which has come out of these studies particularly by studying the Be^{10} in cosmic rays is that the mean cosmic ray residence time in the Galaxy is longer than $\sim 10^7$ years. If it were any shorter the Be^{10} generated as spallation products would not have suffered radioactive decay to B^{10} sufficiently and we would have observed a much larger flux.

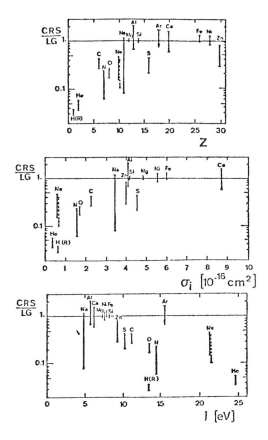

Fig. 2 - Ratio of the relative elemental abundances of cosmic rays at the sources to the local galactic abundances plotted as a function of nuclear charge Z, ionization cross-section σ_i, and the first ionization potential I.

The composition of the nuclei at the sources themselves tell us a lot. First, notice that the differences between the so-called local galactic abundances (Meyer 1979, Cameron 1973) and the composition of cosmic ray nuclei at the sources is within a factor of ~ 10 and selection effects based on atomic properties such as nuclear charge (Cowsik and Wilson, 1973) or first ionization potentials or ionizations cross-sections (Havnes 1971, Kristiansson 1974, Cassé and Goret 1978) could bring the two sets of abundances into correspondence. In Fig. 2 we show these correlations (Meyer, Cassé and Reeves, 1980).

The preliminary results from recent experiments indicate that the correspondance between the cosmic ray an universal abundances continues up to the heaviest elements. In fact these does not seem to be any evidence for usually high abundances of elements synthesised through the R-process thus severing one of the most important connections between cosmic rays and supernovae. In fact there are other factors in the composition of cosmic rays which further weaken the importance of supernova explosions as cosmic ray sources (Cassé and Soutoul 1975). The presence of the Fe-group of nuclei in the Galaxy is believed to be mainly due to explosive nucleosynthesis which provides proton-rich progenitors which decay either by electron capture or β^+ emission. The main decay schemes are

$$^{55}Co \xrightarrow[182h]{ec\ \beta^+} {}^{55}Fe \xrightarrow[2.6y]{ec} {}^{55}Mn$$

$$^{56}Ni \xrightarrow[6.1d]{ec} {}^{56}Co \xrightarrow[77d]{ec\ \beta^+} {}^{56}Fe$$

$$^{57}Ni \xrightarrow[36h]{ec\ \beta^+} {}^{57}Co \xrightarrow[270d]{ec} {}^{57}Fe$$

$$^{59}Cu \xrightarrow[8\,15]{ec\ \beta^+} {}^{59}Ni \xrightarrow[8.10^4y]{ec} {}^{59}Co$$

(18)

The electron capture is strongly inhibited once the nuclei reach high energies, as the atomic electrons would all have been stripped off. Thus the presence of ^{57}Fe and ^{59}Co in cosmic ray sources would indicate that the nuclei had electrons arround them for a considerable length of time before they lost them when they reached high energies. Just from the elemental ratios Fe : Co : Ni one can conclude that there has been a time delay of at least a couple of years between nucleosynthesis and cosmic ray acceleration. In fact if the preliminary indications of finite ^{59}Co in the sources is borne out by further experiments then one may conclude that the delay is longer than $2\times(8.10^4 y)$. In view of such arguments the source of ions for cosmic ray acceleration is the average galactic matter rather than the newly synthesised matter in the supernovae. Since, however the heavy elements in the interstellar matter is depleated on to the grains, the conditions in the cosmic ray sources should be such as to sputter or ablate the grains before acceleration.

There is one important nuclide which does not seem to fit into this general scheme ; it is shown quite definitively that there is an excess of ^{22}Ne at the sources. At the moment the source of this anomaly is not clear though there are many suggestions ; see for example Audouze et al. (1980) and many other papers at this meeting.

V. SOME URGENT QUESTIONS

The analysis presented in the previous section has pointed out two basic questions which need an explanation before we can make further progress in the field. We discuss now these questions in turn.

V.1. Is The Residence Time Of Cosmic Rays In The Galaxy Dependent On Energy ?

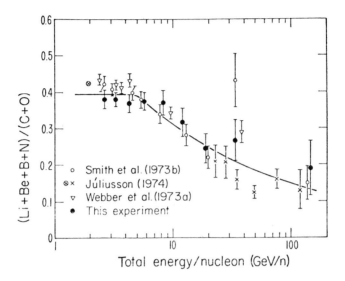

Fig. 3 - Energy dependence of spallation age (from Orth et al. 1978)

The energy dependence of the cosmic ray spallation age is shown in Fig. 3 taken from Orth et al. (1978). The most simple explanation for the fall off at high energies is that cosmic rays of higher rigidity escape from the Galaxy more effectively. An alterative explanation is that the residence time of cosmic rays in the Galaxy is indeed constant but the energy dependence of the L/M ration arises because of more effective trapping of low rigidity particles arround the sources as described in the nested leaky-box model (Cowsik and Wilson 1973,1975). A choice between these two alternate hypotheses is very important and we present here arguments relevant to answering this question.

Let us start with the first explanation in which the residence time of cosmic rays in the interstellar medium can be parametrised as

$$\tau(R) = \tau_0 \text{ for } R < 10 \text{ GV/c} ; \quad \tau(R) = \tau_0 (0.1 R)^{-0.5} \text{ for } R > 10 \text{ GV/c} \quad (19)$$

Obviously this decrease in the residence time cannot continue beyond $\sim 10^4$ GV/c, as it would otherwise lead to too high an anisotropy for the cosmic rays. Thus $\tau(R)$ = constant, for R larger than 10^4 GV/c. Since the spectrum of the observed particles is the product of the injection rate and the residence time ie. $F(E) \sim f(E) \cdot \tau(E)$, the observed spectra of particle would show kinks at ~ 10 GV/c (E= 10 GeV for protons and electrons) unless the source spectra fortuitously have the inverse energy dependence. Indeed the observation of smooth proton spectrum between 1 GeV and 10^5 GeV (Sreekantan 1979) makes the simple explanation unlikely. In the nested leaky box model the more effective trapping of the particles does not change the average source function of the particles and with constant residence time in the galactic volume one expects no kinks in the proton spectrum. Also in this model the cut-off at short path lengths has a very logical and clear explanation. The expected flux of high energy nuclear gamma rays which would be generated if cosmic rays suffer nuclear interactions at the sources is also consistent with the observations (Cowsik and Wilson 1975, Cowsik 1979b). Though for these reasons we prefer the later explanation it is not without its own problems. It seems to predict too low a flux of positrons and antiprotons at high energy. If the preliminary observations of these extremely rare species of cosmic rays are confirmed, it might be the evidence of acceleration of these particles inside the sources.

V.2. Problem Of Sellective Injection Of Particles Into Cosmic Ray Accelerators.

Most processes of particle acceleration that have been discussed in literature rely on the assumption that there is an unknown mechanism that selects some particles from the available bulk matter and gives them an initial energy sufficiently large that their collision cross-section with the matter becomes sufficiently small for the acceleration processes to take over. Since the collision cross-sections drop rapidly with energy, under most astrophysical conditions the minimum injection energy E_i is in the range 1-30 MeV. Beyond this energy the particles are scattered by magnetic irregularities, and hydromagnetic and electromagnetic waves, rather than by the atoms, and so the particles tend to equalise their energy with the bulk motions of the fluid. This process clearly leads to the acceleration of the particles. But despite the fundamental importance of the initial injection process we have very little understanding of the basic mechanisms that are responsible. (Eichler 1979).

VI. THE ELECTRON COMPONENT AND THE MULTIPLICITY OF COSMIC RAY SOURCES.

The propagation of the electron component is very sensitive to the distribution of the cosmic ray sources because of the strong radiative losses suffered by the high energy electrons (Cowsik and Lee 1979, Giler, Wdowczyk and Wolfendale 1980). We may now consider a large number of discrete sources of cosmic rays situated all over the galaxy contributing to the spectrum that is measured. The highest energy electrons come from only the nearest sources as those electrons which leave distant sources with high energies rapidly radiate away their energy through synchrotron emission in the Galactic magnetic fields and by Compton-scattering of the microwave background. The theoretically expected spectral shape for various mean separation between the sources is compared with the observations of Hartmann et al. (1977) in Fig. 4.

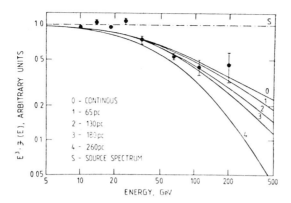

Fig. 4 : The spectrum of electrons summed over all the sources in the Galaxy ; the parameters represent typical spacing between the sources.

It is seen that unless the typical spacing between the sources is less than ~ 180pc the intensity of the high energy electrons cannot be accounted for. With such a typical spacing, within our Galaxy of radius ~ 15kpc, there are at least 3.10^4 sources actively accelerating cosmic rays. Thus there must be other sources of cosmic rays besides supernovae in the Galaxy.

VII. CONCLUSIONS

1. Cosmic ray source composition can be generated by atomic sellection effects operating on normal local galactic abundances.
2. There are probably more than 100,000 sources of cosmic rays in the Galaxy.
3. Cosmic rays suffer nuclear interactions in the vicinity of the sources generating secondary cosmic ray nuclei and gamma rays.
4. The processes of initial injection of selected ions into the cosmic ray accelerators needs critical study.

REFERENCES.
Audouze, J., Chièze, J.P. and Vangioni-Flam, E. 1980, Preprint of Institut d'Astrophysique de Paris.
Axford, W.I., Leer, E. and Skadron, G., 1977, Proc. 15th Internat. Cosmic Ray Conf. 2, 173 (Plovdiv).
Blandford, R.D., and Ostriker, J.P., 1978, Astrophys. J. (Letters), 221, L 29.
Cameron, A.G.W. 1973, Space Sci. Rev. 15, 121.
Cassé, M. and Soutoul, A., 1975, Astrophys. J. (Letters) 200, L75.
Cowsik, R., Yash Pal, Tandon, S.N. and Verma, R.P. 1966, Phys. Rev. Lett. 17, 1298.

Cowsik, R., Yash Pal, Tandon, S.N. and Verma, R.P. 1967, Phys. Rev. 158, 1238.
Cowsik, R. and Wilson, L.W., 1973, Proc. 13th Internat. Cosmic Ray Conf. 1, 500 (Denver).
Cowsik, R. and Wilson, L.W., 1975, Proc. 14th, Internat. Cosmic Ray Conf. 2, 659 (Munich).
Cowsik, R. and Lee, M.A. 1977, Astrophys. J. 216, 635.
    1979, ibid. 228, 297.
Cowsik, R., 1979, ibid. 227, 856.
Cowsik, R., 1979b, Proc. COSPAR Symp. Non-Solar Gamma Ray Astronomy (Bangalore, Pergamon Press ed. Cowsik and Wills).
Cowsik, R., 1980, Astrophys. J. (to be published Nov. 1)
Eichler, D. 1979, Proc. 16th Internat. Cosmic Rays Conf. 2, 61, (Kyoto).
Fermi, E. 1949, Phys. Rev., 75, 1169.
Garcia Muñoz and Simpson, J.A. 1979, Proc. 16th Internat. Cosmic Ray Conf. 1, 270.
Giler, M., Wdowczyk, J., Wolfendale, A.W. 1980, Ast. Astrophys., 84, 44.
Ginzburg, V.L. and Syrovatskii, S.I., 1964, The Origin of Cosmic Rays, (Pergamon Press, New York).
Hartmann, G., Müller, D. and Price, T., 1977, Phys. Rev. Lett., 38, 1368.
Havnes, O., 1973, Ast. Astrophys., 24, 435.
Honda, M. 1979, Proc. 16th Internat. Cosmic. Ray Conf. 14, 159 (Kyoto).
Hundhausen, A.J., Sime, D.G., Hansen, R.T. and Hansen, S.F. 1980, Science, 207, 761.
Jokipii, J.R., Levy, E.H. and Hubbard, W.B., 1976, Astrophys. J., 213, 816.
Jokipii, J.R. 1977, Proc. 15th Internat. Cosmic Ray Conf. 2, 429 (Plovdiv).
Kristiansson, K., 1974, Astrophys. Space Sci., 30, 417.
Meyer, J.P. 1979, Proc. 16th Internat. Cosmic Ray Conf. 2, 115 (Kyoto)
Meyer, J.P., Cassé, M. and Reeves, H. 1980, CENS preprint.
Orth, C.D., Buffington, A., Smoot, G.F. and Mast, T.S., 1978, Astrophys. J. 226, 1147.
Ormes, J. and Freier, P. 1978, Astrophys. J. 222, 471.
Parker, E.N., 1966, Planet. Space Sci., 13, 9.
Silberberg, R. 1980, pvt. comm.
Sreekantan, B.V. 1979, Proc. 16th Internat. Cosmic Ray Conf. 14, 345.
Svalgaard, L. and Wilcox, J.M. 1976, Nature, 262, 766.
Wilson, L.W. 1979, Ph. D. Thesis, Univ. of California (Berkeley).

INTERPRETATION OF COSMIC RAY COMPOSITION: THE PATHLENGTH DISTRIBUTION

R.J. Protheroe* and J.F. Ormes
Laboratory for High Energy Astrophysics
NASA/Goddard Space Flight Center, Greenbelt, MD, U.S.A., and
G.M. Comstock
Cosmo-Science Associates, Stony Brook, NY, USA

The chemical composition of cosmic ray nuclei with $3 \leq Z \leq 28$ between ~ 100 MeV/nuc and a few hundred GeV/nuc are compared with a consistent set of propagation calculations. These include the effects of spallation (energy-dependent cross sections are used), escape and ionization loss in the interstellar medium and deceleration in the solar cavity. This has enabled a consistent study of the cosmic ray pathlength distribution to be made over this entire energy range. Details of the propagation calculation are left to a forthcoming paper.

It has been generally believed that the composition was best explained by a pathlength distribution (PLD) with an absence of short pathlengths (e.g. Shapiro et al., 1973). In an attempt to explain this truncated shape of the PLD, several models have been advanced. Simon (1977) has considered the "two-zone" models or "nested leaky box" models of Cowsik and Wilson (1973, 1975) and shown that in these cases the PLD is the convolution of two exponential distributions. The distribution is uniquely defined by two parameters: the mean pathlength in the source region λ_s, and the mean pathlength in the galaxy λ_b (the mean escape length is $\lambda_e = \lambda_s + \lambda_b$, and the ratio λ_s/λ_b determines the shape of the distribution).

We have used observed and predicted (Be+B)/C and B/C ratios to obtain λ_e. Below 2 GeV/nuc we find $\lambda_e \cong 7$ g/cm^2 for an exponential PLD ($\lambda_s/\lambda_b = 0$) and 5.3 g/cm^2 for $\lambda_s/\lambda_b = 0.5$ (the most extreme two-zone model). Above 2 GeV/nuc λ_e decreases as $E^{-0.4 \pm 0.1}$ (Ormes and Freier, 1978). These values are for an ISM comprising 90%H and 10% He by number.

We have calculated the ratio of Iron secondaries ($21 \leq Z \leq 25$) to Fe and (Be+B)/C as a function of energy for these two cases. The predictions are shown in the figure where we also show the effect of varying the amount of solar modulation. The predictions for the ($21 \leq Z \leq 25$)/Fe ratio may be compared with a survey of the experimental results shown in the figure. From this comparison, it appears that there is no need to

* NAS/NRC Research Associate

invoke a truncated PLD to explain the observed ratio. The data are consistent with the predictions for the exponential PLD; however, because of the large amount of scatter between data points, we cannot rule out the possibility of small values of λ_s/λ_b.

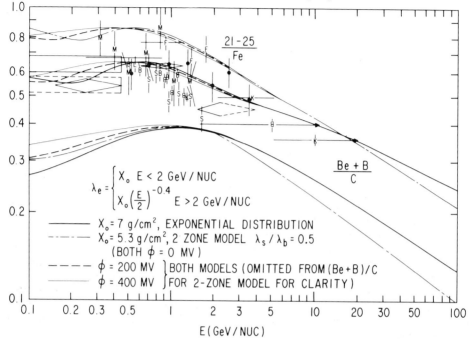

VARIATION OF (Be+B)/C AND (21-25)/Fe WITH ENERGY. M, Maehl et al., 1977; L, Lund et al., 1975; F, Freier et al., 1979; B. Benegas et al., 1975; S. Scarlett et al., 1978; K. Koch, 1980, preliminary HEAO-C data; ●, Lezniak and Webber, 1978; ◇, Israel et al., 1979; ⊂⊃, Garcia-Munoz et al., 1977; ⬜, Garcia-Munoz et al., 1979.

References

Benegas, J.C., et al., 1975: Proc. Munich Conf. 1, 251.
Cowsik, R., and Wilson, L.W., 1973: Proc. Denver Conf. 1, 500.
Cowsik, R., and Wilson, L.W., 1975: Proc. Munich Conf. 2, 659.
Freier, P.S. et al., 1979: Proc. Kyoto Conf. 1, 316.
Garcia-Munoz, M., et al., 1977: Proc. Plovdiv Conf. 1, 224.
Garcia-Munoz, M., et al., 1979: Proc. Kyoto Conf. 1, 310.
Israel, M.H., et al., 1979: Proc. Kyoto Conf. 1, 232 and 13, 402.
Koch, L., 1980: Bull. A.P.S. 25, 563.
Lezniak, J.A., and Webber, W.R., 1978: Astrophys. Space Sci. 63, 35.
Lund, N., et al., 1975: Proc. Munich Conf. 1, 263.
Maehl, R.C., et al., 1977: Astrophys. Space Sci. 47, 163.
Ormes, J.F., and Freier, P.S., 1978: Astrophys. J. 222, 471.
Scarlett, W.R., et al., 1978: Astrophys. Space Sci. 59, 301.
Shapiro, M.M., et al., 1973: Proc. Denver Conf. 1, 578.
Simon, M., 1977: Astron. Astrophys. 61, 833.

ARE STELLAR FLARES AND THE GALACTIC COSMIC RAYS RELATED?

Richard I. Epstein,
NORDITA,
Copenhagen, Denmark.

It has been suggested that the Galactic cosmic rays may be accelerated by a two stage process in which one process, such as stellar flares, inject non-relativistic, super-thermal particles which are subsequently boosted to cosmic ray energies by some other mechanism, perhaps related to supernovae (eg. Cassé and Goret, 1978). Two-stage models in which the injection and re-acceleration processes are uncorrelated are apparently untenable because they cannot fit the observed energy dependence of the LiBeBN/CNO ratio (Fransson and Epstein, 1980). Here it is shown that additional contraints derived by considering the energy losses and nuclear reactions suffered by the super-thermal particles prior to their re-acceleration severely restrict other types of two-stage models.

Two-stage models can be characterized by two important parameters: E_i, the mean energy of the super-thermal particles which ultimately become cosmic rays, and $n\Delta t$, where n is the ambient nucleon density and Δt is the typical time between the injection of the super-thermal particles and their re-acceleration to relativistic energies.

IONIZATION LOSSES. There is an energy, E_t, such that a proton which initially has an energy less than this loses all its excess energy and is thermalized in a time less than Δt.

$$E_t = 50 \; (n\Delta t/5 \times 10^6 \; cm^{-3} \; yr)^{2/3} \; MeV \tag{1}$$

The first point to be made is that for plausible two-stage models

$$E_i \gtrsim E_t \tag{2}$$

There are three lines of reasoning which support this contention.
(a) Composition: Since ionization losses increase with the charge of the ion, if $E_i < E_t$, the higher Z elements would be strongly depleted before re-acceleration. The resulting abundance distribution of the Galactic cosmic rays would then decrease with increasing Z, contrary to observations. This is an important constraint to keep in mind, since

one of the major motivations for appealing to two-stage models is the hope that flaring stars could produce the correct cosmic ray source composition. (b) <u>Energy requirements</u>: If $E_i < E_t$, flaring stars must inject additional super-thermal particles to compensate for those which are degraded; the minimum power which must be emitted in super-thermal particles is $P_{min} \approx \dot{N} E_i (E_t/E_i)^{3/2}$ where $\dot{N}$ is the rate at which the cosmic ray nuclei must be replenished. This requirement is less severe when $E_i \approx E_t$, but even in this case it is likely to be several per cent of the total power supplied to the cosmic rays. (c) <u>Depletion of lower energy particles</u>: If the mean energy spectrum averaged over many flares is less steep than $dN/dE \propto E^{-5/2}$ at energies below E_t, then after a time Δt, the lower energy particles would be largely thermalized. The mean energy of the remaining super-thermal particles would then be greater than $\sim E_t$.

<u>NUCLEAR SPALLATIONS</u>. Super-thermal or cosmic ray CNO nuclei which have energies above the spallation threshold, $E_s \approx 50$ MeV/n, produce light secondary nuclei. The relative abundances of the secondary and primary nuclei in the cosmic rays indicate that the CNO nuclei have traversed only about 5 g cm^{-2} of matter. Since viable two-stage models must not over-produce secondary nuclei, at least one of two conditions must be satisfied: either (a) $m_p n \Delta t v \lesssim 5$ g cm^{-2}, where $m_p v^2/2 \approx E_i$ or (b) $E_i < E_s$. If eq. 2 is valid, then condition (a) is the less stringent one, and it is satisfied only if

$$n \Delta t \lesssim 6 \times 10^6 \text{ cm}^{-3} \text{ yr}. \tag{3}$$

This condition severely restricts possible two-stage models; acceptable models must have the source of the supra-thermal particles and the agency for their final acceleration very closely related. For example, even models in which the flaring stars and the supernovae which re-accelerate the supra-thermal particles are in the same OB association (Montmerle, 1979) appear to violate eq. 3. If one follows Montmerle and takes the density in the OB association to be ~ 100 cm^{-3} and the time before re-acceleration to be of the order of the association lifetime $\sim 10^7$ yr, then $n \Delta t \sim 10^9$ cm^{-3} yr.

References

Cassé, M. and Goret, P.: 1978, Ap.J., <u>221</u>, 703.
Fransson, C. and Epstein, R.: 1980, Ap.J. (in press).
Montmerle, T.: 1979, Ap.J. <u>231</u>, 95.

EXTENDED RADIO SOURCES

Edward B. Fomalont
National Radio Astronomy Observatory
Charlottesville, Virginia 22901 U.S.A.

1. INTRODUCTION

The understanding of the nature of strong radio emission associated with galaxies and quasars has significantly increased in the last five years. A large contribution to this increase has been obtained from the arcsec and milli-arcsec mapping of radio sources which show remarkable features. Because of the recent, extremely good review articles concerning extended extragalactic radio sources (Miley 1980 and Willis 1978), I shall emphasize some new, mostly unpublished works, stemming from VLA observations. This is not to slight European experimental and theoretical work which have led in the formulation of our current thinking, but to try to bridge the inevitable communication gap caused by the Atlantic Ocean and ever higher airfares.

Perhaps, the most important advance in the last five years has been the success of the so-called beam-type models (e.g., Rees 1971, Blandford and Rees 1974) which explain many of the morphological features in radio sources and provide a theoretical framework which has led to a better understanding of the physics associated with galaxies and quasars. The acceptance of this model will be tacitly assumed in the following paper.

2. THE FOUR PARTS OF A RADIO SOURCE

Nearly all extended emission associated with extragalactic objects can be separated into four morphological parts--which also correspond to distinct physical processes occurring in the source. They are: the core, the jet, the hot spots and the diffuse emission. A somewhat simplified picture is given in Table 1 with a contour plot in Fig. 1 of NGC315 (Bridle et al. 1979) which display these features. The following discussion only touches the highlights of the radio emission, is not complete and is biased towards the beam models.

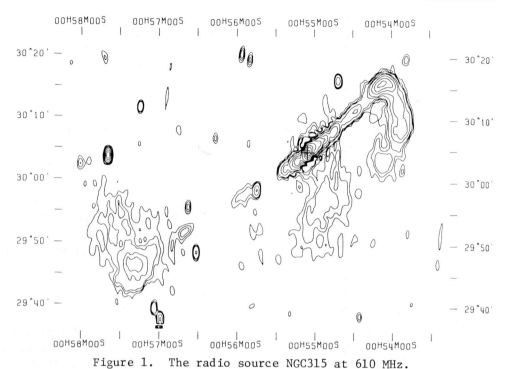

Figure 1. The radio source NGC315 at 610 MHz.

TABLE 1
The Four Parts of a Typical Radio Source

Morph. Part	Linear Size (kpc)	Spectral Index $S \sim \nu^{-\alpha}$	Equip. B Gauss	Percent Polar.	Role in Source
Core	< 0.01	0.0	10^{-3}	< 2	Energy Generation
Jet	2-1000	-0.6	10^{-5}	0-60	Energy Flow
Hot spot	5	-0.6	10^{-5}	15	Energy Conversion
Extended emission	50-1000	-0.9	$10^{-5.5}$	0-60	Energy loss and history

2.1 The Radio Core

Nearly all radio sources contain a small-diameter radio component, often less than 100pc in size, which is coincident with the optical nucleus of the galaxy. The radio strength is often variable over time-scales of months and the core is similar to "isolated" compact sources (Kellermann 1978). The spectral index of most cores is about 0.0 and the spectrum is consistent with self-absorption in some cases. Some cores do have steeper spectra and larger linear sizes but these are probably composed of a radio core and a small radio jet.

The radio properties of cores can only be measured using VLBI techniques and these properties will be described by Pauliny-Toth in a separate article. The distinction of radio sources between those with extended emission and weak radio cores from those with dominant or isolated radio cores is becoming blurred and it is likely that we are dealing with the same class of sources. More on this later.

2.2 The Radio Jets

Many radio sources show linear features which emanate from one side or from opposite sides of the radio core. These jets, as the features are called, often extend hundreds of kiloparsecs and terminate into a hot spot in the radio lobes. The jets are most prominent in the lower luminosity sources; they tend to be one-sided; they often contain gaps, bends and wiggles; and they become wider further from the core. Although most luminous sources do not have prominent jets, the consensus is that beams do exist for these sources and that jets are seen only if some inefficiency in the beam has generated sufficient numbers of relativistic particles which then radiate. It is usually assumed that the visible jets fill the beams and that properties of the beam can be inferred from those of the jets.

The spectral index of most jets is about -0.6. Optical emission has also been seen coincident with the bright features of some radio jets (Butcher et al. 1980) and the spectral index defined from the radio to optical frequencies is also about -0.6. Linear polarization is also seen at optical frequencies and the radiation mechanism is probably synchrotron radiation. Recently x-ray emission has also been detected in jets associated with nearby galaxies.

Non-beam models which explain the existence of jets are not attractive. Plasmon models in which discrete blobs of relativistic and thermal matter are transported from the core to the lobes are not consistent with the generally continuous nature of the jets and the rapid supersonic flow necessary to keep the jets well-collimated. Similarly, the continuous jets do not support models in which the ejection of several large massive objects from the galaxy supply the majority of energy to the lobes. Clearly, the jet morphology is the crucial observational support for the beam-type models.

2.3 Hot Spots in Lobes

Extended radio sources often contain regions of size 5 kpc in the radio lobes which are significantly brighter than the surrounding lobe emission. The hot spots invariably lie near the termination of the radio jet or its extrapolation when one is visible. This morphology strongly suggests that the hot spot is generated by the disruption of the beam with the concomitant generation of relativistic particles. The spectral index of the hot spots is also similar to that of the jets. Occasionally there are several close hot spots in the lobes of luminous sources which complicates this simple interpretation somewhat.

There are several examples of optical emission associated with prominent hot spots in several radio sources (e.g., Simkin 1978, Saslow et al. 1978). As in the case of the jets, the optical/radio spectrum is about -0.6 and radiation at both frequencies is satisfactorily explained as synchrotron radiation.

2.4 Extended Emission

In addition to the hot spots, the radio lobes contain more diffuse emission. The spectral index of this emission is -0.8 and can be as steep at -1.3. In the extended emission the degree of polarization is large, intensity gradients are greatest at the edge of the lobes and the magnetic field structure is somewhat uniform. The morphology and placement of the extended emission with respect to the hot spots and the radio core vary widely.

2.5 The Overall Source Symmetry

The four parts of a typical radio source when taken as an entity do not form a precise linear system but usually show various symmetries and distortions. These have been illustrated by Miley (1980) and include "C"-shaped sources, "S"-shaped sources, curvature in radio jets, reflection symmetries through the core, etc.

These various morphologies are thought to be associated with the velocity and acceleration of the radio source in an external galactic medium, the dynamics and environment found in the galaxy as a whole, and the dynamics and environment of the pc-sized nuclear region. However, these effects do not appear to disrupt the basic energy flow mechanism very much but give the source a non-linear appearance.

3. RECENT OBSERVATIONS AND SPECULATIONS

The rate of production of high resolution radio maps from the VLA, Westerbork, Cambridge and Jodrell Bank, as well as accurate and detailed optical images, is now uncovering important details in radio galaxies. Models are becoming more realistic now that observational data can point out unambiguous features and physical conditions in the radio source.

It is too early to form firm models which fit the recent observations. Rather, I would like to briefly describe recent results and speculations associated with the latest observations (mostly from the VLA) which seem to be telling us something.

3.1 Compact Versus Extended Radio Sources

The separation of extragalactic radio sources into two distinct groups, compact and extended, is becoming blurred and I think one can safely state that 1) nearly all "isolated" compact sources have associated large-scale structure > 10 kpc in extent, and 2) nearly all extended sources have a radio core which may be identical to the so-called "isolated" compact sources. The difference may be only in the relative luminosity of these radio core and extended structure.

First, the recent observational evidence should be discussed. Using the VLA very sensitive and high resolution observations of "isolated" compact radio sources have shown that about half of them in a complete 6cm-survey sample contain extended emission of size about 10 arcsec. At 6cm the extended emission comprise about 5 percent of the total flux density. At 20 cm the percentage rises to about 15 percent. Some examples are given in Figure 2.

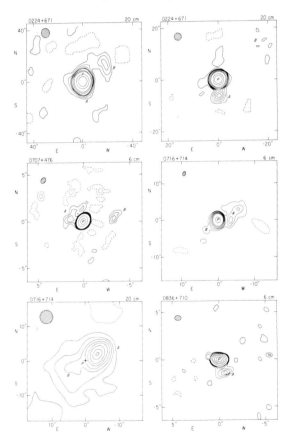

Figure 2. Examples of core-dominated radio sources. The lowest contours are at a level of 0.4% of the peak.

These morphologies are similar to the "D2" class of Miley (1971) or the "C" class of Readhead et al. (1978) in which a faint steep-spectrum component is displaced from a flat-spectrum core. The source 3C273 is a good example. All of these core-dominated sources are luminous radio emitters and are usually identified with quasars or distant galaxies.

Based on maps of about 20 such sources, their radio properties are:
(1) The flux density is dominated by a flat-spectrum core which is coincident with a galaxy.
(2) One or more secondary components are displaced about 10 kpc from the core. These components have spectral indices of about -0.8 and are about 5 kpc in size.
(3) More diffuse emission is usually present in a non-descript morphology and position with respect to the other components. The spectrum of this component is also steep.

This overall morphology does not fit in with the usual structure of extended radio sources. Because of the extreme luminosity in the core the normal collimating and transport mechanisms may not be operating with the result of a smaller, less symmetric radio source.

Recent work on relativistic flow in jets by Scheuer and Readhead (1978) and Blandford and Königl (1978) may provide an explanation of the core-dominated morphologies in the framework of the more typical radio structure. The model suggests that the radiation from a region near the core may be doppler-enhanced if the flow from the core is relativistic and in the direction of the observer. For a flow with $\gamma = 7$ within 8 deg of the line of sight, an enhancement of a factor 500 can be obtained. Any associated extended structure, presumably unaffected by relativistic flow, would be faint compared with the core. Because the source would be viewed nearly along its radio axis, any non-linearities in the source structure would be magnified. Thus, the secondary components seen may be lobe hot spots and the more extended emission would then be the larger-scale lobe emission. A source like Cygnus A, when viewed along the line of sight with a core enhancement of 500, could look similar to the sources shown in Figure 2.

More detailed data are needed to test the above hypothesis. Verification of this model will be an important test in determining if relativistic flow velocities do occur in the inner parts of radio jet in luminous sources.

3.2 The Collimation of Radio Jets

By now the properties of radio jets are well-known. Some examples of sources with these linear features are shown in Figure 1 and 3. Some of the general properties are:
(1) The jets are usually asymmetric in intensity.

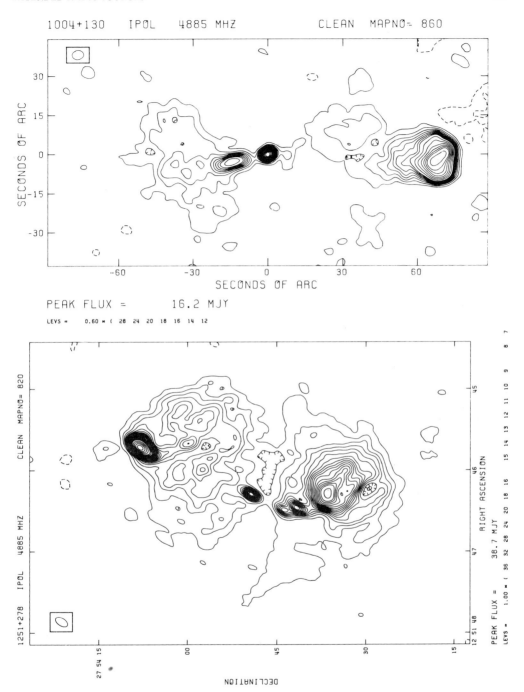

Figure 3. Radio sources with jets.
(a) 1004+130 at 4.9 GHz
(b) 1251+278 (3C277.3) at 4.9 GHz

(2) Often, there are gaps in the jet and most jets do not start until several kpc from the radio core.
(3) There are wiggles, bends and kinks in the jets. Often these features in the jets show a reflective symmetry with respect to the core.
(4) Jets associated with low luminosity sources are generally prominent and have average opening angles of about 10 deg. Jets are not seen often in the higher luminosity sources and when visible have opening angles less than about 3 deg.
(5) The magnetic field direction in the inner region of the more prominent jet is oriented parallel to the jet axis. This orientation flips to perpendicular to the jet axis, the distance from the core where this flip occurs varies directly with the source luminosity. For the luminous sources, the magnetic field remains parallel to the jet axis. The less dominant jet, if visible, has its magnetic field always perpendicular to the jet axis (Fomalont et al. 1980).
(6) Recent dual-frequency observations of the linear polarization in 3C449 (Perley et al. 1979) and NGC6251 (Perley et al. 1980) suggest the presence of significant thermal material.
(7) The opening angle of the jets is not constant but generally decreases with distance from the core.

From point (7) above we see it is apparent that the jets do not expand uniformly as would be expected by a freely expanding supersonic flow. The departures can provide clues to changes in the pressure equilibrium (if the flow is confined) or changes in the effective equation of state (if the flow is free). Recent work by Chan and Henriksen (1980) and Bridle et al. (1980) have investigated the collimation properties in several jets with interesting conclusions.

They adopted the dynamics of a supersonic, but non-relativistic, magnetized beam. They derived a self-similar solution (azimuthally symmetric, radially scale-independent, arbitrary axial dependence) for the magnetic field, flow velocity, specific angular momentum and effective temperature. They also assumed that the jet was heavy; i.e., the relativistic particle energy is only a small fraction of the total energy. This latter assumption seems valid in the light of the depolarization data associated with jets and the requirement of reacceleration in the jets to keep them visible. Most reacceleration mechanisms invoke significant amounts of thermal matter.

One conclusion they draw is that if the energy in the circumferential magnetic field is more than one percent of the kinetic energy in the flow down the jet (somewhat less than the equipartition value one would calculate in the normal way), significant pinching of the beam will occur and radial oscillations would be present. Such oscillations are not observed on the necessary scale and thus rule out the dominance of circumferential magnetic fields in non-relativistic jets (see Fig. 4, curve a and b).

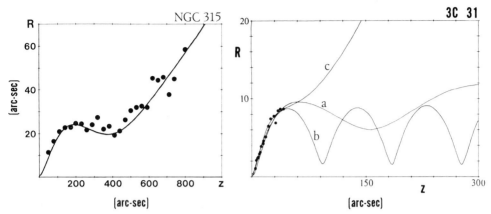

Figure 4. Fits to collimation of NGC315 and 3C31 jets. The dots show the width R of the jet with distance Z from the radio core. The various models are described in the text.

With the x-ray detection of extended (100 kpc), hot (> 10^7 K) gaseous haloes around M87 and Cygnus A (Fabricant et al. 1978, 1979; Fabbiano et al. 1979) it is attractive to consider the possibility that jets in low-luminosity sources are confined by external pressure. A confining atmosphere with a single pressure scale cannot, however, produce decreasing expansion rates such as those observed in most jets. Such a decrease can be effected by a two-component atmosphere. First, the beam is collimated by a dense "nuclear" atmosphere whose pressure decreases rapidly. After some distance this nuclear atmosphere is supplanted by one with a more slowly decreasing pressure. Such a model fit to 3C31 is shown in Figure 4, curve (c); a more impressive fit to the north-west jet in NGC315 is shown in Figure 4. The parameters for this fit were:

Beam radius at sonic point	0.75	arcsec
Scale hight of nuclear atmosphere	5.0	arcsec
Power law of nuclear atmosphere	4.0	
Scale height of second atmosphere	140.0	arcsec
Power law of second atmosphere	2.5	
Height where two pressures are equal	52.0	arcsec
Density where two pressures are equal	1000	cm^{-3} T^{-1}

The decrease in the expansion rate of the jet in the region where the two components of the external pressure are about equal can be explained in the following way. Initially, the pressure balance between the nuclear atmosphere and the relativistic particles sets up a supersonic collimated jet in which the thermal matter inertia keeps the jet collimated with a relatively constant opening angle. This opening angle depends on the rate of decrease of the pressure which varies as Z^{-4} in the model for NGC315, and the internal temperature decreases down the jet. When the outer atmosphere, which decreases as $Z^{-2.5}$ becomes dominant, the external pressure decreases slower than the thermal pressure. The jet width then contracts until the thermal gas heats sufficiently to restore an inertially collimated flow at a new expansion rate. The assumption of this type of galactic atmosphere is ad hoc but at least one can demonstrate that the collimation properties of low luminosity jets can be reproduced from simple pressure confinement with little or no magnetic confinement. The halo densities implied in this model are also reasonable.

The relation between expansion rates and magnetic structure may be different in the more radio luminous jets (Perley et al. 1980, Potash and Wardle 1979). These jets may be highly relativistic and the magnetic field direction is parallel to the jet axis.

3.3 Precession of Jet Axes

The evidence is accumulating that the ejection of energy which operates in the galaxy nucleus precesses or swings significantly. Such a precession produces morphologies in sources with reflective symmetries. The source NGC315 in Figure 1 is a good example of the "S"-shaped structures one obtains. In this and virtually all other sources the jet points to or blends into a hot spot at one edge of the radio lobe. The remaining lobe structure, transverse to the jet axis, is then populated with relativistic particles deposited when the jet pointed in that direction. The length of this transverse lobe region depends on the rate of precession of the jet, the radiating lifetime of the particles, the reacceleration mechanism operating in the lobes and the effective speed of the end of the jet (relativistic time delay effects may be important).

A morphology suggestive of more extreme precession is shown in Figure 5 for the sources 3C315 and NGC326 (Northover 1976, Ekers et al. 1978). These maps were recently made at the VLA. Here the outer lobes are more extended and both show the reflective symmetry necessary for precession. The jet in NGC326 points to the hot spots at the edge of each lobe as expected. Recent high resolution observations at the VLA show that 3C315 does have a short radio jet which also points to a warm spot near the edge of the radio lobe. Distortion of the lobes by orbital acceleration (Blandford and Icke 1978) as in the outer parts of 3C31 and 3C449 cannot explain the morphology in 3C315 when the jet direction is considered.

It seems significant that both radio sources are associated with a close pair of galaxies, indicated by the crosses in Figure 5. Such

EXTENDED RADIO SOURCES 121

close interaction could account for more rapid precession than normal (Rees 1978).

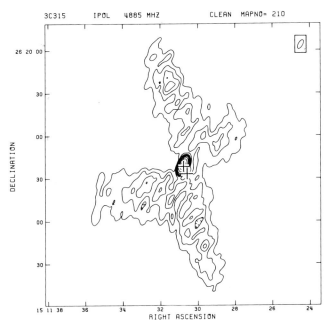

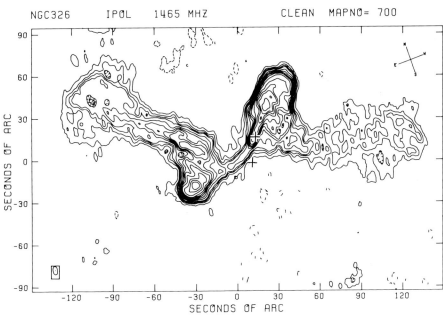

Figure 5. Two sources with rotational symmetry.
(a) 3C315 at 4.9 GHz   (b) NGC326 at 1.5 GHz
The crosses show the galaxies near the core.

3.4 Freaks

Quite often, the study of freak morphologies can better contribute to an understanding of some aspects of radio source evolution. The following two sources may be of interest in this regard.

The source 3C293 is the relatively close, 14-mag galaxy, VV5-33-12 with a peculiar optical structure (Argue et al. 1978). The minor axis of the light distribution is in position angle 60 deg but the galaxy shows several knots on the blue Palomar Sky Survey. The overall radio structure with five arcsec resolution at 20cm is shown in Figure 6 (Bridle and Fomalont 1980). The dashed outline shows the approximate edge of the optical galaxy. There is a dominant core which is steep in spectrum and about 2 arcsec extended in position angle 90 deg. An 0.2 arcsec resolution map at 2cm shows that the "core" is composed of three components and the galaxy nuclei is between two of these components. This is, obviously, not a true radio core. A jet emanates from the core, first in position angle 90 deg, then bends to position angle 180 deg, then bends back to a position angle of about 110 deg before terminating in a radio lobe. Faint emission to the south-west of the core does exist and is better seen on a lower resolution map which is not included here. Thus, 3C293 has a, more or less, typical triple radio structure although the core is very dominant.

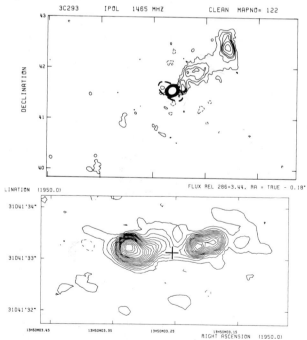

Figure 6. The radio source 3C293.
    (a) 5" resolution at 1.5 GHz with galaxy outlined
    (b) 0".2 resolution at 15 GHz with galaxy nucleus as +

The segmented appearance of the jet is extremely curious and its cause can only be speculated. Precession of the energy source of the jet cannot produce such a morphology. Orbital acceleration of the nuclear engine as it rotates rapidly around another object could produce a flow with the appearance of the north-west jet (Blandford and Icke 1978), but the south-west lobe emission is in the wrong place for this model.

The correlation found by Palimaka et al. (1978) and Guthrie (1978) that the radio axis tends to align with the minor axis of the galaxy suggests that radio jets try to find a path of least resistance out of the galaxy. If elliptical galaxies are oblate rotators, then there may also be a dynamic influence in this minor axis-radio axis correlation. Thus, one might speculate that the jet in 3C293 has taken the path of least resistance and has deflected on two occasions to avoid high density regions. Because of the knotty appearance of the galaxy, such a devious route might be necessary.

The lack of a true, flat-spectrum, radio core in this source is also peculiar. The role of the two arcsec feature near the nucleus is also strange. In some ways this component resembles the nuclear emission associated with 3C236 (Fomalont et al. 1979) which also has a majority of emission in a steep spectrum source near the nucleus of the galaxy.

Fornax A, identified with NGC1316, has recently been observed optically (Schweizer 1980), the large-scale radio emission has been accurately mapped (Goss et al. 1980) and a small radio jet has been detected (Fomalont and Geldzahler 1980). The radio maps are shown in Figure 7. The structure near the nucleus contains three radio components. A core, coincident with the nucleus, with again a relatively steep spectral index of -0.5. A radio jet with spectral index -0.5 emanates from both sides of the core for about 1 kpc in a position angle of about 100 deg. The jet then deflects to an angle of about 145 deg, steepens to a spectral index of -1.2, which is unusual, and disappears after about 2 kpc. This part of the jet points to the outer edges of both radio lobes, but no significant hot spots are apparent. It is suggested that the energy flow in the jet has been recently terminated.

The optical studies of the galaxy show that dust and ionized gas appear to be rotating quickly and their axes are inclined to that of the nuclear stellar system by about 90 deg. It appears that a significant amount of gas is now infalling into the nuclear region and has interrupted energy flow to the outer lobes. Such infall of gas or another galaxy has been suggested as the fueling for strong radio emission (Shklovski 1962, Rees 1978). If so, NGC1316, the third closest strong radio source, is worth detailed study to understand this possible fueling mechanism of radio sources.

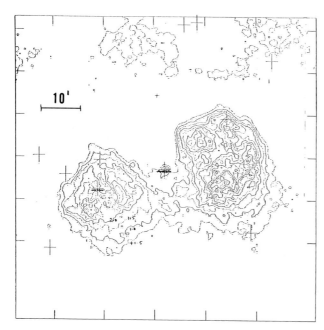 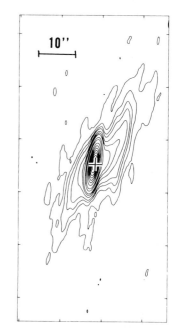

Figure 7. The radio source Fornax A
          (a)   40" resolution at 1.4 GHz
          (b)   3" resolution of central region at 4.9 GHz

4. SUMMARY

It may not be too strong of a statement that for the first time astronomers are getting to the heart of the physical processes which determine radio sources. Continued high-resolution mapping of the details in the sources and more accurate modeling of their physical conditions promise a more complete understanding of one of the most energetic phenomenon in the universe.

REFERENCES

Argue, A. N., Riley, J. M. and Pooley, G. G., 1978, Observatory, 98, 132.
Blandford, R. D. and Icke, V., 1978, M.N.R.A.S., 185, 527.
Blandford, R. D. and Königl, A., 1979, Ap. J., 232, 34.
Blandford, R. D. and Rees, M. J., 1974, M.N.R.A.S., 169, 395.
Bridle, A. H., Davis, M. M., Fomalont, E. B., Willis, A. G. and Strom, R. G., 1979, Ap. J., 228, L9.
Bridle, A. H. and Fomalont, E. B., 1980, in preparation.
Bridle, A. H., Henriksen, R. N., Chan, K. L., Fomalont, E. B., Willis, A. G. and Perley, R. A., 1980, in preparation.
Butcher, H., van Breugel, W. J. M. and Miley, G. K., 1980, Ap. J., in press.

Chan, K. L. and Henriksen, R. N., 1980, Ap. J., in press.
Ekers, R. D., Fanti, R., Lari, C. and Parma, P., 1978, Nature, 276, 588.
Fabbiano, G., Doxsey, R. E., Johnston, M., Schwartz, P. A. and Schwarz, J., 1979, Ap. J., 230, L67.
Fabricant, D., Lecar, M. and Gorenstein, P., 1979, I.A.U. Joint Discussion: "Extragalactic High Energy Astrophysics", ed. H. van der Laan, p. 64.
Fabricant, D., Topka, K., Harnden, F. R. Jr. and Gorenstein, P., 1978, Ap. J., 226, L107.
Fomalont, E. B., Bridle, A. H., Willis, A. G. and Perley, R. A., 1980, Ap. J., in press.
Fomalont, E. B. and Geldzahler, B., 1980, in preparation.
Fomalont, E. B., Miley, G. K. and Bridle, A. H., 1979, Astron. Astrophys., 76, 106.
Goss, W. M. et al., 1980, in preparation.
Guthrie, B. N. G., 1979, M.N.R.A.S., 186, 177.
Kellermann, K. I., 1978, Phys. Scripta, 17, 275.
Miley, G. K., 1971, M.N.R.A.S., 152, 477.
Miley, G. K., 1980, Ann. Rev. Astr. Astrophys., 18.
Northover, K. J. E., 1976, M.N.R.A.S., 177, 307.
Palimaka, J. J., Bridle, A. H. and Fomalont, E. B., 1979, Ap. J., 231, L7.
Perley, R. A., Bridle, A. H., Willis, A. G. and Fomalont, E. B., 1980, Astron. J., in press.
Perley, R. A., Fomalont, E. B. and Johnston, K. J., 1980, Astron. J., in press.
Perley, R. A. and Johnston, K. J., 1979, Astron. J., 84, 1247.
Perley, R. A., Willis, A. G. and Scott, J. S., 1979, Nature, 281, 437.
Potash, R. I. and Wardle, J. F. C., 1979, Astron. J., 84, 707.
Rees, M., 1971, Nature, 229, 312.
Rees, M., 1978, Nature, 275, 516.
Readhead, A. C. S., Cohen, M. H., Pearson, T. I. and Wilkinson, P. N., 1978, Nature, 276, 768.
Saslaw, W. C., Tyson, J. and Crane, P., 1978, Ap. J., 222, 455.
Scheuer, P. A. G. and Readhead, A. C. S., 1979, Nature, 277, 182.
Schweizer, F., 1980, Ap. J., in press.
Shklovsky, I. S., 1962, Sov. Astr. A. J., 6, 465.
Simkin, S. M., 1978, Ap. J., 222, L55.
Willis, A. G., 1978, Phys. Scripta, 17, 243.

COMPACT RADIO SOURCES

I.I.K. Pauliny-Toth
Max-Planck-Institut für Radioastronomie, Bonn, F.R.G.

SUMMARY

The observational data on the radio properties of compact radio sources are presented, including their radio spectra, their occurence, and their structures, as well as the variations of the structures in the "superluminal" sources.

INTRODUCTION

During the past decade, observations made with various antennas, both single dishes and synthesis instruments, have shown that the radio emission from radio galaxies and quasars spans a wide range of dimensions, from a fraction of a parsec in nearby galaxies such as M81, to several million parsecs in the giant radio galaxies such as 3C 236 or DA 240. The range of observed sizes thus extends over at least 9 orders of magnitude. The emission from extended radio sources is dealt with elsewhere (Fomalont, 1980) and we are concerned here with compact sources of radio emission, that is, with emission on linear scales significantly smaller than a kiloparsec, and frequently on a scale of parsecs. The existence of such sources became clear some fifteen years ago, first from the observation that the radio spectra of some objects showed evidence of self-absorption and, second, from the discovery of rapid time-variations in their flux densities (Dent, 1965), both indicating the presence of very small emitting regions.

RADIO SPECTRA

Most of the extragalactic sources found in surveys made at metre wavelengths have radio spectra which can be represented by simple power-laws, i.e. the flux density S depends on frequency ν according to

$$S \propto \nu^{\alpha}$$

where the spectral index, α is typically -0.8. Frequently, the spectrum steepens at short cm wavelengths, that is, α becomes numerically larger. These sources are typically extended, double sources, in which most of the radio emission originates in radio-transparent regions some hundreds of kiloparsecs apart. The spectral index, α, is related to the index of the energy distribution of the relativistic electrons, γ, by

$$\alpha = -(\gamma-1)/2$$

The brightness temperatures of the extended regions at centimetre wavelengths are of the order of 10^4 K, but may reach 10^5 K in bright "hotspots" located at the outer edges of the extended regions. An example of the radio spectrum of such a source is shown in Fig. 1a.

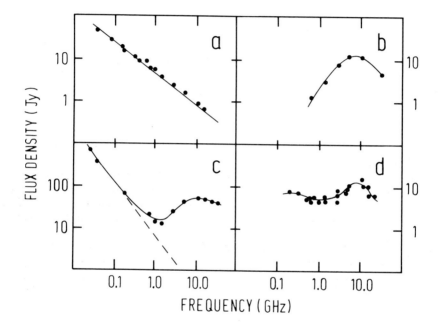

Fig. 1: Four radio spectra: a) 3C 2, a transparent source, b) 2134+004, a self-absorbed, opaque source, c) 3C 84, with both extended, transparent components, and a self-absorbed component, d) 3C 120, containing several self-absorbed components.

The compact sources, on the other hand, are characterised by flat spectra ($\alpha \gtrsim 0$), typically with a cutoff at centimetre wavelengths, below which the source is opaque to its own radiation and the value of α is positive (Fig. 1b). For a homogeneous source and a power-law electron energy distribution, the value of α below the cutoff frequency should be +2.5, but such extreme values have not in fact been observed.

The cutoff frequency, ν_m is related to the flux density S_m (Jy) of the source at that frequency, to the angular size θ (arcsec), and the magnetic field, B (gauss), by

$$\nu_m = f(\gamma) \, S_m^{2/5} \, \theta^{-4/5} \, B^{1/5} \, (1+Z)^{1/5} \text{ MHz}$$

where Z is the redshift of the source, and $f(\gamma) = 34$ for $\gamma = 1.5$ and 33 for $\gamma = 2.5$. The relation can also be written in terms of the brightness temperature T_m at the cutoff frequency as

$$\nu_m \sim 2.3 \times 10^{-17} \, T_m^2 \, B(1+Z)$$

for $\gamma = 1.5$.

A measurement of the cut-off frequency and the angular size can thus be used to determine the magnetic field in the compact sources. For sources having cutoff frequencies in the range $\sim$100 MHz to 10 GHz, the magnetic fields so derived lie in the range 10^{-1} to 10^{-5} gauss and the corresponding brightness temperature between 10^{10} and 10^{12} K (Kellermann and Pauliny-Toth, 1969). No brightness temperatures much in excess of 10^{12} K have been measured directly, and are, in fact not expected to occur, because of the resulting inverse-Compton cooling (Kellermann and Pauliny-Toth, 1969).

In some sources, both extended, transparent components with power-law spectra, and compact, opaque components are present. The resulting spectrum (Fig. 1c) is a power-law well below the cutoff frequency of the compact components, but becomes inverted and reaches a local maximum at the cutoff frequency. If several opaque components, with different cutoff frequencies are present, a complex spectrum results (Fig. 1d)

Some compact sources show flat spectra which are smooth over a wide range of frequencies (Owen et al., 1980). These are probably the result of a gradient in electron density or magnetic field through the source, so that the source becomes opaque over a large frequency range (Condon and Dressel, 1973, Spangler, 1980).

OCCURRENCE

Compact sources are found in both radio galaxies and quasars. Although it is not possible, from the radio spectrum alone, to place the object in one class or the other, statistically, most sources which show flat spectra at centimetre wavelengths are identified with quasars.

The spectral index distribution of radio sources found in surveys at centimetre wavelengths is double peaked: about half the sources have flat spectra ($\alpha \sim 0$) characteristic of compact sources, and half have the steep spectra ($\alpha \sim -0.8$) characteristic of extended sources. Most of the former group are quasars, but a significant fraction ($\sim$16 percent) are identified with radio galaxies (Pauliny-Toth et al., 1978). Optically, such galaxies have bright nuclei and are of the Seyfert or N type. The nucleus contains a compact, opaque radio source which dominates the emission at centimetre wavelengths.

The compact radio sources are opaque and weak at metre wavelengths; sources detected in surveys at these wavelengths are dominated by the emission from the extended regions and have "normal" power-law spectra. Nevertheless, many (perhaps all) of the sources found in such surveys contain compact components with flat radio spectra, located in the nucleus of the optical galaxy, or in the quasar. These "central components" of extended double sources are generally weak at metre and decimetre wavelengths, but become more prominent at short wavelengths. They frequently show time-variations in their flux density (e.g. Hine and Scheuer, 1980) and are in general similar to the stronger compact sources found in short wavelength surveys.

There appears to be a difference in the spatial distribution of sources with flat and steep spectra. For example, quasars with steep spectra ($\alpha < -0.5$) show evidence for strong evolution, in the sense that their space density, or luminosity, was higher at past epochs, whereas quasars with flat spectra ($\alpha > -0.5$) show little, or no evolution (e.g. Schmidt, 1976; Masson and Wall, 1977). The difference is apparent both in the results of the $<V/V_m>$ test (Schmidt, 1968) for the two classes of quasars and in the source counts for sources with flat and steep spectra in centimetre wavelength surveys, where the former show a more rapid convergence at low flux densities than the latter (Kühr, 1980).

SOURCE STRUCTURE

The development of mapping techniques using "closure" phase (Readhead and Wilkinson, 1978) has made it possible to derive reliable maps of compact radio sources on a milliarcsec scale from VLBI data. We may distinguish between sources in which the emission at centimetre wavelengths is dominated by a compact component, and those in which emission from an extended symmetrical double source dominates.

The former class includes the "superluminal" sources and covers a wide range of structures, including core-halo sources, simple doubles, core-jet sources and complex sources. Some typical structures are shown in Fig. 2.

Among core-halo sources are such different objects as the BL Lac-type source OJ 287 and the nucleus of the giant elliptical galaxy M 87. The cores in both have an angular extent of ~ 0.4 milliarcsec (Kellermann et al., 1977), which corresponds to a linear size of ~ 1.5 light-months in the latter. Whereas the flux density of the core varies strongly in OJ 287, no variations have been detected in the core of M 87 (Kellermann et al., 1973). The halo sizes are about 10 times larger than the core sizes in both sources. It is interesting to note that although the structure of M 87 on a scale of arc seconds (i.e. of kiloparsecs) is highly asymmetric, the compact nuclear source shows no significant asymmetry.

Examples of simple double sources are the quasars 4C 39.25 and 2134+004. In both sources, the two components are of comparable strength and show cutoffs in their radio spectra between 5 and 15 GHz. Both sources appear to be "stable": the component separations have not changed significantly over several years, although the flux densities of the components have changed (Shaffer et al., 1977, Bååth et al., 1980, Pauliny-Toth et al., in prep.). In the case of 4C 39.25, which has been observed extensively, the upper limit on the rate of separation is 0.2 c, for 2134+004, where the observations are sparser, the limit is 10 c. The magnetic fields derived from the component sizes and cutoff frequencies are in the range 10^{-1} to 10^{-3} gauss, and the brightness temperatures are near 10^{11} K, close to the "inverse-Compton limit".

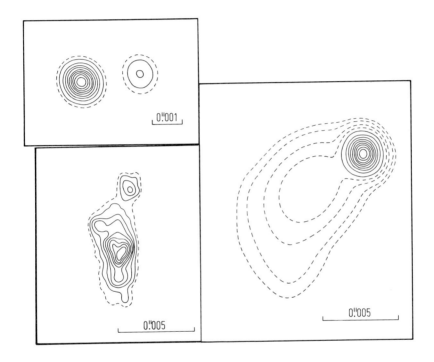

Fig. 2: *Structures of 3 compact sources, where emission from the compact structure dominates: left top, 4C 39.25, a stable double source; right, 3C 454.3, a core-jet source with a variable core; left bottom, 3C 84, a complex, expanding source. Maps are at 6 cm wavelength (Pauliny-Toth et al., in preparation).*

The highly asymmetric "core-jet" structure appears to be common in the sources with "superluminally" separating components, but is not confined to them. Examples are the superluminal sources 3C 345, 3C 273 and 3C 120 (Readhead et al., 1979), but also "stable" sources such as 3C 147 and 3C 380 (Readhead and Wilkinson, 1980). A characteristic of these sources is a curvature of the jet, with a change of P.A. of up to 40° between the innermost part of the jet and the extended structure (Readhead et al., 1978a). The extended ($\gtrsim 1"$) structure in these sources is also one-sided. The curvature of the jets has been explained by Readhead et al. (1978) as the amplification of a smaller curvature by projection effects. A further characteristic of the core-jet compact sources is a systematic variation of the spectral index along the source axis: the spectrum of the core is flat, or inverted, and steepens along the jet (Readhead et al., 1979).

A source with a more complex structure is the nucleus of NGC 1275 (3C 84), which has been shown to consist of three main centres of emission, lying approximately in a straight line, the position angle of which corresponds to that of larger-scale structure (Pauliny-Toth et al., 1976b). The three main regions of emission appear to be stationary relative to each other, with an upper limit of 10^4 km sec^{-1} on the relative velocities, but do appear to be expanding at a rate of about 3×10^4 km sec^{-1} (Preuss et al., 1979). The rate of expansion, as well as the rate of increase of flux density suggest some "initiating event" in the source about 20 years ago. The limit on the relative motion, the source extent of ~ 10 light years and the age of ~ 20 years suggest that particles are injected into, or accelerated in stationary regions.

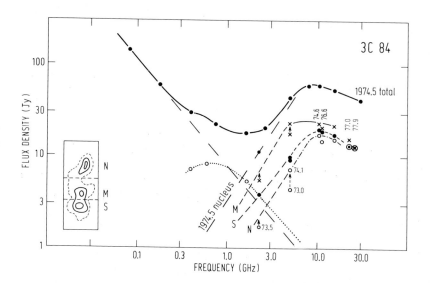

Fig. 3: *Spectrum of the compact source in the radio galaxy 3C 84. VLBI data at several wavelengths give spectra for three main emission regions, labelled N, M, and S in the schematic map at the left.*

Maps of the compact source are available at wavelengths between 13 and 1.3 cm, and permit a crude spectrum of the three main emission regions to be derived (Fig. 3). In contrast to the core-jet sources, the nuclear source in NGC 1275 does not show a systematic change in the spectrum along the axis. All three emission regions have similar spectra and all become self-absorbed, the middle region at a somewhat lower frequency than the outer two, implying a lower magnetic field in the central region. The magnetic fields are in the range 10^{-1} to 10^{-3} gauss.

Maps of the compact central components of a number of symmetric, extended double sources are now available. These include 3C 111 (Pauliny-Toth et al., 1976a), 3C 236 (Schilizzi et al., 1979), NGC 6251 (Readhead et al., 1978b), 3C 390.3 (Preuss et al., 1980) and 3C 405 (Kellermann et al., 1980). Figure 4 shows the structure of the central components in the last two of these sources, together with the extended structure.

In spite of the symmetry of the extended double structure, the central components in all these sources show an asymmetric structure, with an extension towards one of the lobes. The compact structure is aligned with the extended structure to within a few degrees, although in the case of 3C 405 (Cygnus A) the deviation is about $10°$. This is in contrast to the strong core-jet sources referred to earlier. There is as yet no direct evidence that any of these central components show superluminal expansion. The alignment, which extends from scales of ~ 1 pc to $\gtrsim 10^6$ pc shows that the directive mechanism acts over a large range of dimensions and over long times.

SUPERLUMINAL SOURCES

In four radio sources, the quasars 3C 273, 3C 279 and 3C 345, and one radio galaxy, 3C 120, VLBI observations have shown that the components separate at rates which, when translated into linear velocities, appear to exceed the speed of light.

In 3C 345, the compact core consisted of two components, of approximately equal strength at short centimetre wavelengths, in 1974 (Shaffer et al., 1977). VLBI observations between 1970 and 1976 showed that the components separated at a constant rate of 0.17 milliarcsec per year (Cohen et al., 1977), the corresponding linear velocity being 6.7 c (H_0 = 55 km s^{-1} Mpc^{-1}, q_0 = 0.05). There was no dependence of the separation on the wavelength of observation. After 1976, the structure of the source changed, and one of the components became weaker, until, in 1977, the structure was of the core-jet type (Readhead et al., 1979; Cohen et al., 1979) and the rate of separation seemed to have decreased, or stopped. Data obtained at wavelengths of 1.3 cm and 6 cm in 1978/1979 (Bååth et al., 1981; Kellermann et al., 1981) show that the separation of the brightest regions in the source was ~ 1.3 milliarcsec, rather than the 2.2 milliarcsec expected from an extrapolation of the data of Cohen et al. (1977). Possibly, this may represent a new "event" in the source.

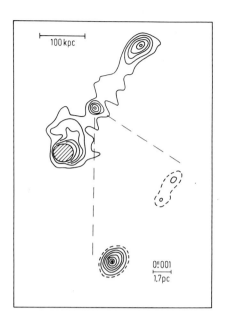

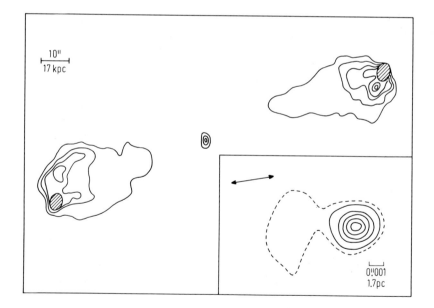

Fig. 4: VLBI maps of the central components in the galaxies 3C 390.3 (top) and Cygnus A = 3C 405 (bottom), shown together with the extended double structure. The extended structure is adapted from Harris (1972) and Hargrave and Ryle (1974); the VLBI maps are at 2.8 cm wavelength, from Preuss et al. (1980) and Kellermann et al. (1980).

3C 273 has a complex structure, consisting of three compact components and a large "jet-like" structure ∿7 milliarcsec (50 pc) away, in the direction of the optical jet. Cohen et al. (1977) derived a separation rate for the components of 0.32 milliarcsec per year, or 4.2 c, from the movement of the first minimum of the visibility function in the Fourier transform plane. This represents the relative motion of the main centres of emission in the source. Again, the structure appears to have become more asymmetric in 1976/1977, although the expansion appears to have continued (Cohen et al., 1979).

3C 120 has shown several outbursts in the total flux density between 1970 and 1978 and at least two clear cycles of expansion. The separation velocities for these were 5 c and 8 c (Cohen et al., 1977). Recently, activity in the source has decreased and the structure is of the core-jet type (Readhead et al., 1979).

3C 279 has not been observed as intensively as the other three sources. Cotton et al. (1979) have analysed data taken between 1970 and 1972 and have shown that they were consistent with a double source, with components separating at a rate of 21 c. Data taken between 1972 and 1974 show that this expansion continued, but that the source was now triple, with the most compact components separating at a rate of ∿40 c (Pauliny-Toth et al., in preparation).

In the cases where more than one "event" or cycle of expansion has occurred (3C 120, 3C 279) the motion has been along the same position angle. The rates of separation for different events in the same source are, however, different. This places restrictions on possible theoretical models for the sources, as discussed elsewhere (Kellermann, 1980).

Further restrictions on theoretical models are placed by the "stationary" sources. For example, the sources 4C 39.25 and 2134+004 have structures and luminosities similar to that 3C 345 had at one stage in its expansion, yet do not expand superluminally.

Yet another problem are the central components of symmetric doubles. They appear to have the one-sided structure typical of the core-jet and superluminal sources. The one-sided structure is unlikely to be the result of relativistic motion nearly along the line of sight, since the compact structure is aligned with the outer lobes, and the latter are not likely to have a preferential orientation along the line of sight. The detection of superluminal motion in such sources would exclude any model which requires a special orientation of the source with respect to the observer.

REFERENCES

Bååth, L.B., Cotton, W.D., Counselman, C.C., Shapiro, I.I., Wittels, J. J., Hinteregger, H.F., Knight, C.A., Rogers, A.E.E., Whitney, A.R., Clark, T.A., Hutton, L.K. and Niell, A.E. 1980, Astron. Astrophys. in press.

Bååth, L.B., Rönnäng, B.O., Pauliny-Toth, I.I.K., Kellermann, K.I., Preuss, E., Witzel, A., Matveyenko, L.I., Kogan, L.R., Kostenko, V. I., Moiseev, I.G. and Shaffer, D.B. 1981 in preparation

Cohen, M.H., Kellermann, K.I., Shaffer, D.B., Linfield R.P., Moffet, A. T., Romney, J.D., Seielstad, G.A., Pauliny-Toth, I.I.K., Preuss, E., Witzel, A., Schilizzi, R.T. and Geldzahler, B.J. 1977, Nature 268, 405.

Cohen, M.H., Pearson, T.J., Readhead, A.C.S., Seielstad, G.A., Simon, R.S. and Walker, R.C. 1979, Astrophys. J. 231, 293.

Condon, J.J. and Dressel, L.L. 1973, Astrophys. Lett. 15, 203.

Cotton, W.D., Counselman, C.C., Geller, R.B., Shapiro, I.I., Wittels, J.J., Hinteregger, H.F., Knight, C.A., Rogers, A.E.E., Whitney, A.R. and Clark, T.A. 1979, Astrophys. J. 229, L115.

Dent, W.A. 1965, Science 148, 1458.

Fomalont, E.B. 1980, this volume.

Hargrave, P.J. and Ryle, M. 1974, Monthly Not. Roy. Astron. Soc. 166, 305.

Harris, A. 1972, Monthly Not. Roy. Astron. Soc. 173, 37.

Hine, R.G. and Scheuer, P.A.G. 1980, Monthly Not. Roy. Astron. Soc. in press.

Kellermann, K.I. 1980, this volume.

Kellermann, K.I., Clark, B.G., Cohen, M.H., Shaffer, D.B., Broderick, J.J. and Jauncey, D.L. 1973, Astrophys. J. 179, L141.

Kellermann, K.I., Downes, A.B., Pauliny-Toth, I.I.K., Preuss, E., Shaffer, D.B. and Witzel, A. 1980 in preparation.

Kellermann, K.I. and Pauliny-Toth, I.I.K. 1969, Astrophys. J. 155, L71.

Kellermann, K.I., Schraml, J., Witzel, A., Pauliny-Toth, I.I.K. and Johnston, K. 1981 in preparation.

Kellermann, K.I., Shaffer, D.B., Purcell, G.H., Pauliny-Toth, I.I.K., Preuss, E., Witzel, A. Graham, D., Schilizzi, R.T., Cohen, M.H., Moffet, A.T., Romney, J.D. and Niell, A.E. 1977, Astrophys. J. 211, 658.

Kühr, H. 1980, Dissertation for the degree of Doctor of Philosophy, University of Bonn.

Masson, C.R. and Wall, J.V. 1977, Monthly Not. Roy. Astron. Soc. 180, 193.

Owen, F.N., Spangler, S.R. and Cotton, W.D. 1980, Astron. J. 85, 351.

Pauliny-Toth, I.I.K., Preuss, E., Witzel, A., Kellermann, K.I. and Shaffer, D.B. 1976a, Astron. Astrophys. 52, 471.

Pauliny-Toth, I.I.K., Preuss, E., Witzel, A., Kellermann, K.I., Shaffer, D.B., Purcell, G.H., Grove, G.W., Jones, D.L., Cohen, M.H., Moffet, A.T., Romney, J.D., Schilizzi, R.T. and Rinehart, R. 1976b, Nature 259, 17.

Pauliny-Toth, I.I.K., Witzel, A., Preuss, E., Kühr, H., Kellermann, K.I., Fomalont, E.B. and Davis, M.M. 1978, Astron. J. 83, 451.

Preuss, E., Kellermann, K.I., Pauliny-Toth, I.I.K., Witzel, A. and Shaffer, D.B. 1979, Astron. Astrophys. 79. 268.

Preuss, E., Kellermann, K.I., Pauliny-Toth, I.I.K. and Shaffer, D.B. 1980, Astrophys. J. in press.

Readhead, A.C.S., Cohen, M.H., Pearson, T.J. and Wilkinson, P.N. 1978a, Nature 276, 768.

Readhead, A.C.S., Cohen, M.H. and Blandford, R.D. 1978b, Nature 272, 131.

Readhead, A.C.S. and Wilkinson, P.N. 1978, Astrophys. J. 223, 25.

Readhead, A.C.S., Pearson, T.J., Cohen, M.H., Ewing, M.S. and Moffet, A.T. 1979, Astrophys. J. 231, 299.

Readhead, A.C.S. and Wilkinson, P.N. 1980, Astrophys. J. 235, 11.

Schilizzi, R.T., Miley, G.K., van Ardenne, A., Baud, B., Bååth, L., Rönnäng, B.O. and Pauliny-Toth, I.I.K. 1979, Astron. Astrophys. 77, 1.

Schmidt, M. 1968, Astrophys. J. 151, 393.

Schmidt, M. 1976, Astrophys. J. 209, L55.

Shaffer, D.B., Kellermann, K.I., Purcell, G.H., Pauliny-Toth, I.I.K., Preuss, E., Witzel, A., Graham, D., Schilizzi, R.T., Cohen, M.H., Moffet, A.T., Romney, J.D., Niell, A.E. 1977, Astrophys. J. 218, 353.

Spangler, S.R. 1980, Astrophys. Lett. 20, 123.

NUCLEI OF GALAXIES : THE ORIGIN OF PLASMA BEAMS

Martin J. Rees
Institute of Astronomy
Cambridge, England

The radio structures mapped by aperture-synthesis techniques with the VLA, and discussed earlier at this conference by Fomalont, are energised by activity in galactic nuclei. The compact radio sources resolvable by VLBI techniques (Pauliny-Toth, these proceedings) are on scales $\sim 10^5$ times smaller; but even this still far exceeds the scale where - according to most theoretical ideas - the primary power production is concentrated. This central "core" - involving a large concentrated mass, probably a black hole of $\sim 10^8$ $M_\odot$ accreting material from its surroundings - is the primary origin of the power for all categories of activity associated with active galaxies and quasars. In this paper, I shall principally discuss how this "core" could give rise to the fast-moving plasma which emerges in directed beams and energises the radio sources; I shall also comment on some interesting features of radio source structure.

1. THE CENTRAL "CORE"

1.1 Some Characteristic Numbers

In any efficient source powered by inflow into a compact object, the primary power output is concentrated within a region only a few times r_s (the Schwarzschild radius) in size: the r^{-1} fall-off in the depth of the potential well means that the gravitational energy released at much larger r is negligible, though of course radiation or particle fluxes generated at small radii can be reprocessed further out.

The following approximate figures are illustrative of the densities, temperatures, etc. in the primary emitting "core". Suppose that the characteristic dimensions are ~ 10 $r_s \simeq 3 \times 10^{14}$ M_8 cm, M_8 being the central mass in units of 10^8 $M_\odot$. The free-fall (or Keplerian) speed throughout this region is $V_{free\ fall} \gtrsim c/3$. The inward drift speed V_{inflow} would be of order $V_{free\ fall}$ for radial accretion; when angular momentum is important, V_{inflow} depends on viscosity. Suppose that the accretion rate yields a luminosity L with an efficiency $\varepsilon \simeq 0.1$. Then characteristic quantities are

Optical depth:
$$\tau_e \simeq 1 \times (L/L_E) \; (V_{free\;fall}/V_{infall}) \tag{1}$$

Particle density:
$$n \simeq 10^{10} \times (L/L_E) \; (V_{free\;fall}/V_{infall}) \; M_8^{-1} \; cm^{-3} \tag{2}$$

The maximum magnetic field, corresponding to equipartition with the bulk kinetic energy, would be

$$B_{eq} \simeq 10^4 \times (L/L_E)^{\frac{1}{2}} \; (V_{free\;fall}/V_{infall})^{\frac{1}{2}} \; M_8^{-\frac{1}{2}} \tag{3}$$

The intensity of the radiation field can be expressed by the ratio

$$\left(\frac{\text{Radiation energy density}}{\text{Rest-mass energy density}}\right) \simeq 0.1 \times \left(\frac{V_{free\;fall}}{V_{infall}}\right)^{-1} \tag{4}$$

This implies that the radiation energy may amount to as much as 100 Mev for each particle in the emitting region, a value not approached in other astrophysical contexts apart from pulsar magnetospheres. Another measure of the radiation intensity is the equivalent black body temperature

$$T_{bb} \simeq 3 \times 10^5 \times (L/L_E)^{\frac{1}{4}} \; M_8^{-\frac{1}{4}} \; K. \tag{5}$$

Any thermal radiation emerging directly from the central "core" would tend to be in the far ultraviolet or soft X-ray parts of the spectrum. Another consequence of (5) is that k T_{bb} is far below the "virial temperature" k $T_{virial} \simeq m_p c^2 (r_s/r)$; gas that has cooled down to thermal equilibrium at a temperature $\simeq T_{bb}$ can therefore only be supported against gravity if radiation pressure exceeds gas pressure by a large factor $\sim (T_{virial}/T_{bb})$.

In the central "core", where the bulk velocities are comparable with c, conditions are exceptionally favourable for the acceleration of ultra-relativistic particles. Indeed, the distinction between "thermal" and "non-thermal" particles becomes rather blurred under these extreme conditions. The energy available for each "thermal" ion is $\sim$ 100 Mev, and at kinetic temperatures corresponding to this we cannot assume that the electron temperature (T_e) and the ion temperature (T_i) are equal, nor that there is any well-defined component of the plasma with a Maxwellian velocity distribution. In general, it is likely that $T_e < T_i$, the electrons being efficiently cooled and maintained at energies $\lesssim$ 1 Mev by synchrotron or compton emission, except for the small proportion that may have just passed through a shock, or a region of magnetic reconnection, where the coupling between electrons and ions is more efficient than that due to ordinary Coulomb encounters.

Physical processes in "transrelativistic" plasmas where $k T_i > k T_e \simeq m_e c^2$ deserve greater theoretical attention, taking account of magnetic fields, pair production effects, weak electron-ion coupling, etc. Such processes must play an essential role in the "core region" of galactic nuclei: this is the region where the observed non-thermal optical continuum and X-rays are probably emitted, and which indirectly energises all the phenomena occurring further out.

There is always the possibility that a small fraction of the particles can be accelerated in shock fronts, or in regions of magnetic reconnection, to very high Lorentz factors. Cavaliere and Morrison (1980) estimate the _maximum_ energy that can be attained by individual particles. At this _maximum_ energy the effects of the most efficient possible acceleration (a 'linear accelerator' with $E \simeq B$) are cancelled by synchrotron and Compton radiation losses. The maximum γ is that for which the gyroradius ($\propto \gamma B^{-1}$) and radiation length ($\propto \gamma_e^{-1} B^{-2}$) are equal. Of course this is very much an upper limit: on overall energetic grounds it is unattainable unless the available power can all be channelled into a very small favoured fraction of the particles (as may happen in pulsar-type 'light cylinder' mechanisms). If the particle acceleration is limited by radiation reaction, note that it is easier for ions than for electrons to attain very high individual energies.

More precise estimates than (1) - (5) for physical conditions in the central core can be made on the basis of specific models (see, for example, Blandford and Rees 1978a, Maraschi et al. 1979, Rees 1980, Cavaliere and Morrison 1980). The requirement that the electron scattering optical depth be ≤ 1 (since otherwise the high polarization observed in the optical non-thermal continuum would be washed out) implies that the amount of material in the emitting core must be $\leq r^2 m_p/\sigma_r$. For $r \leq 10^{15}$ cm this implies a mass $\leq 10^{-3} M_\odot$: in other words, about ten orders of magnitude less than the likely central mass enclosed within r. This occasions no problem in models involving accretion onto a black hole (since the only material that need actually be present around the hole is that required to supply the power for a time $\sim r/c$); but the requirement that the density in the emitting region is 10 orders of magnitude less than the mean density of matter enclosed within it is a severe constraint on spinar-type models where the bulk of the material is postulated to be still in an un-collapsed state.

1.2 Radiation Mechanisms

The optical polarized continuum is most naturally interpreted as synchrotron radiation. The X-rays could be explained similarly, but could alternatively be comptonised bremsstrahlung from plasma with $T_e \simeq 10^9$ °K. For a fuller discussion of continuum radiation from the core, see O'Dell (1979) or Fabian and Rees (1978).

In discussions of non-thermal emission from galactic nuclei, some authors still allude to the so-called "inverse Compton problem" i.e. that the relativistic particles cannot cross the emitting region in their synchrotron lifetime unless B is supposed so weak that the Compton losses becomes overwhelming. While the optical continuum sources are centainly in this parameter range, this difficulty can be evaded if acceleration occurs in a distributed way or in localities (e.g. regions of field reconnection or shock fronts) dispersed through the volume. No location is more propitious for particle acceleration than the "core" with properties described above, throughout which the bulk velocities are $\gtrsim c/3$ and whose field strengths are higher than in any other astronomical context save degenerate stars. The dispersed acceleration needed to avoid the "inverse compton problem" thus seems inevitable rather than an ad hoc contrivance.

A general constraint in the high-luminosity sources, discussed by Blandford and Rees (1978a) and Cavaliere and Morrison (1980) comes from the condition that the saattering optical depth of the emission region must be $\lesssim 1$, other wise the polarization would be reduced by electron scattering. If each relativistic electron injected into the volume were to cool (in $t_{rad} < r/c$) and thereafter remain for $\gtrsim r/c$, this condition would be violated. The conclusion is that such relativistic electrons must be reused several times.

1.3 Coherent Emission?

None of the reliably-detected rediation from galactic nuclei need be coherent. The VLBI radio components studied at $\gtrsim 1$ GHz all lie apparently below the "Compton limit" (Pauliny-Toth, these proceedings); the brightness temperature in the optical and infrared continuum, even if the dimensions are $\sim 10^{14}$ cm, is low enough to be compatible with incoherent synchrotron or Compton emission. At lower radio frequencies (< 1 GHz) variability is now well established in several sources. The inferred brightness temperatures (neglecting relativistic factors) exceed 10^{15} °K, and bulk Lorentz factors as large as ~ 50 may be needed to bring the temperatures below the Compton limit of $\sim 10^{12}$ °K. While this cannot be ruled out, it would be hard to devise a model with plausible efficiency in which the bulk relativistic factor was so large.

The possibility of coherent emission, particularly in the radio band, certainly cannot be ruled out. The field strength in the core being up to $\sim 10^4$ G (cf eqn (3)) the gyrofrequency can be in the GHz band. This field exceeds the estimated value at the light cylinder of pulsars with $\sim$ 1s period: the flow pattern around a black hole in many respects resembles an "inside out" light cylinder (Blandford and Znajek 1977). Thus coherent cyclotron-maser emission cannot be excluded.

But if the radiation is produced deep in the quasar core, it is vulnerable to absorption (the inverse of incoherent emission) at larger radii. Moreover, so high would be the brightness temperature T that induced scattering (which swamps spontaneous scattering by a factor $(k\,T/m_e c^2)$ x (solid angle factors)) could quench the radiation unless there were directions where $\tau_{es} \ll 1$ or the electrons were all flowing out with high Lorentz factors in which case relativistic corrections reduce the scattering (cf Wilson and Rees 1978). Thus, if the sub-GHz variations involve coherence, it is likely to be a mild degree of coherence within a region comparable in size to the radio components already resolved by VLBI, rather than pulsar-type emission from the central "core".

2. OBSERVATIONS OF BEAMS AND JETS

There is now strong morphological evidence that active galactic nuclei give rise to jet-like features, with lengths ranging from a few parsecs to hundreds of kiloparsecs. Phenomena observed to date include:

a) <u>Jets associated with double radio sources</u>. These jets, observed in NGC 6251, 3C 66B, 3C 219 and NGC 315, are linear features with opening angles in the range of 3-15° and are often one-sided. So far, jets have only been reported associated with weaker, comparatively close-by radio doubles, but this may in part be a selection effect.

b) <u>Jets associated with compact radio sources</u>, eg 3C 273, 3C 345, 3C 147 and 3C 380 (see Readhead, 1979 and references cited therein). Recent results from VLBI have shown that compact variable radio sources are usually of the "core-jet" type in which an inhomogeneous jet emerges from an unresolved optically thick nucleus. Apparent superluminal expansion is often observed between prominent features in the jet and the core. This is interpreted as evidence for relativistic motion within the jets (Blandford et al. 1977, Mascher and Scott 1980). Compact jets are usually roughly aligned with larger scale radio features where these exist. However, the jets in the powerful sources are bent through angles of up to 30° as they are observed with progressively larger angular resolution (Readhead et al. 1978). This bending is probably exaggerated because the jet makes a small angle with the line of sight, a condition that can also give rise to superluminal expansion (Scheuer and Readhead 1979, Blandford and Konigl 1979). Extended double radio sources generally contain relatively faint compact central components; the brightest of these (eg 3C 236 and Cygnus A) have also been examined by VLBI and show linear structure. These compact sources are much more closely aligned with the centre radio components, suggesting the absence of a strong projection effect.

c) <u>Optical jets</u>. The most notable optical jets are those associated with M 87 and 3C 273, nuclei in which there is independent evidence of activity. Both of these are also radio jets. Optical emission has also been recently detected from 3C 66 and 3C 31 (Butcher et al. 1980). A jet-like feature displaying emission lines blue-shifted by 3000 km s^{-1} has been associated with DA 240, but this bears no relation to the radio structure (Burbidge, Smith and Burbidge 1975).

d) <u>X-ray Jets</u>. Centaurus A has recently been shown to contain an X-ray jet located a few kpc from the nucleus and apparently feeding the inner radio lobes (Schreier <u>et al</u>. 1979). This may be related to the optical jet reported earlier by van den Bergh.

e) <u>Galactic Jets</u>. The moving emission lines from the galactic object SS 433 have been modelled kinematically as a pair of antiparallel precessing jets (Abell and Margon 1979) with a velocity of $\sim 0.27c$. This is the only case where we actually know the flow speed. There is also evidence for radio jets from SS 433 which may suggest an analogy with Sco X-1.

f) <u>Inferred Directivity on Unresolvably Small Scales</u>. The VLBI data on the "superluminal" compact radio sources suggest that beams exist on scales as small as a few parsecs. This is still several orders of magnitude larger than the compact source from which the power probably derives. There is no real prospect of probing smaller scales by radio techniques. Synchrotron self-absorption suppresses the power from smaller regions; and even if a coherent mechanism operated in some sources earth-based VLBI would have inadequate resolution to resolve them. <u>Indirect</u> evidence for collimation on still smaller scales comes, however, from studies of optical continuum variability. There are some ultraluminous and rapidly variable objects, such as B2 1308+326 and AD 0235+164, whose optical and infrared luminosities would attain values $\sim 10^{49}$ erg s^{-1} if the emission were isotropic. These vary in intensity and polarization on timescales of $\lesssim 1$ week, the degree of polarization being up to $\sim 30\%$. These could be the members of a (more common) population of beamed sources which happen to be pointed almost towards us (Blandford and Rees 1978)). This would obviously ease the energetic problem, as well as the difficulties of accounting for the rapid variability, high polarization, etc. Support for this scheme comes from Angel and Storkman's recent (1980) optical studies of highly polarized and variable objects, which they call "blazars". The most extreme variables (such as the two sources mentioned above) have polarization vectors that vary over all "points of the compass". Contrariwise, the less luminous objects and those associated with extended double radio structure (which would be more nearly transverse to our line of sight) display a preferred long-term polarization orientation.

Compared with a non-relativistic model for a source with a given observed flux and variability timescale, a suitably oriented beam model with bulk Lorentz factor γ_b yields altered estimates for various quantities as follows:

inferred maximum angular size $\propto \gamma_b$;
inferred maximum surface brightness $\propto \gamma_b^{-2}$;
radiation energy density in frame of emitting material $\propto \gamma_b^{-6}$;
overall luminosity of source (integrated over 4π steradians) $\propto \gamma_b^{-2}$.

3. BEAM SPEEDS, AND PHYSICAL CONDITIONS IN BEAM PLASMA

One cautionary comment may be helpful before addressing the physics and collimation of beams:

The beam or jets that concern us in astrophysics are generally neutral. The electrons and the ions (or positrons) which neutralise them have essentially the same density and move with very nearly the same speed. This contrasts with the situation in most laboratory or terrestrial beams (Bekefi et al. 1980). Moreover, for radio source applications gas dynamics is a good first approximation; analogues with hypersonic gas flow (and even wind-tunnel experiments) may prove more helpful in developing our intuition than comparisons with plasma beams. The Debye length of astrophysical beams is vastly smaller than the overall scales involved, except in some extreme models of electrodynamically driven outflow where energy is carried away from a compact object by charges of energies 10^{20} ev (cf Lovelace et al. 1979 and references cited therein). The magnetic field is crucial in making the flow basically fluid-like (cf the solar wind); even though the mean free path for two-particle encounters is very long, the gyroradius is much smaller than the flow scales. Note however that the pressure may become significantly anisotropic if the magnetic field is dynamically important.

3.1 Indirect Evidence for Beam Speeds in Extended Sources

We observe direct and indirect manifestations of beams on scales much larger than those probed by VLBI, but on these scales (where we of course see no direct evidence of rapid motion), evidence on the jet speed V (not to be confused with the expansion speed of the double radio source) is conflicting. Arguments can be given in favour of values ranging from the escape velocity of an elliptical galaxy (~ 300 km s^{-1}) to the speed of light. The existence of superluminal expansion suggests that $V \simeq c$, as might be expected if the jets originate at the bottom of a relativistic potential well. However, there is the problem of "waste energy" first emphasized by Longair et al. (1973). In many sources a lower bound on the momentum flowing through the jet can be obtained by multiplying the minimum pressure in the hot spots by their apparent area. Typical values are in the range 10^{34} - 10^{35} dyne. Further multiplications by V/2 gives the jet power L_0. A large value of V often leads to powers much greater than appear to be dissipated as heat in the radio components. If a jet emerges from a collimation region with a speed $\sim c$ and is subsequently decelerated to a speed $V \ll c$ by entraining surrounding gas, then conservation of linear momentum (approximately valid for a highly supersonic jet) implies that the power in the collimation region exceeds the final power by $\gtrsim (c/V)$. The "waste energy" that is difficult to dissipate invisibly in the radio components has then got to be similarly processed in the nucleus. For a strong source like Cygnus A, a relativistic jet power is 10^{46} erg s^{-1}, much greater than the known power in the nucleus.

There is the opposite problem with the mass flux. If the jet is decelerated by entrainment of surrounding material, then the mass loss must <u>increase</u> by $\sim (c/V)$. Again, using Cygnus A as an example, a jet speed of only 1000 km s^{-1} requires 100 $M_\odot$ yr^{-1}; over the lifetime of the radio source, at least 10^{10} $M_\odot$ of gas would have to be injected into the radio components – much more than is believed to be present within an elliptical galaxy. (Most of this mass would in any case have to be hidden in cool, dense filaments in order not to lead to excessive Faraday depolarization.) So high jet speeds seem to require too much energy and low jet speeds too much mass for the galaxy to provide.

The so-called "head-tail sources" such as NGC 1265 can be interpreted as twin-beam sources where the beams are being curved sideways by the transverse pressure due to the motion of the galaxy through the surrounding intracluster medium (Begelman <u>et al</u>. 1979, Jones and Owen 1979). In this picture, the radius of curvature of the beam (for a given transverse force) tells us the momentum discharge; the energy flux along the beam is then V times the momentum discharge. This line of argument favours a value $V \gtrsim 10^4$ km s^{-1}. If the beams in NGC 1265 were actually moving relativistically, the energy flux would be larger than is required to supply the power input into the tail.

In sources where there is evidence for internal Faraday depolarization within the beam, one can infer the plasma density and thereby estimate the beam speed on the basis of <u>either</u> (a) assuming a value for the power carried by the beam, or (b) by assuming (eg in NGC 6251) that the beam is "free", so that V must exceed the internal sound speed by at least θ^{-1}, where θ is the opening angle of the beam.

With improved and more extensive observations, it should be possible to refine all the various methods of estimating V. At the moment, however, it is an open question whether any of the beams observed on scales >> 10 pc are moving relativistically, and there are good arguments <u>against</u> this view for some relatively weak sources. It is still a plausible conjecture, however, that the beams carrying energy out to the components of <u>strong</u> radio doubles have relativistic speeds. For this reason, and also because of the direct evidence for relativistic jets in components probed by VLBI, it is important to investigate how beams can be collimated in galactic nuclei, and whether relativistic velocities can plausibly be attained.

3.2 Why are the Jets Generally "One-Sided"?

One apparent puzzle is the prevalence of one-sided jets even in sources where the extended structure resembles a symmetric double. There are four classes of explanation:
  (i) <u>Relativistic flow and "Doppler favouritism"</u>. Given the strong evidence for bulk relativistic flow in the variable compact ("superluminal") sources, it is not a great extrapolation to suppose that some larger-scale beams may be moving relativistically – for

instance, those in M 87 and NGC 6251. If two identical jets are observed, pointing in opposite directions along an axis aligned at θ to our line of sight, the relative observed fluxes would be in the ratio $F = [(1 + V/c \cos \theta)/(1 - V/c \cos\theta)]^{2+\alpha}$. (Note that the exponent is $2 + \alpha$ rather than $3 + \alpha$ because we consider the contribution from a given projected length of it, rather than from the plasma ejected in a given interval: each element on the receding side contributes for longer). If $V \simeq c$ the asymmetry is very marked even when θ is not specially small. For random orientation the median value of θ is $60°$; this yields an asymmetry approaching $3^{2+\alpha}$ if V is close to c. The only really compelling arguments against this model would arise if symmetric jets were never seen in double source. Also, if the depolarization suggests a high internal density, this would be incompatible with a relativistic bulk velocity.

[A variant on this is the hypothesis that the relativistic electrons have an anisotopic distribution, even though the field along which they are gyrating is coupled to material that is not moving relativistically].

(ii) <u>Flip-flop behaviour</u>. The one-sidedness may be intrinsic: there may be no beam at all on the other side. But then, to give rise to symmetric double radio lobes, beams must have squirted for comparable times in each direction - averaged over the source lifetime. Moreover, the last injection on the now-defunct side must have been recent enough to generate the highest energy electrons still radiating. But when jets are as long as in NGC 6251 the beam must squirt for $\gtrsim 10^6 (V/c)^{-1}$ yrs between reversals. It is surprising that this interval should be so long. Only a small range of timescales can be squeezed between the two constraints.

(iii) <u>Asymmetric internal dissipation</u>. There are many strong classical doubles where the beams are not directly seen, but can be inferred to be active. In such sources, presumably the kinetic energy of the beam is transported out to the "hot spots" without there being too much internal dissipation or boundary friction along its path. It is unknown what determines the amount of such dissipation - and, in consequence, the amount of radiation from the beam. All that is required is the conversion of a few percent of the kinetic energy. Conceivably this happens to one jet but not to its counterpart on the other side. For instance, the shear across the jet may be larger on one side, due to different conditions near the nucleus, or to effects of the interstellar environment.

(iv) <u>Intervening absorption</u>. A further possibility, which may contribute to the asymmetry at least for compact jets, is that a central gas cloud or disc may obscure the far side of the nucleus. Free-free absorption would be the main mechanism, and this has the ability to absorb at $\lesssim 5$ GHz if $n_e^2 \ell \gtrsim 3 \times 10^7$ (for $T = 10^4$ °K). Values of $n_e \gtrsim 10^4$ cm^{-3} can be expected out to 10pc; this is in itself a possibly sufficient explanation for the absence of counterjets on the VLBI scale. Such absorption cannot, however, prevail out to $10^4 - 10^5$ pc; the thermal plasma that can be expected to pervade a whole galaxy is too rarified to cause absorption at GHz frequencies.

3.3 Inversion Symmetry and Precession Effects

Some extended sources display "inversion" (or 180° rotation) symmetry suggestive of precession in the beams' orientation. (Two good examples are shown in Figure 5 of Fomalont's contribution to these proceedings.) The apparent precession may be a manifestation of complex non-axisymmetric gas flow in the gravitational potential well of an elliptical galaxy. However, if the beams are collimated on a very small scale, near a massive black hole, changes in the hole's spin axis must be involved in the phenomenon.

One possibility (Rees 1978a) is that a hole whose spin axis is misaligned with the angular momentum vector of infalling gas is gradually (on a timescale $\sim M/\dot{M}$) being swung into alignment. The only way a black hole can actually <u>precess</u> at a significant rate would be if it were orbited by another hole of comparable mass. Massive black hole binaries would form if two galaxies, each containing a black hole in its nucleus, were to merge. Mergers between galaxies appear to be frequent. cD galaxies in clusters and small groups were quite probably formed in this manner and there is strong evidence that the nearby active galaxy Centaurus A is a merger product (Tubbs 1980). If massive black holes are present in their nuclei--relics of earlier activity-- then they will settle, under dynamical friction, in the core of a merged stellar system. Many cD galaxies contain double or multiple cores that may have formed in this way. There are other scenarios for binary black hole formation. For instance, a rotating supermassive star may undergo bar-mode instability and fission into two components which both collapse (cf Begelman and Rees 1978).

Binaries involving black holes contract slowly and eventually coalesce. The longest-lived phases, in which such systems stand the biggest chance of being observed, correspond to separations such that (r/r_s) lies in the range $3000 - 10^5$ (corresponding to $10^{17} - 3 \times 10^{18}$ cm for masses $\sim 10^8$ $M_\odot$): when the systems are tighter than this, gravitational radiation causes rapid coalescence; for separations exceeding this range, dynamical friction is efficient. For a given binary, the orbital period varies as $r^{3/2}$ and the precession period as $r^{5/2}$. "Wide" binaries with precession periods $\sim 10^8$ yrs may be responsible for the inversion-symmetric features in large-scale jets and double sources; closer binaries, with orbital periods ~ 30yrs and precession periods $\sim 10^4$ yrs, may produce bending and misalignment in VLBI jets through the kinematic consequences of precession. This scenario has been investigated by Begelman <u>et al</u>. (1980).

3.4 Source Statistics

The naive (but very attractive) twin-jet model for superluminal sources (Blandford and Konigl 1979, Scheuer and Readhead 1979) attributes the range of L_{rad}/L_{opt} in quasars to orientation effects, on the hypothesis that the radio emission from the compact component is relativistically beamed. This model, in principle, offers specific productions about the bivarate radio-optical luminosity function. The luminosity function for the beamed radio continuum (and maybe also for the optical continuum if this is beamed as well) should be broader than the luminosity spread in the emission lines or any other isotropic component. But the failure of the most straightforward predictions need not be fatal to the general scheme – it could mean only that the model is unrealistically oversimplified. Complications such as a spread in Lorentz factor between sources (and even a range of Lorentz factors and ejection angles within each source) should probably be allowed for; when data are available on the distribution of L_{rad}/L_{opt}, the equivalent width of lines, etc. for larger samples of optically-selected quasars (cf Smith and Wright 1980, Condon et al. 1980, Strittmatter et al. 1980) it will be worthwhile exploring these refinements.

Even the present restricted data, however, confront the model with an embarassment: 3C 273, the brightest optical quasar in the sky (in terms of its line and continuum flux), is a superluminal radio variable. This requires an object which is already exceptional by other criteria to have a preferred orientation that has only a $\sim 1\%$ probability ($\sim \gamma_b^{-2}$, where the superluminal behaviour requires $\gamma_b \gtrsim 5$). This betokens the desirability of modifying the simple scheme. There are a number of possibilities, of which the following is one:

The relativistic plasma may emerge over a spread of angles, rather than solely along the beam axis. Retaining the assumption of axisymmetry, we can suppose that the material is ejected over a range of directions making angles ϕ with the symmetry axis, in such a way that $\gamma_b = \gamma_b(\phi)$ and $\mathcal{J} = \mathcal{J}(\phi)$. For illustrative purposes, suppose γ_b is the same for all ϕ and that only the "specific discharge" depends on ϕ. Then the superluminal velocities of order $\gamma_b c$ will be manifested to an observer whose line of sight makes an angle ϕ_0 with the axis of symmetry by portions of the material within the hollow cone defined by the angles $(\phi + O(\gamma_b^{-1})) - (\phi_0 - O(\gamma_b^{-1}))$. The dominant superluminal effects will be on one side of the centre, and will appear – in projection – aligned with the axis on the approaching side, if $\mathcal{J}(\phi_0 - O(\gamma_b^{-1}))$ dominates $\mathcal{J}(\phi_0 + O(\gamma_b^{-1}))$ by a sufficient factor. For instance one possible form for $\mathcal{J}(\phi)$ guaranteeing this would be $\mathcal{J}(\phi) \propto \exp(-K\phi)$, with $K \gtrsim \gamma_b^{-1}$.

To show how this eases the problem of explaining why 3C 273 is a superluminal source, suppose that $\gamma_b \simeq 5$ but that our line of sight makes an angle $\sim 45°$ with the symmetry axis. We would see superluminal effects, although the radio emission would then be only e^{-3} as strong as if we had the optimal alignment. However, the apparent value of L_{rad}/L_{opt} is higher in 3C 279, which is 3 magnitudes fainter than 3C 273 in the optical but has a similar radio flux. Thus, the relatively weak superluminal components of 3C 273 can be explained as the "sidelobes" of the main directed discharge.

4. COLLIMATION OF JETS

Almost all collimation mechanisms involve the preferential escape (or acceleration) of material along "directions of least resistance"; these directions are likely to be related to a rotation axis within the galactic nucleus. One class of possibilities involves the 'twin exhaust' nozzle (Blandford and Rees 1974, 1978b; Rees 1976) in which buoyant material surrounded by a rotating gravitationally bound cloud escapes via two beams. The confinement is provided by a gravitationally bound cloud in the central potential well; within this cloud lies a source of gas which is very buoyant (in the sense that its value of (P/ρ) much exceeds the gravitational binding energy) and may be relativistic. If the gravitationally bound cloud is flattened, owing to rotation and/or the shape of the potential well, the buoyant fluid emerges along the minor axis. If a steady flow pattern establishes itself, its form is calculable provided that the $p(\rho)$ relation for the buoyant fluid is known. A nozzle forms where the external pressure drops to about half the stagnation pressure; further out, the flow is supersonic, and the channel slowly broadens as the external pressure drops off. This general mechanism does not require any pressure anisotropy in the fluids concerned.

The width of the channel adjusts itself so that there is pressure balance across the walls. The channel cross-section is proportional to the energy flux carried by the jets, and varies inversely with the pressure in the external cloud.

4.1 The Properties of the Gravitationally Bound Confining Cloud

The cloud which confines and collimates the jets must be gravitationally bound in a potential well; but its pressure must be sufficient to prevent it from collapsing into the centre (or into a thin disc if it is rotating). Consequently, the value of (P/ρ) for the cloud material must be of the same order as the gravitational binding energy. Blandford and Rees (1974) envisaged that the cloud was confined within a flat-bottomed potential well due to the stars in the inner region of the galaxy, and supported by gas pressure. This required temperatures in the Kev range, implying that the cloud would itself be a source of X-rays.

If the confining cloud were much more compact, and in a deeper potential well, a higher value of P/ρ would be needed to support it. For clouds around massive black holes, where P/ρ may exceed $m_e c^2$, it becomes implausible to suppose that the pressure comes from an electron-ion plasma with $T_i = T_e$: the electrons would then need to be relativistic, and their cooling (via synchrotron and compton processes) would be very rapid. There are two classes of model for a pressure-supported cloud (or thick disc or "donut") in a relativistically deep potential well:

(i) The cloud may be supported by <u>ion</u> pressure, the electrons being cooled by radiative losses to ≤ 1 <u>Mev</u>. (This option is plausible only when the density is low, so that the electron-ion coupling time is long enough to prevent all the ion energy from being drained away too rapidly.) Or (ii) the cloud is supported primarily by radiation pressure. The gas temperature can then be lower by the same factor by which radiation pressure exceeds gas pressure. If the cloud is sufficiently dense and opaque, the radiation will acquire a black body spectrum at temperature T_{rad} and the gas kinetic temperature will also be T_{rad}. If radiation pressure provides the primary support, the leakage of energy must correspond to the "Eddington luminosity" for the central mass.

4.2 General-Relativistic Features of the Flow

Most features of galactic nuclei are probably insensitive to the details of the metric in the relativistic part of the potential well (and therefore are almost independent of which theory of gravity is the right one). But there are two features of the flow pattern around collapsed objects which are distinctively relativistic and play an important role in constraining the behaviour of inflowing gas.

Near a spinning black hole the Lense-Thirring dragging of inertial frames can enforce axisymmetry with respect to the hole's rotation axis, even if the inflowing material possesses angular momentum in an oblique direction. An orbit of radius r and period t_{Kep} precesses on a timescale $t_{prec} = t_{Kep}(J/J_{max})^{-1}(r/r_s)^{3/2}$, J being the hole's angular momentum and J_{max} the maximum angular momentum of a Kerr hole. If the material drifts in only slowly (ie if the viscosity is low enough that $t_{infall}(r) \gg t_{Kep}(r)$) then the precession builds up over many orbits. As Bardeen and Petterson (1974) pointed out, there is then a critical radius r_{BP}, at which $t_{prec} = t_{infall}$. For $r \lesssim r_{BP}$ the gas flow is axisymmetric with respect to the hole irrespective of the original angular momentum. Thus, if the beams are collimated at a radius $\lesssim r_{BP}$, their orientation is, in effect, controlled by the 'gyroscopic' effect of the hole: they are then impervious to jitter resulting from small-scale fluctuations in the flow pattern of the surrounding gas, and can swing or precess only in response to changes in the hole's spin axis.

A second distinctive feature of black holes is that there is a maximum amount of angular momentum which can be swallowed by the hole. This is essentially the specific angular momentum of the circular orbit of zero binding energy (which is located at $r = 4M = 2r_S$ for a Schwarzschild hole). For an axisymmetric flow pattern, we can consequently delineate a 'region of non-stationarity' around the rotation axis. This region, roughly paraboloidal in shape, is bounded by the circular orbits of zero binding energy whose angular momentum equals the critical value. No stationary axisymmetric bound flow can extend into this region. All the material within it must either have positive energy (in which case it will escape) or else has so little specific angular momentum that it falls freely into the hole. Not surprisingly, the 'walls' bounding this "region of non-stationarity" figure in several specific models accounting for the production of collimated jets.

4.3 Thick Discs or "Donuts"

If material with angular momentum is able to cool, it forms a thin centrifugally-supported disc whose structure is relatively simple. But discs becomes more complicated when the thickness $h(r)$ becomes comparable with the radius r. Pressure gradients in the radial direction are then comparable to those in the perpendicular direction; consequently the angular velocity can no longer be assumed exactly Keplerian. The structure of such discs involves more degrees of freedom than for thin discs; the angular momentum is a free parameter, subject to certain constraints. The most thorough investigations of thick discs have been carried out by the Warsaw group (Paczynski, Jaroszynski, Abramowicz, Sikora, Koskowski) is a series of papers, following earlier work by Fishbone and Moncrief (1976) and Lynden-Bell (1978). Attention has been focussed on optically thick discs supported primarily by radiation pressure. The authors of this work have been concerned with application to quasars (and have supposed that the primary photo-ionizing continuum in quasars may be the ultraviolet thermal emission from the outer part of the disc, radiating at the Eddington luminosity). These specific models do however help us to develop a feel for thick discs and rotating clouds in general, and their role in the production of beams.

The angular momentum distribution is subject to some constraints. For instance, stability may require that the specific angular momentum increases outwards; if viscosity is to cause material to drift inward, then the <u>angular velocity</u> must <u>decrease</u> outwards. If the inner edge of the disc in the equatorial plane lies at r_i, then r_i must lie between r_o, the radius corresponding to a circular orbit of zero bounding energy, and r_{ms}, the radius of the minimum stable Keplerian orbit.

The following comments can be made about thick discs in general, whether their pressure support comes from trapped radiation, or from ions with high kinetic temperature.

(i) The shape of the disc is determined by the surface distribution of angular momentum. Once this is specified, the binding energy over the surface is determined. In the limiting case where the angular momentum is constant, and the binding energy at r_i tends to zero (i.e. r_i is close to r_o) the 'donut' would be bounded by the paraboloidal "region of non-stationarity".

(ii) The shape of the isobars within the 'donut' depends on the internal equations of state, and on the internal distribution of angular momentum. (The arbitrariness of the latter can be reduced by imposing the "von Zeipel condition" for stability of a barytropic rotating fluid - the relativistic generalisation of the requirement that the angular velocity is constant on cylinders around the rotation axis. However, there is no obvious reason for requiring barytropicity, nor for excluding convective or circulatory motions). The pressure maximum must always occur in the equatorial plane at a radius outside r_{ms}, where the rotation velocity is Keplerian. Outside this radius the rotation rate is slower than Keplerian because the pressure gradient is directed outwards. In the limit of very thick and extensive 'donuts', where the angular momentum is low, the pressure-density relation can be almost isotropic except in a narrow "funnel" near the rotation axis.

The 'donuts' supported by radiation pressure would have approximately $\rho \propto r^{-3}$ away from the axis. They can have a large overall extent (i.e. a "photospheric radius" $r_{phot} \gg r_i$) only if they are very optically thick in the inner parts. This implies that the viscosity, parameterised by the α-parameter introduced by Shakura and Sunyaev (1973), must be very low ($\alpha < (r_{phot}/r_s)^{-2}$).

These 'donut' configurations obviously have the right shape to aid the collimation of energy released near a central black hole.

4.4 Narrowing of the Beam Angle .

The "twin-exhaust" configuration (or its generalisation to the 'donut' geometry) is naturally able, by purely hydrodynamical processes, to produce an initial bifurcation of the outflow into two beams along the rotation axis. Another rather different possibility utilises the anisotropic stress from an ordered magnetic field. If such a field is initially aligned along the rotation axis of a disc, and is then sheared (and rendered almost axisymmetric) by strong differential rotation, it can exert a strong torque and generate outflow along the field lines which will be (to some extent) directed along the rotation axis (Blandford 1976).

But even when the outflow has become sufficiently collimated to be supersonic (i.e. a directed velocity $> c/\sqrt{3}$ if the beam consists of ultrarelativistic matter) its collimation can be improved if it propagates outward under conditions of transverse pressure confinement. A beam emerging from the nucleus of a galaxy may pass through a region of steadily decreasing ambient pressure for up to several tens of kiloparsecs. Even if the beam is confined, and its flow obeys Bernoulli's equation, this decrease in pressure will result in an increase in cross-section; but the degree of collimation can still increase, if the diameter of the channel d increases less rapidly than the distance from the nucleus, r. This is well outside the region where supersonic flow begins, so the flux of bulk kinetic energy does not change significantly in this region if there is no entrainment. For an equation of state $p \propto \rho^{\Gamma}$ (not necessarily adiabatic) and a pressure run $p \propto r^{-n}$, the opening angle of the jet $\theta \sim d/r$ varies as

$$\theta \propto r^{(n - 2\Gamma)/2\Gamma} \qquad (6)$$

for a non-relativistic equation of state in the jet, and $\theta \propto r^{(n-4)/4}$ for $c_s \simeq c/\sqrt{3}$. The Mach number varies as $r^{(\Gamma - 1)n/2\Gamma}$ and $r^{n/4}$, respectively. Collimation increases with decreasing n and increasing Γ, i.e., with a harder equation of state. The jet composed of ultra-relativistic fluid is easiest to collimate, since the formula for θ corresponds to setting $\Gamma = 2$ in equation (6), although the equation of state in the comoving frame has $p \propto \rho^{4/3}$. If resistance to sideways compression is dominated by conserved angular momentum about the jet axis or a longitudinal magnetic field, then the effective Γ also equals 2.

The confining cloud is likely to extend to large r, but with density decreasing outwards. For a cloud with an equation of state $p \propto \rho^{\Gamma_c}$ surrounding a point mass, $n = \Gamma_c/(\Gamma_c - 1)$. Thus, a jet of nonrelativistic gas with $4/3 < \Gamma < 5/3$ will be decollimated unless Γ_c exceeds 10/7. A radiation-dominated isentropic atmosphere ($\Gamma_c = 4/3$) cannot enhance collimation, although the jet may have originated in a thick disk of radiation-dominated gas. Where convergence does occur within a hydrostatic atmosphere it is in general rather slow.

Semi-empirical estimates of the mean pressure run over larger distances offer greater hope for effective pressure collimation. The pressure in the vicinity of a 10^8 $M_\odot$ black hole ($\sim 10^{14}$ cm) is unlikely to exceed 10^8 dyne cm^{-2}, while the pressure at ~ 1 pc is typically of order 10^{-2} dyne cm^{-2} for VLBI jets such as NGC 6251 as well as for the broad emission line regions of quasars. This gives a mean value for n of about 2, resulting in considerably better collimation than one obtains with the $n = 5/2$ hydrostatic atmosphere which corresponds to $\Gamma_c = 5/3$. In the five decades of radius between the black hole and where we see VLBI jets, more than 20-fold collimation can occur for $\Gamma > 4/3$, with as much as 300-fold collimation possible for a jet with an ultrarelativistic equation of state. This is more than adequate to explain the apparent opening angles, $\theta \sim 3^\circ - 10^\circ$, which have been observed in VLBI jets.

If cooling occurs in the beam material, this aids collimation by reducing the internal pressure; conversely, viscous effects cause the beam to broaden. A further interesting possibility is that the magnetic field in the beam may establish a transverse self-focussing pattern.

The run of pressure with radius is of course unlikely to follow a smooth power law over the many decades of r which are relevant to the radio observations. If p becomes very small, or falls off suddenly with increasing radius, the beam may become a "free jet" expanding sideways (in its own frame) at its internal sound speed. Free jets must have Mach numbers exceeding θ^{-1}; they are of course not so vulnerable to Kelvin-Helmholtz instabilities.

It is possible that beams produced deep in a galactic nucleus can be destroyed by instabilities, and then re-collimated on a larger scale. In this case the orientation of the larger-scale beams would reflect the overall shape of the galaxy, rather than its nucleus;

characteristic differences between the radio morphology of spirals and ellipticals could be a consequence of the different gaseous environments in these systems (Shu and Sparke 1980).

4.4 Stability, etc.

Even if equilibrium flow patterns exist that can give rise to beams, they may be subject to serious instabilities. Given the difficulty of predicting the stability of terrestrial and laboratory fluid flows, one should be cautious about attaching too much weight to model calculations of stability, when they are applied to the flow of magnetised (possibly relativistic) plasma flowing through a medium of uncertain properties. Kelvin-Helmholtz instabilities can occur anywhere along the beam's path but even more serious is the possibility that Kelvin-Helmholtz and Rayleigh-Taylor instabilities (both of which could in principle occur in the nozzle region) can prevent the collimated flow from ever being set up.

Several authors (e.g. Wiita 1978, 1980, Smarr et al. 1980) have attempted 2-D numerical calculations of how the 'twin exhaust' flow pattern might be established. A gravitationally bound cloud is taken to be already established in a flat-bottomed potential well; the central source of 'hot' plasma is then switched on; it inflates a bubble, which finally breaks out along the minor axis; in some cases, a quasi-steady flow pattern may then be set up. the most recent calculations along these lines are those of Smarr et al. (1980). They find that the twin-exhaust pattern seems stable only if the width of the nozzle is comparable with the scale height - i.e. for a limited range of energy fluxes (the external pressure being given). While these calculations are the best we yet have, they are based on assuming simple equations of state for each fluid. Furthermore, they use a 2-D code rather than full 3-D hydrodynamics. Thus the 'instabilities' are due to ring-shaped protrusions around the flow boundary. It is not obvious whether this underestiamtes or overestimates the instability - these axisymmetric perturbations are harder to excite than a localised perturbation; on the other hand, once they are formed, their effect on the flow is more drastic.

If the collimation occurs close to a massive black hole - especially if it occurs in the domain where Lense-Thirring precession can enforce axisymmetry - the confining cloud lies in a potential which is approximately of the form r^{-1}, rather than being flat-bottomed. As Blandford and Rees (1974) already recognised, this may make it harder to establish a stable twin exhaust geometry. The pressure in the cloud would fall off as $\sim r^{-4}$ (if supported by radiation pressure) or as $\sim r^{-5/2}$ (if supported by ion pressure); in either case, this fall-off is so steep that a <u>homologous</u> perturbation of an equilibrium flow may lead to instability. If, however, the confining cloud possesses some angular momentum, so that its pressure distribution has the form expected in the 'donut' configuration, the prospects for stability are much improved.

This is because the pressure can in principle rise by several orders of magnitude as one moves out from the 'zone of non-stationarity'. Even a drastic change in the power of the central source need therefore cause only a small enlargement of the channel. In computing such flow patterns, rotational effects in the jet material itself should also be taken into account.

If the collimation is established on scales $\lesssim 10^2 \, r_s$, it ceases to be realistic to regard the confining cloud and the expelled jet material as separate entities: each is part of a rotational flow pattern in the gravitational potential well of a relativistic central body. Conditions in the cloud can change on the same timescale as the jet flow itself; the flow may involve continuous gradations in (P/ρ) rather than two distinct components.

5. PRODUCTION OF BEAMS WITH $\gamma_b \gtrsim 5$

5.1 Electron-positron pairs

The VLBI data imply that, in the cases of the superluminal sources at least, γ_b is in the range 5 - 10. (As explained in section 3, however, the evidence on beam speeds in other objects is ambiguous). For an electron-proton plasma to attain this energy, each proton must acquire 5 - 10 Gev, which is $\gtrsim 20$ times the maximum mean energy per particle that can be made available in an accretion process. (If the jet production involves a transition from subsonic flow through a nozzle, then the material must start off with a <u>thermal</u> energy per particle of this order). This is a general difficulty with purely gas-dynamical processes; it suggests that a different process must be invoked. Two possibilities are:

(i) Hydromagnetic mechanisms which channel most of the energy of infalling matter into a small fraction of the particles. This sort of process is envisaged to occur near the light cylinder of pulsars. In extreme cases (Lovelace 1976, Blandford 1976) beams with $\gamma_b \gtrsim 10^{10}$ could be generated.

(ii) The beams may consist of $e^+ - e^-$ plasma rather than containing ions; $\gamma_b \simeq 10$ can then be attained with only ~ 10 Mev rather than ~ 10 Gev per electron.

These two possibilities are not disjoint: some pulsar models involve the production of cascades of $e^+ - e^-$ pairs, and similar processes can happen around black holes (Blandford and Znajek 1977).

There are, however, more mundane ways of generating $e^+ - e^-$ pairs: indeed their creation is unavoidable in a compact region containing high energy particles, emitting X-rays and gamma-rays. If the X-ray luminosity (characteristic photon energy ε) is L_X, then the optical depth of the core region to a gamma-ray of energy $> 2(m_e c^2)^2/\varepsilon$ is

$$\sim 10^2 \, (L_X/L_E) \, (\varepsilon/m_e c^2)^{-1}$$

Unless the X-ray luminosity is very low, any radiation process in the core that generates gamma-rays automatically yields $e^+ - e^-$ pairs.

An obvious constraint on electron-positron beams is that their constituent particles must not annihilate. Defining γ_{random} as the mean Lorentz factor of the particles, measured in a comoving frame, the energy flux is

$$\left(\frac{\pi d^2}{4}\right) m_e c^2 \, n_e \, c \, \gamma_b^2 \, \gamma_{random}$$

The annihilation timescale, measured again in the moving frame, is $\sim (n_e c \, \sigma_T / \gamma_{random}^2)^{-1}$. If the jet originates at a very small radius, this implies that it cannot carry a large energy discharge <u>unless</u> γ_b is large; otherwise it will annihilate. (As a parenthetical remark, note however that a gamma-ray beam resulting from such annihilation is an efficient loss-free mechanism for transporting energy out of a strongly magnetised core region. The gamma-rays could eventually reconvert into pairs after propagating for a few parsecs by interacting with X-rays in the galaxy, thus providing, in effect, an "in situ" acceleration mechanism for $e^+ - e^-$.)

5.2 Radiation Pressure

Radiation pressure on free electrons balances gravity if the luminosity L equals the well-known "Eddington luminosity" $L_E = 4\pi \, Gm_p Mc/\sigma_T$ (This expression for L_E assumes that the Thomson cross-section is applicable, and that the material is primarily hydrogen; modifications to allow for other possibilities are straightforward). If the effective luminosity exceeds L_E (or if, even for lower L, the appropriate cross section σ exceeds σ_T) radiation pressure accelerates material outwards. In the non-relativistic approximation, the terminal speed of a particle released from radius r_1 is of order $[(L - L_E)/L_E]^{\frac{1}{2}} (GM/r_1)^{\frac{1}{2}}$. There is therefore a chance of attaining relativistic velocities if the acceleration starts from a compact object, with r_1 only a few times larger than r_s. Even if L exceeds L_E by orders of magnitude, there is no great likelihood of achieving <u>very</u> high Lorentz factors. This is because, when the velocities get relativistic, the terminal γ_b rises only rather slowly with L/L_E. If the source of radiation is extended, the acceleration saturates at a value of γ_b such that aberration shifts some of the radiation (in the electron's frame) into the forward direction. (These effects are still important even in an electron-positron plasma, where the inertia per unit cross-section is lower by ~ 1840). Relevant calculations are presented by Jaroszynski <u>et al</u>. (1980). A specially interesting possibility arises in the cases of thick 'donuts' supported by radiation pressure. The radiation emerges anisotropically, being especially intense along directions close to the rotation axis. This is because centrifugal effects greatly enhance the effective gravity on the walls of the 'funnel' around this axis. Accurate estimates of this effect, even in idealised models, are complicated because of reflection effects in the 'funnel' (cf Abramowicz and Piran 1980). However, Sikora (1980) finds that, for a 'donut' whose photosphere extends out to 500 r_s, the specific intensity in directions within 15° of the axis is $\gtrsim 70\ L_E$. While this implies that any material expelled by radiation pressure would preferentially lie in jets along the rotation axis, the gain in terminal speed resulting from the collimation is meagre: the high ambient density of radiation within the funnel provides a compton drag which prevents the attainment of relativistic speeds until the material reaches the outer radius of the donut; thick donuts permit large values of (L/L_E) along the axis, but this is compensated by the fact that the effective r_1 is then large.

6. SOME FURTHER QUESTIONS REGARDING THE PHYSICS OF BEAMS

The obvious key issues concern the flow pattern in the region where the beams are formed, the beam speed, the nature of the collimation mechanism and the stability. Collimation may occur in the relativistic domain near a black hole, but it is unclear whether it is basically a fluid-dynamical process or whether a crucial role may be played by electromagnetic effects, anisotropic radiation pressure, etc. The jet could conceivably comprise electron-positron pairs rather than ordinary matter; it may contain a magnetic field, or even carry a current (Lovelace 1976, Benford 1979). The radio emission -- even the milli-arcsecond structures seen by VLBI techniques - involves scales vastly larger than the primary energy source. If some of the extreme cases of optical variability involve directed relativistic outflow (Blandford and Rees 1978a, Angel and Stockman 1980) they may provide clues to jet behaviour on rather smaller scales.

Even though the jets may be moving at a high (or even relativistic) speed the material in them could cool down to a low temperature, yielding a high Mach number and the possibility of unconfined "free jets". A dense jet emanating from near $r \simeq r_s$ with an energy flux of order L_E could certainly cool to a non-relativistic value of T_e. But such a jet would only be directly observable if:

(i) internal magnetic field energy were dissipated and partially converted into relativistic electrons;

(ii) friction and entrainment at the jet boundaries (or interaction with "obstacles" in the jet's path) dissipated some of its bulk kinetic energy; or

(iii) the ejection velocity were variable, and thus led to the development of internal shocks, which accelerated relativistic electrons. (cf Rees 1978c).

The fact that the jet in, for instance, NGC 6251 is detectable at all, means that some re-randomisation of kinetic energy and reacceleration of particles must be occurring along its length to counteract radiation and adiabatic losses. Even the thermal component of the beam material may have a large enough emission measure to be detected (cf the X-ray jet in Cen A, though this could alternatively involve non-thermal emission).

The radio observations of curvature in jets indicate that the physics can be further complicated by such effects as transverse motion of an external medium, buoyance, misalignment of the galaxy with the spin axis of its central black hole, and possible precession of the hole itself.

The range and complexity of realistic beams would much surpass the simple idealised models discussed above. For instance, a beam whose volume is filled predominantly by relativistic plasma may contain cool material embedded in it, owing to

(i) injection of such material in the nucleus

(ii) entrainment from the surrounding medium

(iii) obstacles (e.g. supernova remnants, planetary nebulae) in the path of the jet.

or (iv) cooling of ejected material (via bremsstrahlung etc.), leading to thermal instability, if $t_{cool} \lesssim t_{outflow}$ just outside the nozzle.

For any of these reasons, one may expect cool, relatively dense clouds embedded in the beam plasma, in pressure balance with it, and coasting out with speed $\sim V$, (a scaled-up version of the gas in the SS 433 jets that produces the emission lines). This may be relevant to the large double sources such as DA 240 where there seem to be several hot spots. Smith and Norman (1980) attribute these to separate blobs ploughing supersonically through the intergalactic medium and being decelerated at different rates.

An extreme possibility is that the blobs condensing out may become self-gravitating: having been accelerated to a high (even relativistic) speed, some beam plasma could cool and condense into stars or supermassive objects. This requires rather extreme conditions; it illustrates, however, the wealth of possibilities that can be incorporated within the beam model. If the evidence for anomalous redshifts, "quasars" shot out of galaxies, etc. were ever to become compelling, then these possibilities should be explored before feeling driven to invoke new physics.

7. ACCRETION FLOWS, COOLING, AND SCALING

7.1 The Cooling Timescale Versus the Collapse Timescale

If material falls radially into a black hole, then the efficiency (i.e. the fraction of the rest mass radiated before being swallowed) is small (cf Shapiro 1973, Rees 1978b). This is because, for radial infall, $t_{cool} \propto \dot{M}^{-1}$ whereas t_{infall} is independent of $\dot{M}$. It is perhaps less evident that this low efficiency cannot necessarily be 'cured' by introducing angular momentum. Interestingly, however, this proves to be the case.

Suppose that an accretion disc bulges up, owing to internal viscous dissipation, until its thickness h is of order the radius r. The inward drift timescale $\sim t_{Kep}(h/r)^{-2} \alpha^{-1}$ is then of order $\alpha^{-1} t_{Kep}$, α being the usual viscosity parameter. If the primary agent for viscosity is a tangled-up magnetic field, then the effective α will depend on the time-averaged field geometry and the rate of reconnection (which determine the ratio of Alfven and Keplerian speed). There is, however, no reason why the effective α arising from magnetic viscosity should depend at all sensitively on $\dot{M}$.

This means that as $\dot{M} \rightarrow 0$, t_{inflow} becomes a fixed multiple of t_{Kep}; but the cooling time becomes arbitrarily long in this limit. In other words, angular momentum can be redistributed much faster than the binding energy is radiated away. When $\dot{M}$ is sufficiently low, what must therefore happen is that the inner parts of the disc resemble a 'donut' extending in almost to the location, r_o, of the orbit of zero binding energy. Material that loses enough angular momentum to reach such an orbit can spill into the hole without having radiated significant binding energy.

This flow pattern may well be the norm in galactic nuclei containing massive black holes ('dead quasars') but where the fuelling rate has now fallen very low, to a value corresponding to $\dot{M}/\dot{M}_{crit} \ll 1$. The only perceptible radiation emitted from such objects would be due to a non-thermal tail of electrons (accelerated behind shocks, or by processes related to the magnetic reconnection and viscosity). The pressure distribution of the hot thermal plasma (in which $T_i > T_e$) - apart from a 'funnel' along the rotation axis - would be approximately of the form $p \propto r^{-5/2}$ at large distances from the hole. This cloud may be responsible for the collimation of jets and beams from such systems. The high-entropy plasma escaping in the beam could be produced by exotic processes near the hole (maybe involving Blandford-Znajek (1977) electromagnetic effects).

An example of an active galactic nucleus with $\dot{M} \ll \dot{M}_{crit}$ is M 87. If it indeed contains a black hole of $5 \cdot 10^9 \, M_\odot$, as has been claimed (Young et al. 1978), then the level of activity in the nucleus yields only 10^{-4} of the critical luminosity. This nucleus certainly cannot contain a radiation-supported 'donut' emitting at the Eddington luminosity; on the other hand, provided magnetic (or any other) viscosity yields $\alpha \gtrsim 10^{-2}$, the cooling timescale for thermal plasma accreted onto it could still be longer than the inward drift timescale, allowing the possibility of a cloud supported by ion pressure. The compact radio source in the nucleus of M 87 displays no superluminal effects; nor is it apparently aligned along the jet. Non-thermal emission from a quasi-spherical cloud could account for this emission. Over a range of radii, one would expect $B \propto r^{-5/4}$; the observed radiation would be the sum of contributions from different radii, the self-absorption turn-over occurring at higher frequencies for smaller r.

7.2 Scaling Laws

The flow pattern when accretion occurs is determined by the values of the parameters L/L_E (which determines the relative importance of radiation pressure and gravity) and the ratio t_{cool}/t_{inflow}. If $\dot{M}$ is scaled in proportion to M, there is no reason why accreting objects of very different mass should not involve similar flow patterns: t_{cool}/t_{inflow}, α, the radiative efficiency, and L/L_E would then be independent of M.

So the apparent analogy between stellar-scale phenomena (SS 433, Sco X1 etc.) and active galactic nuclei may indeed reflect a close physical similarity. The relevant parameter is $\dot{M}/\dot{M}_{crit}$ ($\propto L/L_E$ for a given efficiency):

	$L/L_E \simeq 1$	$L/L_E \ll 1$
$M = 10^9 M_\odot$	Quasars	M 87
$M = (1 - 10) M_\odot$	SS 433	"radio stars" γ-ray sources

Just as M 87 and similar low-luminosity active nuclei may display the most purely non-thermal emission (because the cooling time for non-relativistic gas is too long); so these objects may have small-scale counterparts in our Galaxy, detectable only in the radio (and maybe also the gamma-ray) band.

8. CONCLUDING REMARKS

This is a cosmic ray conference. I should therefore attempt, before concluding, to relate galactic nuclei and radio sources to the theme of cosmic rays.

(i) The central cores of galactic nuclei, where irregular bulk velocities comparable with c are inevitable, are the most efficient sites (apart possibly from pulsar magnetospheres) for the production of fast particles. <u>All</u> the particles may be at least mildly relativistic; a variety of electromagnetic processes can generate some particles with high γ.

(ii) The central cores, and the beams emerging from them, could conceivably be composed of electron-positron plasma.

(iii) There must be regions within extended radio sources - in the "hot spots" now resolvable by the VLA - where the bulk kinetic energy of beam plasma is being randomised, and particle acceleration is occurring by processes qualitatively similar to those encountered in more local contexts such as the solar wind.

(iv) The cumulative debris from all extended radio sources must have built up a relativistic particle flux pervading the entire intergalactic medium. Electrons with $\gamma > 200$ lose their energy within the Hubble time via the inverse Compton effect, but the ions will accumulate. Even if (as now seems unlikely) most of the X-ray background came from these intergalactic electrons, the associated ions would contribute a universal flux of only a few times 10^{-3} ev/cc (Rees and Setti 1968) even if we assume the traditional 100:1 factor favouring ion acceleration over electrons. Any intergalactic magnetic field is likely also to be primarily the relic of radio source activity.

(v) Cosmic rays of the very highest energy ($\gtrsim 10^{19}$ ev), commonly thought to be of extragalactic origin, could have been accelerated in the lobes of extended nearby double sources: the product of length scale and field strength in these lobes is large enough that statistical acceleration mechanisms could operate at these energies (Cavallo 1978). The central cores of active galaxies are less promising sites because of the high photon density associated with them.

To sum up: it seems that the phenomena of active galactic nuclei are beginning to fit into a pattern. It is at least clear what physical processes are the main ingredients of a model and should be explored further. In particular, the formation of beams seems a generic property of the flow pattern around collapsed objects, where there is a relativistically deep potential well. These beams are relevant to extended radio sources, "superluminal" variations in compact sources, and the injection of high-energy plasma into the intergalactic medium.

ACKNOWLEDGEMENTS

I acknowledge helpful discussions with many colleagues, especially M. Begelman, R. Blandford and L. Smarr.

REFERENCES

Abell, G.O. and Margon, B.: 1979, Nature, 279, p. 701.
Abramowicz, M. and Piran, T.: 1980, preprint.
Angel, R. and Stockman, P.: 1980, Ann. Rev. Astr. Astrophys. (in press).
Bardeen, J. and Petterson, J.A.: 1974, Astrophys. J. (Lett), 195, L.65.
Begelman, M.C., Blandford, R.D. and Rees, M.J.: 1980, Nature (in press).

Begelman, M.C. and Rees, M.J.: 1978, Mon. Not. R. astr. Soc., 185, p.847.
Begelman, M.C., Rees, M.J. and Blandford, R.D.: 1979, Nature, 279, p.770.
Bekifi, G., Feld, B.T., Parmentola, J. and Tsipis, K.: 1980, Nature, 284, p.219.
Benford, G.: 1979, Mon. Not. R. astr. Soc., 183, p.29.
Blandford, R.D.: 1976, Mon. Not. R. astr. Soc., 176, p.465.
Blandford, R.D. and Konigl, A.: 1979, Astrophys. J., 232, p.34.
Blandford, R.D., McKee, C.F. and Rees, M.J.: 1977, Nature, 267, p.211.
Blandford, R.D. and Rees, M.J.: 1974, Mon. Not. R. astr. Soc., 169, p.395.
Blandford, R.D. and Rees, M.J.: 1978a, in "B.L. Lac Objects" ed. A. Wolfe (Pittsburg).
Blandford, R.D. and Rees, M.J.: 1978b, Physica Scripta, 17, p.265.
Blandford, R.D. and Znajek, R.L.: 1977, Mon. Not. R. astr. Soc., 179, p.433.
Burbidge, E.M., Smith, H.E. and Burbidge, G.R.: 1975, Astrophys. J. (Lett), 199, L.137.
Butcher, H., Van Breugel, W. and Miley, G.: 1980, Astrophys. J. (in press).
Cavaliere, A. and Morrison, P.: 1980, Astrophys. J. (Lett), 238, L.63.
Cavallo, G.: 1978, Astron. Astrophys., 65, p.415.
Condon, J.J. et al.: 1980, Astrophys. J. (in press).
Fabian, A.C. and Rees, M.J.: 1979, in "X-ray Astronomy" ed. W.A. Baity and L.E. Peterson (Pergamon).
Fishbone, L.G. and Moncrief, V.: 1976, Astrophys. J., 207, p.962.
Jaroszynski, M., Abramowicz, M.A. and Paczynski, B.: 1980, Astron. Acta, 30, p.1.
Jones, T.W. and Owen, F.N.: 1979, Astrophys. J., 234, p.818.
Longair, M.S., Ryle, M. and Scheuer, P.A.G.: 1973, Mon. Not. R. astr. Soc., 164, p. 243.
Lovelace, R.V.E.: 1976, Nature, 262, p.649.
Lovelace, R.V.E., MacAuslan, J. and Burns, M.: 1979, in Proceedings La Jolla Workshop on Particle Acceleration Mechanisms (A.I.P.)
Lynden-Bell, D.: 1978, Physica Scripta, 17, p.185.
Maraschi, L., Perola, G.C., Reina, C. and Treves, A.: 1979, Astrophys. J., 230, p.243.
Marscher, A.P. and Scott, J.S.: 1980, PASP, 92, p.127.
O'Dell, S.L.: 1979, in "Galactic Nuclei" ed. C. Hazard and S. Mitton (CUP).
Readhead, A.C.S.: 1979, in "Objects of Large Redshift", ed. G.O. Abell and P.J.E. Peebles (Reidel).
Readhead, A.C.S., Cohen, M.H., Pearson, T.J. and Wilkinson, P.N.: 1978, Nature, 276, p.768.
Rees, M.J.: 1976, Comm. Astrophys. & Sp. Phys., 6, p.112.
Rees, M.J.: 1978a, Nature, 275, p.516.
Rees, M.J.: 1978b, Physica Scripta, 17, p.193.
Rees, M.J.: 1978c, Mon. Not. R. astr. Soc., 184, 61P.
Rees, M.J.: 1980, in "X-ray Astronomy" ed. R. Giacconi and G. Setti (Reidel).
Rees, M.J. and Setti, G.: 1968, Nature, 217, p.362.
Scheuer, P.A.G. and Readhead, A.C.S.: 1979, Nature, 277, p.182.

Schreier, E. et al.: 1979, Astrophys. J. (Lett), 234, L.30.
Shakura, N.I. and Sunyaev, R.A.: 1973, Astron. Astrophys., 24, p.337.
Shapiro, S.L.: 1973, Astrophys. J., 180, p.531.
Shu, F. and Sparke, L.S.: 1980, Astrophys. J. (Lett) (in press).
Sikora, M.: 1980, Mon. Not. R. astr. Soc. (in press).
Smarr, L., Wilson, J. and Norman, R.: 1980, preprint.
Smith, M.D. and Norman, C.A.: 1980, Astron. Astrophys., 81, p.28.
Smith, M. and Wright, A.E.: 1980, Mon. Not. R. astr. Soc., 191, p.871.
Strittmatter, P.A., Hill, P., Pauliny-Toth, I.I.K., Steppe, H. and Witzel, A.: 1980, preprint.
Tubbs, A.: 1980, preprint.
Wiita, P.: 1978, Astrophys. J., 221, p.436.
Wiita, P.: 1980, preprint.
Wilson, D. and Rees, M.J.: 1978, Mon. Not. R. astr. Soc., 185, p.297.
Young, P.J. et al.: 1978, Astrophys. J., 221, p.721.

THE COMMON PROPERTIES OF PLERIONS AND ACTIVE GALACTIC NUCLEI

N. Panagia
Istituto di Radioastronomia CNR
Bologna, Italy, and
K.W. Weiler
National Science Foundation
Washington, D.C., U.S.A.

Plerions, i.e. supernova remnants resembling the Crab Nebula, are characterized at radio wavelengths by having: 1) a centrally peaked brightness distribution, 2) flat radio spectrum ($\alpha > -0.3$, $S_\nu \propto \nu^\alpha$), 3) high linear polarization, and 4) a highly ordered magnetic field (Weiler and Panagia, 1978, 1980). At higher frequencies (10^{11}-10^{13} Hz depending upon the age) the spectrum turns over and attains a slope of $\alpha \sim -1$. In particular, for the Crab Nebula the turnover frequency

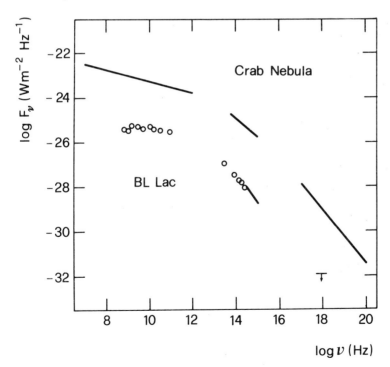

Electromagnetic spectra of the Crab Nebula (upper curve) and BL Lacertae (lower curve).

occurs at $\nu_c \sim 3\times 10^{12}$ Hz (cf. Fig. 1). Moreover, the optical spectrum displays a slope of -0.9 and the radiation is linearly polarized. Non-thermal emission is also detected in the X-ray and γ-ray domains (Fig.1). Similar spectral characteristics are found for other bona fide plerions. As discussed by Weiler and Panagia (1980), the plerion phenomenon is determined by the presence of a highly energetic and active central object (a fast spinning neutron star) which is able to both accelerate and inject continuously relativistic electrons into the remnant with a typically flat energy distribution roughly proportional to E^{-1}.

It is interesting that this picture, which was originally devised to explain the properties and the evolution of plerions, can apply equally well to the case of compact nuclei in many extragalactic radio sources. In fact, we can recall that BL Lacertae objects are characterized by having, in addition to variability, a flat radio spectrum, steep optical and X-ray spectra (cf. Fig.1) and high integrated linear polarization at high radio frequencies and possibly in the visual. Similar properties are also found for the compact cores of extragalactic radio sources, which indeed may be different from BL Lac objects only in the fact that the latter ones are the "naked" version of the former ones (cf. Weiler and Johnston, 1980). The close similarity between plerions on the one side and BL Lac sources, as well as active nuclei, on the other side suggests the operation of comparable energy supply mechanisms in both classes, i.e. efficient conversion of rotational (as in the Crab pulsar) and/or gravitational (as may be the case of W50 - SS433) energy into relativistic electrons and magnetic fields, on vastly different scales. The properties of compactness and activity of the source then determine the characteristic acceleration process which is inherent to the source in that it must occur at, or very near to, the source itself and is independent of the conditions of the surrounding medium. We note that this would provide a more natural explanation for the flat radio spectrum of compact extragalactic radio sources than the presently accepted idea of "onion" layers of synchrotron self-absorbed components.

In conclusion, it appears that two types of acceleration processes are present in non-thermal sources:
a) A plerion-type mechanism which takes place in compact sources. It is essentially the same phenomenon at all scales and produces an electron spectrum $N(E) \propto E^{-1}$. This mechanism operates in plerions and compact nuclei of extragalactic sources.
b) A second acceleration process arises from the interaction of relatively low energy plasma with a large scale ambient medium. This process, which produces steep energy spectra $N(E) \propto E^{-2} - E^{-3}$, operates in shell type SNR and big lobes of extragalactic sources.

References

Weiler, K.W. and Johnston, K.J.: 1980, M.N.R.A.S. 190, 269.
Weiler, K.W. and Panagia, N.: 1978, Astron. Astrophys. 70, 419.
Weiler, K.W. and Panagia, N.: 1980, Astron. Astrophys., in press.

CENTRAL RADIO SOURCES IN GALAXIES

E. Hummel
Kapteyn Astronomical Institute, University of Groningen
Postbus 800, 9700 AV Groningen, The Netherlands

From radio continuum observations it is known that the central regions of galaxies can be an important source of cosmic rays, in particular relativistic electrons. In order to get some insight in the properties of the central sources and their dependence on optical properties, especially morphological type, a large sample of galaxies (∼ 400) was observed with the Westerbork Synthesis Radio Telescope at 1.4 GHz. The sample definitions, observations, reduction, survey characteristics and the analysis of the data are given in Hummel (1980, 1980a). Here we will concentrate on the results obtained for the central sources (< 20").

The observational results can be summarized as follows:
 i) The distribution of the spectral index[1] for elliptical and lenticular galaxies has a median of $\alpha = -0.1$, indicating compact, nuclear sources. Central sources in spirals have a steeper spectrum ($\alpha \simeq -0.7$) and usually have rather complex structures.
 ii) The central sources in ellipticals and early type spirals are at 1.4 GHz a factor 10 stronger than those in lenticulars. 10% of all the early type spirals, S0/a, Sa, Sab, have $\log P \geqslant 21.6$ WHz^{-1}.
 iii) For the spiral galaxies there is on average a linear proportionality between the radio power of the central source and the total optical luminosity, but there is certainly no one to one correlation.
 iv) The median radio power of the central sources in spirals decreases by a factor 10 from early type spirals (S0/a, Sa, Sab) to late type spirals (Scd, Sd, Sdm, Sm).
 v) Central sources in barred spirals are on average a factor 2 stronger than those in non-barred spirals. About 10% of all barred spirals have $\log P \geqslant 21.2$ WHz^{-1}.
 vi) The central sources in interacting spiral galaxies are on average a factor 2 to 3 stronger than those in isolated spiral galaxies. 10% of all interacting galaxies have $\log P \geqslant 21.3$ WHz^{-1}.

---

[1] spectral index α defined by $S(:)\nu^{\alpha}$

The higher radio power of the central sources in ellipticals compared to lenticulars suggests that the radio power of a compact, nuclear source is related to the mass and the rotation of the bulge component. This agrees with the fact that spirals in general do not have compact nuclear sources which are stronger than those in lenticulars (Hummel and Kotanyi, 1980). Also the change in radio power of the central sources in spirals with morphological type indicates that central sources are related to the mass distribution. Spirals which have the more centrally concentrated mass distribution (early type spirals, barred spirals), as inferred from studies of the dynamics of spirals (Roberts, 1978, Rubin et al. 1978) and from the light distribution, have on average stronger central sources.

Steep spectrum central sources are rarely detected in lenticular galaxies (Hummel and Kotanyi, 1980) but these sources are very common in spirals. The much lower gas content in lenticulars compared to spirals suggests that the gas content has some influence on the occurrence of these steep spectrum central sources. For spirals it is known that the gas content increases with morphological type (Roberts, 1975, Shostak, 1978) while the radio power of the central sources decreases. This suggests that there is a threshold in the gas content below which no central source can be maintained and above which there is hardly any dependence of the steep spectrum central sources on gas content.

The enhanced radio emission from the central sources in barred spirals might be (partly) related to non-circular motions of the gas in the region of the bar and the shocks induced by this. The kinematics of the gas may also be of importance to explain the enhanced emission in the interacting spiral galaxies. The stronger central sources in interacting galaxies can be explained by the gravitational interaction or by possible differences in initial conditions at the formation of double systems. Since there is a trend indicating that the central sources get stronger when the distance between the galaxies of a pair decreases (Hummel, 1980, Stocke, 1978) and since central sources in general are short lived phenomena we favour the gravitational interaction hypothesis. Several authors (cf. Condon and Dressel, 1978) have proposed that due to the interaction matter is falling toward the central region and this activates the central source. However, here the main question is whether the infalling matter can lose its angular momentum fast enough (cf. Gunn, 1979).

References

Condon, J.J., Dressel, L.L.: 1978, Astrophys. J. 221, 456
Gunn, J.E.: 1979, in "Active Galactic Nuclei", Cambridge University Press
Hummel, E.: 1980, Astron. Astrophys. Suppl. in press
Hummel, E: 1980a, Ph.D. Thesis, University of Groningen
Hummel, E, Kotanyi, C.G.: 1980, Astron. Astrophys. submitted
Roberts, M.S.: 1975, in "Stars and Stellar Systems Vol. IX" Univ. of Chicago Press
Roberts, M.S.: 1978, Astron. J. 83, 1026
Rubin, V.C., Ford, W.K., Thonnard, N.: 1978, Astrophys. J. 225, L107
Shostak, G.S.: 1978, Astron. Astrophys. 68, 321
Stocke, J.T.: 1978, Astron. J. 83, 348

A SELFCONSISTENT MULTIPLE COMPTON SCATTERING MODEL FOR THE X AND γ-RAY EMISSION FROM ACTIVE GALACTIC NUCLEI

L.Maraschi, R.Roasio and A.Treves
Istituto di Fisica dell'Università and
L.F.C.T.R. - C.N.R. - Milano, Italy

We refer to a model of active galactic nuclei where a massive black hole ($M \simeq 10^8 M_\odot$) is accreting spherically (1-3). Turbulent dissipation of the magnetic field produces an efficient heating. For low Thomson optical depth the main cooling mechanism is selfabsorbed cyclo-synchrotron (c-s) radiation from thermal electrons. The energy transfer of the thermal electrons to the c-s photons via Compton scattering (comptonization) becomes increasingly important for increasing optical depth. In ref. 2 the temperature profile of the inflowing gas was calculated balancing the heating and cooling mechanism but neglecting comptonization due to the difficulty of an analytic estimate of this process.

Extensive numerical computations by Pozdniakov, Sobol and Sunyaev (4) show that soft photons of energy $h\nu_s$, injected at the centre of a hot homogeneous sphere, emerge with a power law energy distribution from $h\nu_s$ to $h\nu = 3$ kT, the spectral index of which is given by the approximate expression, valid also in the semirelativistic regime:

$$\alpha = \left\{ -\lg \tau + 2/(n+3) \right\} / \lg (12 n^2 + 25 n) \tag{1}$$

where $n = kT/m_e c^2$.

Assuming that at each radius in the flow the spectral index $\alpha(r)$ can be evaluated from (1) with the optical depth and temperature at that radius, the Compton losses can be estimated integrating the power law normalized to the synchrotron flux at low frequencies up to $h\nu = 3kT$. This yields a Compton cooling rate

$$\Lambda_c = \frac{3}{r} \frac{2\pi kT}{(1-\alpha)c^2} \nu_s^{*(2+\alpha)} \left[\nu_{max}^{(1-\alpha)} - \nu_s^{*(1-\alpha)} \right] \tag{2}$$

with α given by (1). Here ν^* is the frequency at which the plasma becomes transparent to c-s radiation and $\nu_{max} = 3kT/h$. The determination of ν^* is one of the technical difficulties of the calculation, since in the semirelativistic regime a finite expression of the c-s absorption coefficient is not available.

In fig.1 the temperature profiles obtained with and without in-

Table 1. M = black hole mass; $\dot{M}$=accretion rate; τ =Thomson optical depth; L_{tot}=total luminosity; α =energy spectral index of Comptonized emission.

M (M_0)	$\dot{M}$ (g/s)	τ	L_{tot} erg s^{-1}	α
10^7	10^{23}	0.04	1.5×10^{43}	1.08
10^8	10^{24}	0.04	1.5×10^{44}	1.04
10^9	10^{25}	0.04	1.5×10^{45}	1.0
10^7	10^{24}	0.4	1.5×10^{44}	0.83
10^8	10^{25}	0.4	1.5×10^{45}	0.80
10^9	10^{26}	0.4	1.5×10^{46}	0.78
10^7	3×10^{24}	1.2	4.5×10^{44}	0.72
10^8	3×10^{25}	1.2	4.5×10^{45}	0.69
10^9	3×10^{26}	1.2	4.5×10^{46}	0.67

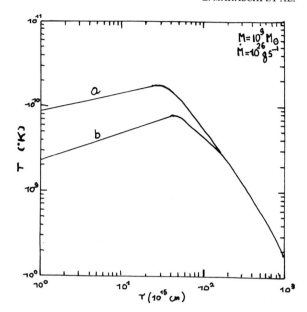

Fig.1. Temperature profile without (a) and with (b) comptonization losses.

clusion of comptonization are compared. In the cases examined the spectral index of the inner shells ($r \lesssim 30 r_s$) is practically constant and is given in Tab.1 for a range of parameters. The important result which is apparent from Tab.1 is that by varying the central mass and accretion rate by two orders of magnitude the spectral index is made to vary from 1 to 0.7. This can be understood as due to the strong dependence of the Compton cooling Λ_c on α and may represent a step in the explanation of the strong correlation of optical and X-ray luminosities in Seyfert galaxies and quasars (5,6).

References

1. P.Meszaros, 1975, Astr.Ap. 44, 59
2. L.Maraschi, G.C.Perola, C.Reina and A.Treves, 1979, Ap.J. 230, 243
3. L.Maraschi, G.C.Perola, A.Treves, 1980, Ap.J., in press
4. L.A.Pozdniakov, I.M.Sobol, R.A.Sunyaev ,1979, Sov.Astr.Lett. 5, 149
5. H.Tananbaum et al., 1979, Ap.J. 234, L9
6. M.Elvis et al., 1978, M.N.R.A.S. 183, 129

SYMMETRY BREAKING AND INVARIANT MASS APPROACH TO THE SPIRAL
STRUCTURE OF GALAXIES

K.O. Thielheim
Institute of Pure and Applied Nuclear Physics
University of Kiel
Federal Republic of Germany

The physical conditions in the centre of spiral galaxies as our own galaxy still are not very well known. They offer themselves to more or less exotic speculations, some of which involve the possible existence of black holes and mechanisms for the production of high energy cosmic rays. In view of such argumentations I feel it worthwhile to look in more detail into recent progress obtained in the understanding of the dynamics of the central region of spiral galaxies which turns out to be intimately related to the galaxy as a whole, especially its spiral structure.

The persistence of galactic spiral structure in the presence of differential rotation is explained in terms of a spiral shaped density wave rotating at a certain angular velocity inside the stellar disk (B. Lindblad, 1941). This notion still is the basis of the understanding of the dynamics of galactic spirals. Still the question of the physical mechanism underlying this density wave is not finally answered. According to "gravitational" density wave theory (Lin and Shu, 1964) adopted by many astronomes today, the density wave is produced primarily by the mutual gravitational attraction of stars inside the galactic disk.

One way of testing this hypothesis is by performing N-body simulations of differentially rotating stellar disks. Typical results (Hohl, 1971) show a short living 's'-shaped transient spiral and, finally, the formation of an oval shaped configuration rotating around its smallest axis without essentially changing its structure. This object is surrounded by a disk shaped low density distribution of stars, rotating differentially around the same axis. So far, these N-body simulations have not reproduced long-living spirals. Recently, the density distribution of stars inside the   disk has been investigated carefully (Berman and Mark, 1979; Sellwood, 1980) and long-living spiral shaped trailing density waves have been found. These results obviously seem to confirm the basic assumption of "gravitational" density wave theory, namely that the spiral pattern is generated by the mutual gravitational interaction of stars inside stellar disks. As a further test of this assumption we have performed a different type of N-body simulations (Thielheim and Wolff, 1979). In these, no mutual gravitational interaction was provided for between stars inside the surrounding disk. Nevertheless, long-living spiral shaped trailing density waves were found again. These results therefore suggest, that the mutual gravitational attraction of stars inside the galactic disk might not be the mechanism which is primarily responsible for the existence of spiral density waves. Instead they suggest that the generation of a spiral pattern in a differentially rotating stellar disk primarily is a response to a central, rotating oval shaped mass distribution which may be understood as an equilibrium configuration growing "adiabatically" by accretion of stars from the surrounding disk. This interpretation implies that the ground state of a differentially rotating, self gravitating stellar system is (under certain conditions) not rotationally symmetric (but only mirror symmetric). This symmetry breaking may be understood as a consequence of the non-linearity of the dynamics governing the central object. The dynamics of the surrounding disk is essentially linear. But at this stage mirror symmetry is broken through the time dependence of the gravitational force exerted by the central object onto the stars in the disk, thus producing a spiral shaped, trailing density wave (Thielheim, 1980). The form of the spiral is determined by the distribution of mass in the central oval object, its (adiabatic) increase with time and by its rotational pattern velocity on the one hand and by the mean distribution of mass inside the surrounding disk on the other.

We are thus led to the conjecture that the oval shaped equilibrium configuration found in N-body simulations is what is otherwise observed as elliptical galaxies. During the formation of an elliptical galaxy essentially all the mass available is successfully incorporated into this stable distribution. In other cases part of the total mass was left outside as a disk. According to this line of thought each spiral galaxy in its centre houses an elliptical galaxy, the adiabatic increase of which produces a spiral density wave in the disk, the form of which would be influenced by gravitation among stars in the disk and interaction with the viscous interstellar gas.

References:
Berman,R.H. and Mark.J.W.-K. (1979), Astron.Astrophys. 77, 31
Hohl,F. (1971), in A. Hayli (ed.), Proc. IAU Symp. 69, 349
Lin,C.C. and Shu,F. (1964), Astrophys.J. 140, 646 and subsequent work
Lindblad,B. (1941), Stockholm Obs. Ann. 13, No. 10 and subsequent work
Lindblad,P.O. (1960), Stockholm Obs. Ann. 21, No. 4
Miller,R.H. and Prendergast,K.H. (1968), Astrophys.J. 151, 699
Sellwood,J.A., private communication
Thielheim,K.O. and Wolff,H. (1979), Astron. Mitt., in press
Thielheim,K.O. (1980), Astron. Mitt., in press

NEUTRINO EMISSION FROM GALAXIES AND MECHANISMS FOR PRODUCING RADIO LOBES

R. Silberberg and M. M. Shapiro
Laboratory for Cosmic Ray Physics
Naval Research Laboratory
Washington, D. C. 20375 U.S.A.

ABSTRACT

In the context of the proposed DUMAND-type detector for high-energy neutrinos, it is important to estimate the neutrino fluxes from powerful radio galaxies. This would indicate whether a detectable event rate can be expected, and would help choose among possible mechanisms for generating the radio lobes associated with many such galaxies. Among the models that have been proposed are: (a) a beam from the galactic core that interacts with the extragalactic medium; (b) ejection of clouds of plasma, particles, and magnetic fields into the lobes, either in an explosive burst, or by diffusive escape; (c) black holes or spinars ejected from the galactic nucleus by a gravitational slingshot mechanism.

Many large elliptical galaxies have a prolific central energy source that gives rise to a pair of radio lobes. The total energy estimates for the relativistic particles and magnetic fields in these lobes are $\approx 10^{56}$ to 10^{61} ergs. Various models have been proposed for the generation of radio lobes:
(a) A relativistic beam of particles emitted along the axis of rotation (Scheuer 1974; Blandford and Rees 1974; Wiita 1978). A beam model invoking a massive black hole as the power source (Lovelace 1976).
(b) Two oppositely directed clouds of thermal plasma, relativistic particles and magnetic fields that are confined by ram pressure (DeYoung and Axford 1967; Sturrock and Barnes 1972; Sanders 1976). Two special cases of this model have been suggested, one in which a large amount of energy builds up at the galactic nucleus and is then ejected in explosive bursts, and another in which the particles within the plasmons are emitted diffusively from the galactic nucleus. (c) Massive condensed objects are ejected from the galactic nucleus by a gravitational slingshot mechanism (Saslaw, Valtonen, Aarseth 1974). We shall here explore how observations of neutrinos can help to discriminate among the various models.

As an example we calculate the neutrino flux from Cen A. We shall try the alternative proton energy spectra characterized by exponents $\alpha = 2.35$ and $\alpha = 2.0$ at energies > 1 GeV. For each of these power laws, we shall assume alternative values of 1 and 10 for the ratio q_{pe} of energy input into protons to that into electrons. For confinement prior to the outburst, we adopt the conservative value $\bar{x}$ = 4 g/cm^2. This could be considerably higher.

A value of $\bar{x} \approx$ 4 g/cm^2 can be inferred in the case of diffusive escape of protons. DeYoung (1976) reports that the extended radio sources have separation velocities of $\sim$ 0.1. Assume that the outward diffusion of the cosmic-ray gas from the galactic center proceeds with the same velocity, though the emission velocity of the radio lobes vs. distance from the galaxy is model dependent. We adopt 0.2 g/cm^2 as the path length for <u>straight-line</u> escape (Sanford and Ives 1976), a value based on observed X-ray absorption. Then we get a mean path length $\bar{x}$ = 4 g/cm^2. For relatively free escape, we assume a path length of 0.4 g/cm^2, twice the value for linear traversal from the source to our galaxy.

Table 1 shows how observations with DUMAND could help differentiate between models of proton confinement with explosive bursts (or with diffusive escape).

Table 1.

Annual Event Rates of Neutrinos (E > 4 x 10^{12} eV) from Cen A for Two Models* (I and II) of Radio-Lobe Production

Parameter Values		Neutrino Events per Year	
α	q_{pe}	I: ($\bar{x}$ = 4 g/cm^2)	II: ($\bar{x}$ = 0.4 g/cm^2)
2.35	1	$\sim$2	$\sim$0.2
2.35	10	$\sim$20	$\sim$2
2.0	1	$\sim$100	$\sim$10
2.0	10	$\sim$1000	$\sim$100

*Model I: Confinement and explosive burst, or diffusive escape.
Model II: Free escape in a beam. (The detector is assumed to have an effective volume of 10^{10} tons of water for observing ν_μ and $\bar{\nu}_\mu$, and a volume of 10^9 tons for ν_e and $\bar{\nu}_e$).

REFERENCES

Blandford, R.D. and Rees, M.J.: 1974, M.N.R.A.S. <u>169</u>, 395.
DeYoung, D.S.: 1976, Ann. Rev. Astron. Astrophys. <u>14</u>, 447.
DeYoung, D.S. and Axford, W.I.: 1967, Nature <u>216</u>, 129.
Lovelace, R.V.E.: 1976, Nature <u>262</u>, 649.
Sanders, R.H., 1976: Ap. J. <u>205</u>, 335.
Sanford, P.W. and Ives, J.C., 1976: Proc. Roy. Soc. A350, 491.
Saslaw, W.C., Valtonen, M.J. and Aarseth, S.J., 1974: Ap. J. <u>190</u>, 253.
Scheuer, P.A.G., 1974: M.N.R.A.S., <u>166</u>, 513.
Sturrock, P.A. and Barnes, C., 1972: Ap. J. <u>176</u>, 31.
Wiita, P.J., 1978, Ap. J. <u>221</u>, 41.

PARTICLE ACCELERATION BY PULSARS

Jonathan ARONS*
Service d'Electronique Physique - Section d'Astrophysique
Centre d'Etudes Nucléaires - Saclay
and
Department of Astronomy and Space Science Laboratory
University of California - Berkeley**

ABSTRACT

The evidence that pulsars accelerate relativistic particles is reviewed, with emphasis on the γ-ray observations. The current state of knowledge of acceleration in strong waves is summarized, with emphasis on the inability of consistent theories to accelerate very high energy particles without converting too much energy into high energy photons. The state of viable models for pair creation by pulsars is summarized, with the conclusion that pulsars very likely lose rotational energy in winds instead of in superluminous strong waves. The relation of the pair creation models to γ-ray observations and to soft X-ray observations of pulsars is outlined, with the conclusion that energetically viable models may exist, but none have yet yielded useful agreement with the extant data. Some paths for overcoming present problems are discussed.

The relation of the favored models to cosmic rays is discussed. It is pointed out that the pairs made by the models may have observable consequences for observation of positrons in the local cosmic ray flux and for observations of the 511 keV line from the interstellar medium. Another new point is that asymmetry of plasma supply from at least one of the models may qualitatively explain the gross asymmetry of the X-ray emission from the Crab nebula. It is also argued that acceleration of cosmic ray nuclei by pulsars, while energetically possible, can occur only at the boundary of the bubbles blown by the pulsars, if the cosmic ray composition is to be anything like that of the known source spectrum.

---

* John Simon Guggenheim Memorial Foundation Fellow.
** Permanent address.

I - INTRODUCTION

In this paper, I will (1) discuss the existing evidence that shows pulsars accelerate relativistic particles, (2) describe the theory of such particle acceleration, paying close attention only to internally consistent scenarios and models, and (3) use the models to assess the possibility of pulsars being the origin of cosmic rays. In § II, I summarize the evidence, emphasizing the importance of γ-ray observations as our only probe of energetic significance into the basic workings of the rotating magnetosphere. In § III, I discuss (a) the theory of pulsar spin down, (b) the scenario of particle acceleration in "strong" waves in the "wave" zone of the magnetosphere, and (c) theories of electrostatic acceleration of ultrarelativistic beams and the creation of pairs in the near zone. My conclusion is that extant models for acceleration in the near zone imply the spindown occurs through the formation of a dense, relativistic magnetohydrodynamic wind ; the strong wave model as commonly used in the cosmic ray literature does not apply. I also show that some aspects of these models may be testable by soft X-ray observations, and that no acceleration model has yet succeeded in accounting for the observations of pulsed γ-rays. I then apply these models to the problem of cosmic ray origins in § IV. I argue that direct acceleration from the surface of the neutron star, or from the near or wind zones of the magnetosphere, do not contribute to the nucleonic component cosmic rays, but I do show that observations of the positronic component of the galactic cosmic ray flux and of the 511 keV recombination line from the interstellar medium may put useful constraints on the pulsar models and/or may be explained by pulsars as the primary source of the positrons.

I also argue that the boundary layer where the relativistic winds from young, rapidly rotating pulsars interact with the surrounding interstellarmedium/supernova remnants is an energetically viable site for the acceleration of the cosmic ray nuclei.

II - OBSERVATIONS OF DIRECT RELEVANCE TO PARTICLE ACCELERATION BY PULSARS

Alsmost all pulsars are known as radio sources (see the review volume by Manchester and Taylor, 1977). Because of the short duration and low frequency of the pulses and the galactic distances to the sources, one readily concludes that a collective emission mechanism is required. For example, some pulsars broadcast essentially all their radio emission in "micropulses" of duration < 100 μsec, from which one can conclude that brightness temperatures in excess of 10^{28} K probably are required (see Cordes, 1979 for a review). Since the emission is collective, there is no requirement that the emitting medium be relativistic ; beamed, high brightness temperature radio emission occurs in the terrestrial magnetosphere with surprisingly high efficiency, for example, without involving any relativistic particles. However, the optical and X-ray synchrotron emission from the Crab nebula points obliquely toward relativistic particles being associated with pulsars, while the observations of γ-ray pulses (hν > 30 MeV) from several radio pulsars implies ultrarelativistic particles exist quite near the neutron stars.

(a) Crab Nebula

The fact that the nebular X-ray source requires continuous acceleration of relativistic electrons in well known, as is the fact that the pulsar's total rotational energy loss is sufficient to power the nebular emission (see for example, IAU Symposium No. 46 for reviews of the situation 10 years ago). The assumption of synchrotron emission was confirmed by the detection of polarization in the nebular X-rays (Novick et al., 1972 ; to prove the synchrotron hypothesis would require detection of circular polarization, impossible in practice in the X-ray region, and perhaps impossible in principle if the emitters in the X-ray emission zone are $e^{\pm}$ pairs, as is quite likely). The nebula requires at least 10^{38} erg/s to explain the photon emission ; the pulsar's spin down supplies $4\pi^2 I \dot{P} P^{-3} = 5 \times 10^{38}$ $(I/10^{45}$ g.cm$^2)$ ergs.s^{-1} with I the moment of inertia, $P = 0^s.033$ the pulsar rotation period, and $\dot{P} = 4.3 \times 10^{-13}$ s/s the spindown rate. Thus the pulsar supplies the energy needed in some unknown form, mostly to be absorbed by the nebula on the south west side of the pulsar (Ricker et al., 1975 ; Giacconi, 1979). The fact that $I \sim 10^{45}$ g.cm^2 is needed to give the requisite output is the strongest support of the rotating neutron star hypothesis. The fact that the excitation of the nebula is so one sided may be a peculiarity of the pair creation by the pulsar, to which I return below.

However, this observation does not tell us whether the pulsar puts its energy into $\sim 10^{38}$ electrons/sec with energy ~ 1 TeV, then shoots them into the nebula, or if the pulsar emits the energy in some lower entropy form which is reprocessed in the nebula into relativistic particles. Put differently, we don't know if the nebular electrons are injected or are accelerated in situ. There are many handwaving models of both types, all unsatisfactory to my way of thinking. One way of proceeding is to see if we can understand the pulsar a bit better. For this, the γ-ray observations of pulses are invaluable.

(b) Pulsed γ-rays

Satellite observations (Thompson et al., 1975 ; Bennett et al., 1977 ; Kanbach et al., 1980) have clearly established the emission of pulsed γ-rays from the Crab and Vela pulsars with luminosities $\sim 10^{35}$ erg.s^{-1} and $\sim 10^{33.5}$ erg.s^{-1} respectively. In addition, 3 σ results and/or upper limits have been reported for several other pulsars (Ogelman et al., 1976 ; Thompson et al., 1976 ; Kanbach et al., 1977 ; Mandrou et al., 1980). I will use only the Crab and Vela in my discussion, as only here are there clearly established results.

For the Crab pulsar, the essential facts are that the optical, X- and γ-ray pulses are coincident to within the timing errors, and coincide with the main radio pulse and interpulse. 15 μsec time resolution is needed to resolve the peak of the optical pulse (Smith et al., 1978) ; the X-ray pulse is unresolved at 60 μsec. The total 0, X and γ output is about 0.03 % of the total spin down luminosity (this is still orders of magnitude brighter than the radio emission), with a spectral slope in the X-rays and γ-rays somewhat flatter than the nebular emission. The optical emission is polarized, indicating an emission mechanism involving a magnetic field, and its intensity level is appropriate for the optical emission to be the Rayleigh-Jeans tail of the X-ray spectrum (Shklovsky, 1970). This fact strikes me as strong support for the optical emission being incoherent radiation from relativistic particles (these are needed since the brightness temperature in the optical is above 10^{11} K), despite a number of well known difficulties with this interpretation ; coherent mechanisms (eg, Sturrock, Petrosion and Truk, 1975) explain the coincidence between the optical spectrum and the RJ tail of the X-rays as a coincidence.

Granted the incoherent interpretation of the optical flux, one can make some simple inferences, all of them old (Goldreich et al., 1972 ; Epstein and Petrosian, 1974). Since the optical intensity must not exceed the black body limit, the area of the optical emission zone must be much larger than that of a neutron star. Therefore, the emission region must be well above the star. Since all of the pulses are coincident (to within the timing errors), the emission region must be at high altitude for all frequencies (radii r >> 10 km). This argument does not apply to the radio precursor, of course. If the emission region is fixed with respect to the star, so that pulses are due exclusively to stellar rotation plus some sort of beaming, the angular and radial extent of the emission region must be small, otherwise sharply cusped pulses cannot be generated. Furthermore, the emission region cannot fluctuate by more than a few % in beaming angle or radius, otherwise the averaged pulses (all that is observed at high frequency) would not show cusped structure. Finally, and obviously, the emission zone must contain relativistic particles of energy > 10 GeV, since the observed photon spectrum extends to these energies and a coherent process is exceedingly hard to contrive for wavelengths less than the Compton wavelength. The particles may or may not be accelerated within the emission zone itself. Ultra high energy γ-rays (hν > 500 GeV reported by Grindlay et al., 1976) most likely refer to a different region, since the rotation phase of this emission differs from all the rest.

Unfortunately, the Vela pulsar is different. There is no X-ray pulse known (see Kanbach et al., 1980, for a review of the observational situation), the single radio pulse looks like the Crab's precursor, while the twin optical pulses lie between the two γ-ray pulses. This morphology might be viewed as similar to the Crab, if the Crab's main radio pulse and interpulse are regarded as unusual, while the Crab's precursor is the normal radio pulse (quite possibly true, based on comparison of the radio spectra to other pulsars). The remainder of the transformation of the Crab into Vela comes from supposing the optical and γ-ray pulses to be emitted with a hollow cone pattern with beaming along dipolar field lines (see below), with optical emission in Vela coming from a substantially smaller radius than applies to the γ-ray emission, while for the Crab, the radii of emission are much closer together. The physics behind this may be that the emission is synchrotron emission, radiation reaction is stronger at low altitude, where the magnetic field is stronger, which may allow higher particle energies and γ-ray emission only at high altitudes. In the case of the Crab, the much stronger energization might lead to the regions being closer together, so that they look like a single zone, within the resolution of γ-ray waveform. This idea predicts that the centroids of the optical pulses in the Crab should have a slightly smaller phase separation than the centroids of the γ-ray pulses, something that has not been carefully studied, so far as I known. The lack of an X-ray pulse in Vela is the main embarrassment for this picture, and probably can be explained only by a rather strong dependence of excitation of the particles' gyrational energy on period and field strength. As I will discuss, the efficiency of acceleration <u>along B</u> does have a very strong period dependence in the models I favor.

Vela converts ∼ 0.15 % of its rotational energy loss into γ-ray emission, to give a γ-ray luminosity in the vicinity of 10^{34} erg.s^{-1} with a power law γ-ray spectrum somewhat flatter than the Crab pulsar's. There is no <u>direct</u> necessity to place the emission region at high altitudes, but I will show that conversion of this much energy into accelerated particles probably forces one to altitudes large compared to 10 km. Like the Crab, the γ-ray pulses in Vela are separated by about 140° of longitude. Such pulse-interpulse structure is usually thought to mean that the emission comes from field lines connected to opposite magnetic poles. However, an alternative, and I think better, interpretation comes from the radio astronomers and the theorists. It has proved possible to organize the vast zoo of pulsar light curves ("waveforms") with a simple kinematic model (Backer, 1976), in which the emission comes from a <u>hollow</u> cone rigidly rotating with the star, as shown in Figure 1. Emission is hypothesized to arise only in the annular rim of the cone. An observer whose trajectory is as shown in A then sees a double pulse, while a trajectory like B produces a single pulse. Note that in this picture, sharp pulses are produced because of strong localization of the beaming in <u>angle</u>, a property of an emission region contained in a right circular cone.

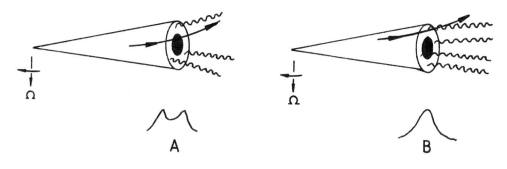

Figure 1. The hollow cone model.

In the original work, the opening angle was picked to accomodate the known separation seen in pulsars with double pulses ($\sim 20°$ is typical). Manchester and Lyne (1977) pointed out, however, that if the half angle of the cone is large (70°, for example), the 140° phase separation of the Crab pulsar could be understood within the same context. Some justification for this comes from the polar beaming model, in which pulsar emission is due to relativistic particles streaming out relativistically along approximately dipolar field lines whose opening angle increases in proportion to $r^{1/2}$, as shown in Figure 2. Note that if radiation is beamed along field lines of the

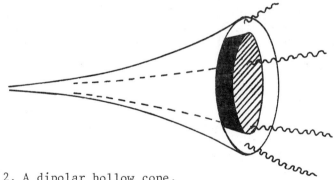

Figure 2. A dipolar hollow cone.

annular rim of this widening cone, an observer sees radiation from different radii at <u>different</u> phases, in contrast to the right circular cone of Figure 1. Therefore, an observer with a broad band detector (such as a spark chamber) will receive sharp pulses only if the emission region is localized in radius, as well as being localized to the annular rim of the polar field lines A narrow band detector, such as a radio receiver, will see a sharp pulse if the emission is also narrow band with each frequency associated with a given radius even if the bolometric radio emission is not radially localized, <u>or</u> if all the radio emission is radially localized . If such localization occurs at high altitude (r a fraction more than 0.1 of $cP/2\pi$ = 48 000 P km), half angles as large as 70° are quite possible. For more discussion of these ideas, including the statistical evidence for the applicability of very wide hollow cones and the fragments of theoretical foundation for the idea, see Manchester and Lyne (1977), Manchester (1978) and Arons (1979).

III - THEORY

The γ-ray observations certainly do show that relativistic particles are present in some region near the pulsar where the magnetic field can confine the acceleration and emission to a restricted region. (I will assume the existence of pulses is not due to a relativistically expanding sheet or sphere, where time compression can make the emission appear pulsed. This idea, first proposed by Michel and Tucker (1969), has various virtues and problems discussed by Michel (1971) and Arons (1979) which are not worth addressing here). The fact that pulsars lose angular momentum strongly suggests the presence of large scale electric fields in the magnetosphere. The fact that particle are accelerated to ultra relativistic energies near the pulsar probably requires the existence of substantial components of E parallel to B. This systematic acceleration of particles in $E_{||} \equiv \underline{E}.\underline{B}/B$ actually has been discussed mostly on theoretical grounds, well in advance of the γ-ray observations, and remains the main possibility for explaining the particles which emit gamma rays. As we will see, this mechanism may be energetically sufficient, but the resulting models have not (yet) succeeded in explaining the data. By contrast, almost all work on cosmic ray acceleration by pulsars has invoked the electric fields of "strong waves", which I think are a failure.

(a) Pulsar spin down

Pulsar wave forms define an excellent clock with characteristic period P. In all well studied cases, P slowly increases at the rate $\dot{P}$. Typically, $\dot{P} \sim 10^{-15}$ s/s, but a substantial number of pulsars have large $\dot{P}$ ($\sim 10^{-13}$ s/s) and small $\dot{P}$ ($\sim 10^{-17}$ s/s). There is no correlation of $\dot{P}$ with P ; there exist a few very short period pulsars with very small $\dot{P}$, and there is a substantial group of long period objects with large values of the spin down rates. Thus evolution in the $\dot{P}$-P diagram along evolutionary tracks starting from the same initial conditions seems unlikely, although the certainty of any interpretation is poor because of the uncertain selection effects in the published data (see Manchester and Taylor, 1977), a situation which will change upon completion of the Molonglo-NRAO survey.

The fact that pulsars keep time so well and slow down steadily has been the main support of the rotating magnetized neutron star hypothesis. The reason is that a massive flywhell keeps excellent time and, when endowed with a suitable electromagnetic field, has an average spin down like that of the observed stars. An observation of P and $\dot{P}$ yields the total energy loss :

$$\dot{E}_{rot} = \frac{d}{dt}(\frac{1}{2} I\Omega^2) = -4\pi^2 \frac{I\dot{P}}{P^3} = 3 \times 10^{32} \frac{I}{10^{45} \text{ g.cm}^2} \frac{\dot{P}}{10^{-15}} (\frac{0.5}{P})^3 \text{ erg.s}^{-1}$$

$$= 5 \times 10^{38} \frac{I}{10^{45}} \text{ erg.s}^{-1} \quad \text{(Crab)}$$

The only theory used here is the theoretical number for I = moment of inertia. The spin down is explained by assuming the object has an intense magnetic field, of strength so high that at radii $\sim$ vacuum wavelength $\lambda = c/\Omega = cP/2\pi = 48\,000$ P km, the energy density in particles is still less than the total electromagnetic energy density (in most ideas, much less). At radii $r \gg R$ = stellar radius, it suffices to assume the magnetic field is given by the dipole component, which is a good description of the field out to radii $\sim \lambda$. To make a pulsar, the dipole must be inclined with the rotation axis. The torque on the star is explained by assuming sufficient current flow exists in the dipole field at $r \sim \lambda$ to induce a toroidal field $B\phi$ $(r \sim \lambda) \sim B_{dipole}$ $(r \sim \lambda) \sim \mu/\lambda^3$. This requires currents of magnitude $\sim B/P$; whether they are displacement currents or conduction current does not matter. Simultaneoustly, the rotation of the magnetic field induces a poloidal electric field $E \sim (r/\lambda) B_{dipole} \sim \mu/\lambda^3$ at $r \sim \lambda$. Then $\underline{E} \times \underline{B}$ includes an outward poynting flux which, when summed over the sphere of radius $r \sim \lambda$, gives a total electromagnetic energy loss rate

$$\dot{E}_{Em} \sim 4\pi \lambda^2 \frac{c}{4\pi} (E_\theta B_\phi)_{r \sim \lambda} = c\frac{\mu^2}{\lambda^4} = \frac{\mu^2 \Omega^4}{c^3} \quad (3)$$

With the replacement $\mu^2 \to 2\mu^2 \sin^2 i/3$, i = angle between angular velocity and magnetic moment, (3) yields the familiar expression for vacuum dipole radiation, the only "successful" theory of pulsar spindown! In this case the current is entirely displacement current. However, the argument that led to (3) shows that all one needs are currents $\sim B/P$ and relativistic velocity of energy flow, and indeed the relativistic mhd wind models, where the currents are all conduction, of Michel (1969) and of Kennel et al. (1979) have exactly the same energy loss as in (3), to within factors of order 1. Goldreich and Julian's (1969) scenario for a charge separated wind in the aligned rotation almost certainly does not involve current of this magnitude (Jackson, 1976, 1980 ; Scharlemann et al., 1978 ; Mestel et al., 1979 ; Michel, 1980), thus making the aligned rotator an unlikely setting for understanding even basic pulsar physics. In the oblique rotator (i $\gg 0°.6$ $P^{-1/2}$) conduction currents of magnitude B/P are quite likely (Scharlemann et al., 1978 and below) in addition to displacement currents of comparable magnitude, thus making a complete theory of spindown quite inaccessible to detailed solution. However, the order of magnitude of μ and of the magnetospheric electric field can be estimated by equating (3) to the observed spin down energy loss.

In the case of the Crab pulsar, it is possible that the spin down proceeds in proportion to $\Omega^{3.5}$ (Groth, 1975) ; however, it is also possible that the true spindown rate $\alpha\ \Omega^4$ is masked by the random walk of $\dot{P}/P^2$ (Cordes, 1980).

The main point I want to draw from (3) is that the magnetosphere has a large scale magnetic field $B \sim 10^{12}\ (R/r)^3$ Gauss and a large scale electric field $E \sim 6 \times 10^8\ P^{-1}\ (R/r)^2$ V/m for $r < \lambda$; the "exact" value of course depends on the object. At distances greater than λ, the magnetic field is expected to be predominantly toroidal, with magnitude $B_\phi \sim 4\ (\mu/10^{30}\ \text{G.cm}^3)\ P^{-2}\ (10^5\ \text{km}/r)$ G and $E \sim B$, independent of whether the out flow is a "vacuum" wave or a relativistic wind. If particles accelerate in these electric fields, relativistic energies are easily achieved. However, in the near zone ($r \sim \lambda$), the strong magnetic field provides a two dimensional insulator by confining particles to motion along B, while in the far zone ($r \sim \lambda$), the nature of the acceleration depends on whether the electric field is like that of a vacuum wave or like that of a wind.

(b) Acceleration in strong waves

A lot of work on particle acceleration has followed the suggestion by Gunn and Ostriker (1969), who noted that one test particle injected into the wave zone of a vacuum magnetic rotation (where $E/B = 1$) would be accelerated to extreme relativistic energies (monoenergetic electron distributions of energy $\sim 10^{7.5}\ m_e c^2$ looked possible for Crab pulsar parameters). I only want to point out that this idea really doesn't work, once even the slightest attempt at self-consistency is made. They correctly pointed out that vacuum-like propagation of the wave requires the wave frequency = rotation frequency to exceed the effective plasma frequency. For strong waves, this requires the number density to satisfy

$$n\ (r \sim \lambda) < B\ (r \sim \lambda)/2\ \text{Pec} \sim 4\pi^3\ \mu/ec^4 P^4 = 0.3\ (\mu/10^{30}) P^{-4}\ \text{cm}^{-3} \quad (4)$$

Physically, all that is said here is that the current nec induced in the plasma by the wave's electric field must be less than the spindown current $\cong$ displacement current $\sim B/P$. If the inequality is reversed, one finds the sufficient conditions for mhd theory to apply (Michel, 1969). If the number density is phrased in terms of a particle loss rate, the total loss rate must be less than

$$\dot{N}_c \cong 6 \times 10^{30}\ \frac{\mu}{10^{30}\ \text{G cm}^3}\ P^{-2}\ \text{s}^{-1} \quad (5)$$

if the particle flux from the star covers all 4 steradians. Modern work suggests $\dot{N} \gg \dot{N}_c$ for observed and observable pulsars, thus implying a hydromagnetic wind as the relevant zeroth order picture (see below). However, a particle flux of magnitude

$$f_c = (4\pi^2 \mu/ec^3 P^4)(cP/2\pi r)^2 = 10^{10}(\mu/10^{30} \text{ G.cm}^3) \times P^{-4}(cP/2\pi r)^2 \text{ cm}^{-2} \text{ s}^{-1} \tag{6}$$

as Ostriker and Gunn (1969) suggested might be present. This leads to complete disruption of their mechanism. The reason is that the basic mechanism relies on the electric field of the wave to accelerate the particle to speed = c in the direction orthogonal to the propagation direction and to the wave magnetic field $\vec{B}$. Then the strong V x B force accelerates the particle in the radial direction to speed = c also, so that in the guiding center frame, the particle sees the V x B force as almost static and moves to regions of weaker acceleration (larger radius) before the force reverses. Even a little plasma (flux << f_c) thoroughly destroys this phase matching by changing the wave phase speed to velocities > c with E/B > 1 (Akhiezer and Polovin, 1956; Max and Perkins, 1971; Max, 1973). Indeed, in the extremely strong wave limit $eB/m_e c \Omega \gg 1,836$ Z/A, where Z = nucleonic charge and A = mass number, applicable to radii somewhat greater than $cP/2\pi$ but much less than the radius of the bubble blown by the star's Poynting flux, steady travelling wave solution show that the wave demands a flux = f_c and that the energy is equipartitioned between particle and field (e.g. Kennel et al., 1973; Kennel and Pellat, 1976), but that individual particle energies are far less than estimated by Gunn and Ostriker. Of course, these plane wave results implicitly assume that the star is willing to supply this flux, which may or may not be true. In fact, the local critical flux required by these solutions is proportional to field strength and therefore declines α r^{-1} in a spherical wave, while conservation of number implies average flux must decline in proportion to r^2, indicating that the equipartition solutions cannot be extended to spherical geometry. Recent work by Llobet (1980) indicates a lack of such maximal flux solutions in the spherical case.

While I will argue below that the total flux probably greatly exceeds f_c, we don't know this to be true in any specific case. Therefore I think a more powerful limit is set by the results of Asseo et al. (1978), who show that Non-Linear Inverse Compton (= Synchro-Compton) radiation damping in an otherwise stable strong wave would turn the wave into γ-rays in a quite short distance, even for flux << f_c. In addition, a number of powerful instabilities have been uncovered, most of which assume the plasma zero temperature but some of which include the possibility of relativistically high temperatures, which occur even when f_c << actual flux (Max and Perkins, 1972; Arons et al., 1977; Romeiras, 1979; Asseo et al., 198L) and all of which are very likely to heat the low density plasma until radiation reaction can take the energy out of the system, most likely at γ-ray energies. Since in general pulsars are far less than 100% efficient at producing γ-rays (Ogelman et al., 1976; Kanbach et al., 1977), we can fairly safely conclude that particle losses in sufficient number to make the objects interesting cosmic ray sources are not consistent with strong wave acceleration. Physically, all this happens because E/B > 1 in these waves. Then the particles always move so as to make the field approximately a pure electric field in the rest frame of the particle's oscillation center, with very large V.E work done on the particle. By contrast, in mhd winds, the guiding centers move so that E = 0 in the comoving frame, with obvious differences in the V.E work that can be done in accelerating particles, creating instabilities and radiating the energy. Since I think the particle flux from a pulsar is at least f_c and probably is $\gg f_c$ I think any correspondence between particle acceleration models which rely on strong waves and observation is purely coincidental, resulting from the fact that the gross electromagnetic energy loss rate is independent of the specific form taken by the Poynting flux. For various examples of strong wave models applied to in a variety of circumstances, see Rees, 1971, Kulsrud et al., 1972, Rees and Gunn, 1974, Arons et al., 1975, Gaffet, 1976, Weinberg, 1980 and Kundt and Krotscheck, 1980.

(c) Acceleration in large scale E_{11}

Most modern work on the pulsar itself suggests fluxes vastly exceed f_c, thus implying an mhd wind. Several distinct lines of thought suggest that above the surface of the star, along polar field lines, there are charge separated zones within which large scale parallel electric fields E_{11} = E.B/B $\neq$ 0 exist, creating linear (electrostatic) accelerators and which also extract currents from the star of magnitude B/P which flow along B. The motion of the accelerated charges along B leads to the emission of γ-rays at low altitude, whose magnetic conversion leads to the formation at a dense electron-positron plasma. In all of the consistent theories so far investigated, the formation of this plasma limits the acceleration to a small fraction of the ultimate energy available.

(i) Starvation electric fields

Since neutron stars have intense gravitational fields (g $\sim 10^{14}$ cm.s^{-2}) and low surface temperatures (T* < few x 10^6 K, Giacconi, 1979), the gravitational scale height of any atmosphere is tiny, leading to an electrical vacuum above the surface (Pacini, 1967). But, the rotationally generated vacuum field includes components parallel to B whose force vastly exceeds surface gravity (Deutsch, 1955; Goldreich and Julian, 1969), leading to the <u>electrical</u> extraction of a charge separated plasma and (possibly) current from the star (in this respect, pulsars differ from planetary magnetospheres, whose plasmas are not supplied by electrically driven flow along B for the most part). Particle fluxes comparable to (6) are just what one requires to poison E_{11}, and one expects the extraction to continue until such poisoning occurs.

To see why this is so, consider a star whose electric field is static in the corotating frame (this includes vacuum dipole radiation), an approximation which very likely applies to many processes which occur on time scales $\ll P$ (Arons, 1979) and may apply globally. Then there exists a potential Φ such that the electric field in the local Lorentz frame corotating with the star is a potential field, or

$$E + \frac{1}{c} (\Omega \times r) \times B = - \nabla \Phi \qquad (7)$$

(Schiff, 1939; Mestel, 1971; Fawley et al., 1977). The interpretation of (7) is simple. At low altitude, the magnetic field is so strong that $E + c^{-1}(V \times B) = 0$ is a good approximation, where V is the velocity of any charged particle perpendicular to B. Solving for V with (7) for E yields

$$V = (\Omega \times r)_\perp + c \frac{B \times \nabla_\perp \Phi}{B^2} \qquad (8)$$

If $\Phi = 0$, the particles corotate (true for the rigid crust). However, if $\nabla \Phi \neq 0$, the particles don't corotate; in this context, this is possible only if $\nabla_\parallel \Phi \neq 0$, since the crust is an excellent conductor. Since $\nabla_\parallel \Phi = - E_\parallel$, the lack of corotation is intimately associated with particle acceleration, and we need to ask, how big is Φ ?

To study this question, we relate electric fields to the particles through Gauss's law (Poisson equation). With (7) for E, this yields :

$$- \nabla^2 \Phi = 4\pi (\eta - \eta_R) \qquad (9)$$

where η = charge density and

$$\eta_R = - \frac{\Omega \cdot B}{2\pi c} + \text{small terms} \qquad (10)$$

The potential Φ can be estimated by noting that

$$\Phi = 4\pi (\text{length})^2 (\eta - \eta_R) \qquad (11)$$

with the length chosen to be the smallest in whatever geometry is considered. Of course, Φ can be calculated by the usual machinery of potential theory, if $\eta - \eta_R$ is known. The question then reduces to how much "charge separation" $\eta - \eta_R$ exists. Note that the actual plasma can be completely charge separated with $\eta = Ze\eta$ as one expects if the magnetospheric charges are supplied by electric force, yet as far as acceleration and non-corotation are concerned, the rotating system acts as if it is charge neutral if the charge density <u>exactly</u> equal to η_R is supplied. In magnitude

$$\frac{\eta_R}{e} = 7 \times 10^{10} \frac{B(r)}{1 \text{ TG}} P^{-1} \text{ cm}^{-3} \qquad (12)$$

Multiplied by c, this density yields the flux f_c referred to the stellar surface along dipole field lines (flow tube area $\propto r^3$).

The vacuum $\eta/\eta_R \ll 1$ yields potentials over the magnetic pole $\Phi > 10^{13}$ V; I assumed the polar cap has a radius $\sim 10^4 P^{-\frac{1}{2}}$ cm in estimating this voltage, with resulting vacuum $E_{//} \sim 10^{11}$ V/m. This Φ so vastly exceeds the gravitational binding energy that all binding seems unlikely, except in very unusual circumstances (as we will see, the unusual may be usual, in limited regions). Goldreich and Julian (1969) suggested that flow off the stellar surface continues until the match between η and η_R is excellent, so close that the residual electric field just balances gravity (in electron zones, where $\eta_R < 0$, this means $(\eta - \eta_R)/\eta_R < 10^{-9}$, typically). This results in either non relativistic flow or (more likely) no flow at all; see the references on the aligned rotator above. However, if for some reason, a zone is even a little starved of particles, electric fields much larger than the Goldreich-Julian estimate, yet smaller than vacuum fields, might occur. If such "starvation" can be maintained in a manner <u>consistent</u> with the processes which supply particles and relax $E_{//}$ to its "quasi-neutral" value, the resulting combination of electric field, rotational EMF and plasma is a good candidate for a relativistic particle accelerator. Note that if starvation is the basic reason for strong $E_{//}$, the essential physics is still the same as the reason for non zero V.E in strong waves, namely, the vacuum rotational EMF is incompletely cancelled by the conducting plasma. In the near zone, approximately electrostatic fields under consideration here, the phase velocity effects which are so important to strong wave acceleration do not cause any problem, leading to <u>some</u> possible models which have yet to run into definitive observational or theoretical difficulties.

(ii) Discharge models

Sturrock (1970, 1971) suggested the first, electrostatic acceleration model. I will not review his scheme since when viewed from the demands of constructing a <u>consistent</u> acceleration theory, he vastly overestimated the electric potential available at the stellar surface, essentially because by neglecting η_R compared to η, he implicitly assumed the presence of an "anode" somewhere above the freely emitting stellar surface (= "cathode"), which could support the surface charge needed to give a net accelerating electric field when combined with the space charge of the accelerated stream (see Fawley et al., 1977 and Arons, 1979, for more comment; the model of Kennel et al., 1979, suffers from the same defect). Given his potential, Sturrock suggested a number of interesting consequences, especially copious pair creation and curvature γ-ray emission, which have survived into later work. He did not make any realistic attempt to account for the tendency of the pairs to poison this electric field, and as remarked by Arons (1979), the unsteady creation of relatively low energy pairs (E $\sim$ 1-100 GeV) in regions whose voltage drop (e $\Delta\Phi \sim 10$-10^3 TeV) vastly exceeds the initial energies of the secondaries almost certainly lends to very intense "auroral" bombardment of the polar cap with X-ray emission vastly in excess of past and present observations, an interesting constraint which will be of some importance below. Nevertheless, his results are still used (e.g. Sturrock and Baker, 1979). I regard his work as a fascinating and innovative piece of order of magnitude phenomenology; if one does not ask questions about the physics of the acceleration and simply accepts his estimates of particle output as given, a large number of plausible estimates of various pulsar phenomena can be generated. However, despite much propaganda to the contrary, no specific, convincing model of pulsar pulses has been constructed on his basis which covers more than a small fraction of the known facts. Since this problem plagues all models, I prefer to stick to those which are not internally contradictory, to which I now turn.

Ruderman and Sutherland (1975) proposed a model which preserves many of the features of Sturrock's yet does give a reasonable if crude account of the accelerating voltages that might be available, under one set of assumption that lead to starvation. They hypothesized that currents of magnitude B/P exist, so that the polar field lines continuously lose plasma to infinity (the polar field is assumed to be "open"). They also hypothesized that the surface has very high work function, so high that particles of the same sign as η_R cannot be extracted from the star even by the vacuum electric field. Typically, this requires the work function to exceed 5 keV. Since energies this high may be typical of the binding energy of a perfect, uniform lattice of pure cold iron immersed in a superstrong magnetic field (Ruderman, 1971; Flowers et al., 1972), Ruderman and Sutherland (= RS) hypothesized that those stars for which $\Omega \cdot \mu < 0$ and $\eta_R > 0$ would not be able to emit charges to replace plasma flowing off to infinity. Then as plasma leaves the poles, a starvation zone is opened up (a "gap") in which $\eta = 0$ and the full vacuum potential becomes available. The polar cap size is roughly given by the area (Goldreich and Julian, 1969)

$$A_{cap} \sim 7 \times 10^8 \, P^{-1} \, (R/10 \text{ km})^3 \text{ cm}^2 \tag{13}$$

and since the field lines are assumed to be good conductors, (11) implies the maximum potential, if the whole polar zone of volume $A_{cap}^{3/2}$ is evacuated, is

$$\Phi_{max} \stackrel{\sim}{=} 7 \times 10^{12} \, \frac{B}{1 \text{ TG}} \, P^{-2} \, (\frac{R}{10 \text{ km}})^3 \text{ Volts} \tag{14}$$

The voltage drop is about equally along and across B. This voltage is the same as that assumed to be present by Sturrock.

However, just as Sturrock pointed out, voltages of this magnitude are unstable. Any stray γ-ray that gets into the polar region (for example, from the interstellar medium), is absorbed in the magnetic field and creates an $e^{\pm}$ pair. Once created in the vacuum zone, the $e^{\pm}$ accelerate to energies $> 10^{12}$ eV and radiate more γ-rays of energy ~ 1 GeV, since the $e^{\pm}$ move along curved field lines. The secondary γ-rays are absorbed within the polar cap region, thus generating more pairs in the region where $E_{\parallel} \neq 0$, more γ-rays, etc... An avalanche results, which grows at each point until a charge density $\sim \eta_R$ is created, at which point the accelerating E is assumed to be shut off. The result is a discharge composed of an outwardly moving positron beam and an inwardly moving electron beam, each of total number flux $f \sim \eta_R c/e = 2 \times 10^{21}$ (B/1TG) P^{-1} cm^{-2} s^{-1}. Because the electric field is assumed to be quenched when $f \sim \eta_R c/e$, the voltage is limited to values much below Φ_{max}. Since many discharges occur in the cap at the same time under most circumstances, they argued that the outwardly moving positron beams and further pair creation by the γ-rays emitted by the beams would keep the voltage limited in a quasi-steady manner. Then the polar gap is supposed to be like Figure 4, with $E_{\parallel} \neq 0$ on the upper surface because of the mobile pair plasma appearing at and above this point while $E_{\parallel} \neq 0$ on the stellar surface because of the strong ion binding in the crust.

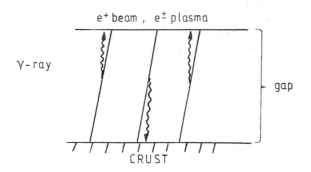

Figure 3. RS gaps

RS went on to elaborate this scheme into a complete scenario for pulsar emission, about which I have made various remarks elsewhere (Arons, 1979). The main aspect that concerns me here is whether this picture has enough energy in accelerated particles to explain γ-ray pulsars. The answer is no, if their 1-dimensional view of electric field poisoning is correct. The total J.E work done in their zone is

$$E_{RS} = 10^{30} \left(\frac{B}{1\ TG}\right)^{6/7} P^{-15/2} \text{ erg.s}^{-1} \qquad (15)$$

Most of this energy is in the positron beam, whose energy goes mostly into creating more pairs above the gap in the case of the Crab and Vela pulsars. Then the gamma ray luminosity is less than 10% of (15) in each of these cases, while the particle luminosity (15) is already a factor of 40 below the γ-ray luminosity of the Crab pulsar and a factor of 10 below the gamma-ray luminosity of the Vela pulsar. Nevertheless, the properties of this accelerator are interesting for other reasons. The output of each polar cap, in addition to the dissipation rate (5), is a positron beam of initial energy 2 $(B/1\ TG)^{-1/2}\ P^{-1/2}$ TeV. For pulsars with large P the positron energy decays by γ-ray emission to a final energy ∼ 30% of its initial value at the top of the gap. The average particle loss in each beam is about N_c per polar cap, with N_c given by (5). The pair plasma created in large P cases has a flux at least $10^3 N_c$, giving a lower bound to the loss rate for the Crab pulsar ∼ 3 x 10^{36} s^{-1}; another factor of ∼ 50 is quite likely, from synchrotron emission by the first generation of pairs and the magnetic conversion of the gamma rays. These pairs are injected into the outer magnetosphere and into the wind with relatively low energy, typically ∼ 100 MeV to 500 MeV (the exact value depends on the paper containing the estimate, and almost all estimates have ignored the broad momentum dispersion resulting from this injection mechanism).

A fairly strong prediction of the model is that an amount of energy equal to (15) is put into the polar cap by the downward flux of electrons and γ-rays. Here the dense stellar material easily thermalizes the energy giving rise to thermal photon emission with effective temperature

$$T_{cap}\ (RS) \stackrel{\sim}{=} 400\ \left(\frac{B}{1\ TG}\right)^{3/21} \left(\frac{0.089}{P}\right)^{2/7}\ eV \qquad (16)$$

and a soft (0.1 - 2 keV) X-ray flux at the earth ∼ 1 Uhuru count x (500 pc/distance)^2 for the Vela pulsar. Fluxes of this magnitude would be easily observed in the Einstein observations of Vela, since the stellar rotation would make the caps appear as a modulated source (other cases are possible too, but Vela is the most favourable). Cyclotron scattering might lead to a change in the modulation pattern, if the closed zone of the magnetosphere has a density greatly in excess of η_R/e, but is unlikely to change the photons' energy (Arons, 1979), so this prediction is a strong test of the model.

So far, such X-rays have not been observed. One possibility is that the whole idea is wrong. However, the use of the vacuum approximation is an inconsistent assumption. The reason is that the temperature (16) is high enough to lead to free <u>thermionic</u> emission, even from a perfect metallic surface. Other effects (impurities, lattice defects, ridges, ...) all suggest that the effective work function is lower than the calculated binding energy anyway. When ions are freely available, the charge density of the extracted ion current is very close to η_R and $\Phi \ll \Phi_{max}$ near the star, where the magnetic opacity is large, even without pairs present to poison the potential. Nevertheless, the potential may be barely large enough to allow e± discharges to occur in the midst of the ion stream (Cheng and Ruderman, 1977, Fawley et al., 1977; Jones, 1978). Because $\eta - \eta_R \ll \eta_R$ even before the discharge occurs, the electric field <u>may</u> be quenched with a lower flux created in the electron and positron beams and therefore less polar cap heating. Whether or not this effect is real, and whether or not the resulting scheme violates the X-ray data, is unknown; a real answer requires following the system of discharges as each enters the regime of non linear growth (see Arons, 1979), a task only addressed for a single spark in a vacuum field by Fawley's (1978, 1980) one dimensional particle-in-cell simulations.

My opinion is that these phenomena might occur in the manner outlined, as far as the internal consistency of the theory is concerned, with the question of observational tests unresolved. If these discharges do occur (with free emission of ions), the plasma density in the outer magnetosphere probably greatly exceeds η_R/e, $N \gg N_c$ and the energy loss is then through an mhd wind, not a superluminous strong wave. But, the energy put into particle acceleration by these surface gaps cannot explain the high frequency emission from the Crab and Vela pulsars; the upper limit (15) to the energy available is too small.

To find this energy, the discharge theorists have turned to "outer gaps" where the same sort of spark phenomena are supposed to occur. These are regions near the null surface where $\Omega \cdot B = 0$, on one side of which the magnetosphere demands a negative charge density, while on the other side, a positive density is required. If these particles can be supplied only by the flow along field lines from the star of a charge separated plasma, most geometries have a problem in that positive particles cannot get to the required region by flow through the negatively charged zone, and vice versa, a difficulty realized and swept under the rug by Goldreich and Julian. Holloway (1973) pointed out that the magnetosphere might form a vacuum zone near this surface, as the particles on open field lines streamed away from the star (if such streaming is allowed by the global electric field). Cheng, Ruderman and Sutherland (1976) took advantage of this idea to develop a rough discharge model at high altitude, a picture elaborated by Cheng and Ruderman (1977b) into a complete scenario for the Crab pulsar and revised by them (1980) and elaborated by Ayashli and Ogelman (1980) into a scenario for all γ-ray emitters. The basic physics is the same (except that in the 1980 paper of Cheng and Ruderman, pair creation occurs through the collision of curvature γ-rays with thermal photons from the heated polar caps, instead of magnetic, which may allow some models to produce acceptably small thermal X-ray emission). Because the opacity at high altitude is much lower than at the surface, the emission of magnetically convertible photons requires higher particle energy, larger voltages and larger J.E work done in accelerating the e+ and e- beams in the gap. The result (for the curvature emission-magnetic conversion model) is a particle acceleration luminosity

$$\dot{E}_{gap} \sim 3 \times 10^{34} \left(\frac{B_{surface}}{1 \text{ TG}}\right)^{4/5} \left(\frac{33 \text{ msec}}{P}\right)^{3/5} \text{erg.s}^{-1} \quad (17)$$

This dissipation rate is in the right ball park, especially since the particles in the beams radiate away most of their energy as soon as they leave the gap in the short period models. However, most of these gamma rays are converted into pairs, so the models are not 100% efficient in putting energy into γ-emission, but they are

∼ 50% efficient in converting the luminosity (197) into soft X-ray emission from the polar caps. Auroral bombardment by one of the beams and its secondary plasma creates an X-ray luminosity $\sim \dot{E}_{gap}/2$ with effective temperature

$$T_{cap} \text{ (outer gap)} = 3.0 \left(\frac{B_{surface}}{1 \text{ TG}}\right)^{1/5} \left(\frac{89 \text{ msec}}{P}\right)^{3/20} \text{ keV} \quad (18)$$

An X-ray flux of this magnitude and colour would be easily observable from the nearer pulsars, including Vela. Furthermore, the energization of the cap cannot be reduced by unknown amounts by assuming the existance of a space charge limited flow of ions from some other source, as is possible in the surface gap case. Therefore, I think outer gaps in their original form are in direct conflict with X-ray observations of pulsars. In fact, I don't think outer gaps exist, since injection of dense pair plasma from the poles in either the surface gap model or the slot gap model to be outlined next would short out the electric field at the $\Omega \cdot B = 0$ surface, and plasma circulation and radial diffusion into the closed zone (Arons, 1979; Kennel et al., 1979) can easily keep the null surface shorted out in the closed zones. In addition, there are many problems, common to all of these theories, with the ability of outer gaps to actually produce the spectrum and pulse shape of γ-ray pulses, as I'll mention below.

(iii) A steady flow theory : Slot gaps:

This theory was proposed by Arons and Scharlemann (1979) = AS, based on previous work on the acceleration problem by Fawley et al. (1977) and Scharlemann et al. (1978); see Arons (1979) and Scharlemann (1979) for qualitative discussion. My remarks here are based on the quantitative work by Arons (1980 a). It does not suffer from the problems I have described in the discharge models and does (more or less) have enough particle energy to explain γ-ray pulses, but does not, in its present level of development, produce sufficient hard photon emission that looks like a pulsar. Since it is a steady flow theory, a fairly simple quantitative expression of ideas is possible.

Like the surface discharge models, it is based on there being a starvation zone over the pole, but of a more subtle type. Consider a polar flux tube whose form is shown in Figure 4, whose sense of curvature is <u>toward</u> the rotation axis (this was called "favourably curved" by AS). Assume $\Omega \cdot \mu > 0$. Then the electric field extracts electrons from the thin atmosphere at the stellar surface with $E_{\parallel} = 0$ in the atmosphere. The acceleration mechanism also applies to ion zones, but the consequences for pair creation are then quite different. Assume (as is required by the energetivs of spin down) that the polar flux tube is bounded by a region of $\Phi = 0$; this boundary condition may identify the transition field lines where return currents flow and pairs precipitate from high altitude. Finally, assume the potential is monotonic with height above the star;

Figure 4. A favourably curved flux tube.

the opposite hypothesis, needed in Mestel et al's (1979) nonrelativistic models of the aligned rotator, very likely leads to sufficient dissipation to relax any other structure to the maximal, relativistic acceleration found here. Then the flux tube shown is a starvation zone in spite of the free availability of the electrons. The reason is that at low altitude $(r - R \ll cP/2\pi)$ the relativistic electron flow is entirely parallel to B. Then $\eta = J_{||}/c \propto (\text{Area})^{-1} \propto B$, since the polar flow tube is a magnetic flux tube. To have $E_{||} = 0$ at the surface, one must have $\eta = \eta_R$ at the surface, to a high degree of accuracy. But, $\eta_R \propto \underline{\Omega} \cdot \underline{B} \propto B \times \cos$ (angle between $\underline{\Omega}$ and $\underline{B}$) and this cosine increases with increasing altitude above the star in favourably curved flux tubes. Thus, above the surface, $\eta_R/\eta > 1$ and the tube is starved with the fractional charge separation increasing in proportion to r^2-R^2. This configuration is possible only in the oblique rotator; for this reason, my opinion is that studies of the aligned rotator are irrelevant to active pulsars.

It's easy to show that this starvation mechanism leads to large voltages fairly near the star. Since the flux tube is long and narrow, the length that enters in (11) is the local width of the tube, with the result

$$\Phi \sim 2 \times 10^{12} \frac{B}{1TG} \left(\frac{0.2 \text{ s}}{P}\right)^{5/2} \left(\frac{\text{dipole radius of curvature}}{\text{actual radius of curvature}}\right) \left(\frac{r}{R} - 1\right) V \quad (19)$$

in the region tube width $\ll r-R \ll R$. For detailed theory see Arons and Scharlemann (1979) and Arons (1980 a). Voltages of this magnitude cause electrons to radiate GeV γ-rays by the curvature process, just as in Sturrock and RS models, except that now the characteristic heights are more typically $\sim 1-2$ km instead of 10-100 m. The mean free path for these γ-rays is exponentially dependent on the height of the emitting particles, in the form mfp $\propto \exp(\text{constant}/(\text{height})^4)$. Thus pairs appear abruptly, at and above a well defined surface, whose location is about where the mfp = height. The γ-rays from lower altitude electrons are almost all of lower energy, and thus create pairs above the surface (called the "pair formation front" = PFF by AS), or escape. The pair creation rate at the PFF is quite large, $\sim 10^2$ pairs km^{-1} per primary electron, and the pairs' relatively low energy and free mobility along B result in the PFF being a surface of $E_{||} = 0$. This results in very few of the newly created positrons being caught in the electric field and accelerated down onto the star so that pair creation does not disrupt the starvation electric field but merely terminates it continuously at the PFF. The PFF is rather like a detonation wave in a chemically reacting flow; below it ("upstream") the polar flux tube is starved and supports a fairly strong $E_\parallel$ (10^{10} V/m is typical), but at the PFF, the partial vacuum "explodes", a dense pair plasma appears and travels outwards ("downstream") along with the original electron beam. The region below the bottom point on the PFF is called the "diode" zone by AS, in analogy to space charge limited relativistic diodes.

There is an additional feature to this scheme. Near the "walls" of the polar flux tube, $E_\parallel$ is weak and Φ is small in the high opacity region r < 2R. Then pairs are not created at the outer rim of the primary electron beam. This means the pair plasma is surrounded by a thin annular starvation zone which extends to high altitude, a "slot" as illustrated in Fig. 5. Within the unshaded region, which extends at least to radii $\sim cP/2\pi \gg R$, $E_\parallel$ is not shorted out and high voltages are achieved (Even though $E_\parallel \sim 10^6$-10^7 V/m in the center of the slot is small compared to $E \sim 10^9$ - 10^{11} V/m typical of the center of the diode, the much larger length along field lines allows the voltage to accumulate). The minimum model for particle acceleration is then diode + slot region, assuming no further energy is generated within the pair plasma or at the boundary layer between the slot and the pair plasma (the dynamics of the PFF were studied by AS, who showed this thin layer contributes miniscule dissipation and also allows the plasma to adjust to $E_\parallel = 0$ with only $\sim 10^{-4}$ positrons being accelerated toward the star for each primary electron accelerated away from the star). The total luminosity in particle acceleration depends on the volume of the diode + slot, which is found by solving the appropriate nonlinear free boundary problem (Arons, 1980 a). The result for the total $J \cdot E$ work on the electron beam in the diode + slot can be put in the form

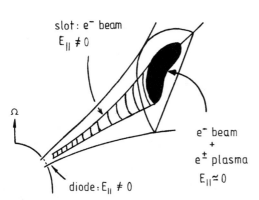

Figure 5. Diode + slot model of starvation zone.

$$L_P = \sim \frac{\mu^2 \Omega^4}{c^3} \left[1.5 \times 10^{-2} P^{-1/2} E_D (P/P_D) + E_S (P/P_D) \right] \quad (20)$$

where P_D = period when the opacity and voltage are just sufficient for the creation of <u>one</u> pair per primary e- along the central field line of the flux tube, with $P_D \sim 0^s.3$-3^s the typical values depending precisely upon B(R), R and the radius of curvature of the surface magnetic field (the exact expression is given by Arons, 1980a). Since $\mu^2 . \Omega^4/c^3$ is the spin down luminosity, $1.5 \times 10^{-2} P^{-1/2} E_D$ and E_S are the efficiency of conversion of spindown energy lost into an accelerated electron beam in the diode and slot regions respectively, with

$$E_D \overset{\sim}{=} 0.09 \left(\frac{P}{P_D}\right)^{2.24} \qquad (21)$$

and

$$E_S = 0.5 \left(\frac{P}{P_D}\right)^{5.08} \qquad (22)$$

The properties of the e± plasma are straight forward consequences of the generation scheme. I find (Arons, 1980 b) $\sim 10^3$ pairs per primary beam particle (for an object like the Crab, the number is more like 5×10^4) with a broad dispersion momenta parallel to B on most field lines of the plasma flow tube, in the range ~ 10 MeV/c to ~ 1 GeV/c. The primary electron beam on field lines that cross the PFF puts most of its energy into creating pair plasma in pulsars with large P, while in those with small P or long P, the beam escapes to high altitude with most of its energy intact. In either case, the total energy put into this part of the beam is given by the first part of (20), and the energy per beam particle at the PFF is ~ 3 TeV to 10 TeV, depending on the exact location. In the slot, acceleration continues to higher energy, with the exact asymptotic final energy unknown because it depends on the unknown structure of the outer magnetosphere and on other high altitude physics (expression 20 is calculated assuming acceleration continues out to $r \sim cP/2$ in a dipole field, where it ceases). In total, the beam has a flux f_c. Therefore, when $P/P_D < 1$, this model, like the RS picture, puts out a plasma with total flux $\gg f_c$, indicating spindown is basically by an mhd wind. The presence of special low density zones should not be ignored, however.

Application of (20) to Vela shows that more than enough $\underline{J \cdot E}$ work is done in the model to explain this pulsar's γ-rays, while for the Crab (20) can be adjusted to be within 1/2 of the required γ-ray luminosity. This is encouraging, since (20) is a minimum to the work done (see below), and in principle particle energies can be radiation reaction limited so that 100% of the J.E work goes into radiation (just as stars put essentially 100% of their nuclear energy generation into radiation). However, in the specific application of the pure slot gap model to pure curvature emission from the Crab and Vela, I find such radiative efficiency is not so; in both cases, $L_\gamma/L_p < 0.1$. For this acceleration model, this negative result is a good thing. Just as in all other curvature emission theories with more or less monoenergetic beams as the γ-ray emitters at each altitude, the γ-ray spectrum is too flat (cf. Massaro and Salvati, 1979, and Hayvaerts and Signore, 1980, for general phenomenological studies). Also, as in other curvature emission theories so far proposed, there is insufficient localization of the source to give a sharp pulse. Unlike other models, the slot gap theory does localize the acceleration at high altitude and its emission to a thin sheaf of field lines, but since the field lines diverge with curvature, an observer sees beamed radiation from different altitudes at different rotational longitudes, giving a broad pulse (see § II above). These diseases are typical of all models employing laminar acceleration and pure curvature emission, starting with Sturrock's and ending with the model described here.

Despite my obvious prejudice, I think the diode + slot gap scheme is capable of overcoming its present problems. It has a number of attractive features. It produces an interestingly large total acceleration energy at high altitude without having manifest internal contradictions and without creating thermal emission from the poles in excess of current observations (see AS for the minimal estimates). The relatively large total acceleration is produced with localization of the acceleration zone to a thin sheaf of polar flux surfaces, which is 1/2 of the localization needed to make a successful model of pulses in the higher energy photons. It automatically yields a single pole scheme, since an observer oriented at random with respect to the rotation axis can see outflow and resulting beamed radiation from the favourably curved part of only one polar flux tube. Even if the magnetic field is completely symmetric, the observer's line of sight looks into unfavourably curved regions when the star rotates by $180°$ (see Fig. 11 of Arons, 1979), a zone where strong outward acceleration, pair creation and (presumably) beamed photon emission do not occur. Another, new argument in favour of this scheme is that such single pole models may explain the gross assymetry in the excitation of the Crab nebula evident in the HEAO-B images of the nebula. If one pole of the star has magnetic structure like that of figure 4, while the other pole is much less able to create outflow and pairs (as can be true for example, in the dipole + axisymetric quadrupole model of Barndard and Arons (1980) when the magnetic axes are aligned with each other but not aligned with the rotation axis; see Fig. 7 of Arons, 1979), the composition of the stellar wind is grossly asymmetric with respect to the rotational equator, with much larger particle flux in one hemisphere. This is probably what is needed to understand the morphology of the nebular emission, if, for example, the acceleration behind the nebular X-ray emission occurs in a shock wave as the wind collides with the inner edge of the nebula (see below), and is likely only if the pulse-interpulse morphology of the stellar emission arises in a single pole, as is possible in the slot gap scheme.

The basic problems in the current form of this model are (1) its lack of radial localization of high energy emission, (2) a mild lack of total energy for application to the Crab, (3) the inapplicability of curvature γ-ray spectra from essentially monoenergetic beams to the observed data, and (4) the inability of laminar, ultrarelativistic flow of an electron beam along curved field lines to create high brightness temperature radioemission. All of these problems are traceable to the ideal conductor approximation used to represent the boundary between the slot gap and the pair plasma. This is a good approximation at low altitude, but becomes poor at large radii where

relative streaming between the plasma components in the much weaker
magnetic field may give sufficient collisionless dissipation to
enhance the acceleration efficiency and create collective radio
emission (I favour the electromagnetic form of the E x B shear flow
instability described by Arons and Smith (1978), but this is by no
means the only possibility). More efficient emission of high energy
photons by the synchrotron process may occur from excitation of pairs
in the boundary layer to finite Larmor orbits by the relevant form of
a beam cyclotron instability. Because of the finite threshold for this
type of instability, radial localization of the synchrotron emission
becomes possible, in addition to the angular localization already
imposed by the pair creation model of the boundary layer.

It is not known now whether these effects are sufficient to
give an explanation of the photon data, within the context of the
consistent dynamical model outlined above. However, recently
Machabelli and Usov (1980) have shown how some aspects of such effects
might lead to an interesting model of high frequency emission from
pulsars. They adopted Tademaru's (1973) elaboration of Sturrock's high
voltage scheme and show the highest energy plasma component in
Tademaru's model has a resonant beam cyclotron instability at radii
$\sim cP/2$, when applied to the Crab pulsar. They use quasi-linear theory
to argue that the resulting acceleration of the pairs into finite
gyrational states might give rise to X-ray and γ-ray synchrotron
emission with a spectrum like that observed. They are unable to reproduce the optical spectrum, nor, given the lack of angular localization
in their emission region, can they get a sharp pulse (they do have
some radial localization of the emission because of the finite
threshold of the instability). As basic theory of particle acceleration, this model is still plagued with the problem recognized by
Tademaru in his particle acceleration paper, namely, that the creation
of the dense pair plasma is not consistent with the high voltage assumed at the polar cap. Nevertheless, if viewed as a semi-phenomeno-
logical model of photon emission, I think Machabelli and Usov's work
reveals the kinds of effects needed in order to make a successful
theory.

IV - COSMIC RAYS

Cosmic rays are basically a spectrum of relativistic nuclei.
The models I have outlined produce mainly pairs. Therefore, one may
well ask, what do pulsars have to do with cosmic rays?

(a) Pairs and positrons

Both surface gap and slot gap + diode models, when applied
to an object like the Crab pulsar, produce a minimum of a few x 10^{36}
electron-positron pairs/sec. In addition, further pairs are produced
by the synchrotron photons emitted by preceding generations of pairs
can enhance this number by another factor $\sim$ 50-100 in the slot gap +
diode case (Arons, 1980 b); this is probably true in the surface gap

case also, but I haven't studied this. The work of Alber et al. (1975) suggests that the Sturrock-Tademaru view has similar results, when done with some attention to consistency.

These rates of particle supply are sufficient to explain the Crab nebula (where the density is too low to yield a 511 keV annihilation line), but further acceleration beyond that supplied by the starvation zones in either model is needed to explain the energetics of the nebular X-ray emission (see (15) and (20). It is obvious that the composition of these model pulsars' output is nothing like the observed cosmic ray source spectrum (cf. Cassé, 1979). Therefore, the only role observed pulsars might play as direct cosmic ray sources is in providing positrons and electrons.

To estimate the significance of this possibility, I assume the number of pairs created per primary beam particle is a constant $\sim 10^{+3}$, independent of magnetic field and P (in fact, it is a function of P/P_D, but I stick to simple estimates here). I also assume the angular momentum loss is given by (3). Then the total number of pairs ejected by a pulsar is

$$N_{\pm} = \chi \frac{Ic^2}{e} \ln(T/\tau) = 10^{49} \frac{\chi}{10^3} \frac{I}{10^{45} \text{ g.cm}^2} \frac{10^{30} \text{ G.cm}^3}{\mu} \frac{\ln(T/\tau)}{7} \quad (23)$$

where T = pulsar lifetime (= time to cessation of pair creation?). $\sim 10^6$ y (Gunn and Ostriker, 1970; Fujimura and Kennel, 1980; Barnard and Arons, 1980), τ = initial spindown time = $Ic^3/\mu^2\Omega_i^2$, Ω_i = initial angular frequency. Assume the pulsar birthrate is $R_p^1 \sim 10^{-10}$ pc^{-2}yr^{-1} (corresponding to $\sim 1/25$ yr in a galactic disk of radius 10 kpc, roughly in accord with Taylor and Manchester's (1977) birthrate), while the confinement time is $\tau_{CR} \sim 2 \times 10^7$ y in a cosmic ray disk of scale height $H_{CR} \sim 500$ pc (Cesarsky, 1980). Then the total isotropic flux of electron-positron pairs one expects in the solar neighbourhood is

$$J_{\pm} \stackrel{\sim}{=} 32 \frac{N_{\pm}}{10^{49}} \frac{R_p}{10^{-10} \text{ pc}^{-2} \text{ y}^{-1}} \frac{\tau_{CR}}{2 \times 10^7 \text{y}} \frac{500 \text{ pc}}{H_{CR}} \text{ m}^{-2} \text{ster}^{-1} \text{s}^{-1} \quad (24)$$

If no further acceleration of the pairs occurs after they are created, the particles in the total flux (24) all have energy << 1 GeV, because of adiabatic losses in the expanding wind. On the other hand, most of each pulsar's rotational energy loss goes into the wind; only a small fraction is used creating the pairs. Then further acceleration as the winds encounter their surroundings could reaccelerate the pairs (see below). Since the observed positron flux at energies ~ 1 GeV is about a factor of 5 below the total flux (24) (Fanselow et al., 1969; Buffington et al., 1975; Golden et al., 1979), the observed positron flux might yield a useful constraint on the abilities of pulsars and their environs to accelerate particles, although I think a lot more will be learned by understanding the accelerator in the Crab nebula.

The flux (24) might have interesting implications for interstellar γ-ray emission. Suppose cosmic ray particles diffuse into molecular clouds with a very short mean free path. Then all of the positrons in the flux (24) incident on the cloud surface annihilate, giving rise to a maximum 511 keV line flux at the earth

$$f_{511} = 2\pi J_{\pm} \left(\frac{R}{D}\right)^2 \sim 2 \times 10^{-6} \frac{N_{\pm}}{10^{49}} \frac{R_P}{10^{-10}} \frac{\tau_{CR}}{10^{7.3}} \frac{500}{H_{CR}} \left(\frac{R}{10 \text{ pc}}\right)^2 \left(\frac{1 \text{ kpc}}{D}\right)^2$$

$$\text{cm}^{-2}.\text{s}^{-1} \qquad (25)$$

for each cloud of size R ∼ 10 pc and distance D ∼ 10 kpc. Such fluxes are too small to observe now but may be accessible with the GRO. However, very small diffusion mean free paths are thought to be unlikely and are excluded if cosmic ray ionization is the source of the free electrons in molecular clouds (Cesarsky and Völk, 1979). If the positrons stream freely through the clouds, the expected annihilation line from the flux (24) is many orders of magnitude smaller than (25), since the clouds have insufficient (by ∼ 10^4) grammage to slow the positrons to nonrelativistic velocity. In special place however, where R_ρ is high, the pulsars might all spin rapidly and if the total density is high, annihilation might lead to currently observable line emission for which the obvious candidate is the 511 keV line from the galactic center (Leventhal et al., 1979). See Bussard et al. (1979) for an analysis of the emission and various possible positron sources. I only point out here that pulsars are a good source, since they are prone to make pairs without making relativistic nuclei, thus avoiding some energy problems pointed out by Audouze et al. (1980). Sturrock and Baker (1979) have claimed that only Sturrock's model produces enough pairs per pulsar. I don't agree but further discussion of this point depends on better modelling and is left for elsewhere.

(b) Nuclei

In the surface gap scenarios, which allow free ion emission, some ions are accelerated outwards. However, the number accelerated is too small (by ∼ 6 orders of magnitude), the energy spectrum in the models is monoenergetic, not a power law, and the composition of the ion beam is usually said to be pure iron. This last statement probably is not true, since the top ∼ 100 g.cm^{-2} of the neutron star's crust is exposed to continuous bombardment by TeV electrons and GeV γ-rays, with the result that the accelerated beam could contain a whole spectrum of the spallation products of iron. However, I see no possibility for the composition that results from grinding up iron being a decent model for cosmic rays.

(c) Energetics

All theories of particle acceleration and pair creation developed so far make use of only a small fraction of the electromagnetic energy loss (3); see (20) for the most explicit example. While the particles supplied by the pulsar itself cannot be the cosmic rays, in these models, the unused energy might be reprocessed by the surroundings where the composition is more normal into a more acceptable spectrum of particles. This sort of "planetary nebula" idea was first suggested by Kulsrud, Ostriker and Gunn (1972), who used strong vacuum waves to calculate particle spectra, an approximation forbidden by modern models. However, the general idea might be OK, if the pulsars have enough energy.

If the relativistic energy loss rate applies all the way back to time 0, the Crab pulsar initially had $\sim 4 \times 10^{49}$ ergs of rotational energy. This is too small, by about one order of magnitude, to explain cosmic rays, assuming the birthrate = supernova rate, 100% efficiency of converting rotational energy into cosmic rays in the initial spindown, and all pulsars started out like the Crab. Note that this is a semi-theoretical argument; one has to use the relativistic spin down rule of thumb (3) right back to the beginning.

For other pulsars, we have no knowledge of the age in specific cases, therefore we don't know how to integrate (3). Furthermore, $P/\dot{P}$ ages don't tell us much, since the presence of too many short period pulsars with small $\dot{P}$ clearly shows that $\dot{P}$ is not a simple function of P; pulsars are not all on the same evolutionary track. The large space motions and small scale height of pulsars does show that the true age of the population does not exceed a few x 10^6 years (cf. Manchester and Taylor, 1977, for a review of this topic).

The loss rate (3) is then consistent with large initial angular velocities (e.g., P = 5 msec initially implies 2×10^{51} ergs of rotational energy), but there is no way to infer a specific value of the initial energy. Thus, all one can say is that the pulsars might have energy; whether they do is an open question.

If one accepts the hypothesis that they do, one still has to face the mechanisms by which the particles are accelerated. For reasons outlined above, I doubt that strong waves are present. Instead, one has to deal with acceleration of nucleons at the boundary of the wind from the pulsar. If the wind passes through the fast mode critical points, it must decelerate through a shock wave as it collides with the surrounding nebula/interstellar medium (Rees and Gunn, 1979, discuss this possibility in their mixed model). Such an interior relativistic shock in the electron-positron flow may be an excellent model for the excitation region of the Crab nebula (the "wisps"), especially since the single pole model outlined in the last paragraph may explain the gross asymmetry of the energy injection into the nebula. However, it is obviously not a good site for accelerating nucleons, since none is in the flow from the pulsar. If the pulsar drives the supernova (Ostriker and Gunn, 1972; Bodenheimer and Ostriker, 1975; Gaffet, 1976), the exterior shock of the supernova remnant might accelerate particles in the now quite popular scenario of Axford et al., 1977; Bell, 1978 and Blandford and Ostriker, 1978. Here, it makes no difference if the energy source is neutron star rotation (currently unfashionable) or neutron star "bounce" (currently fashionable), since the particle acceleration has nothing to do with the pulsar as such. A final possibility is at the tangential discontinuity separating the shocked wind from the pulsar from the surroundings (be it supernova envelope or shocked interstellar medium). Because the shocked wind stores much of its energy in the magnetic field, dissipation of the magnetic field in the boundary layer might accelerate the nuclei without running into energetic difficulties. If the subsequent adiabatic losses are not too severe, this environment might be a cosmic ray source peculiar to pulsars. In general, however, I doubt that there is sufficient magnetic energy storage to make this idea viable.

My opinion is that pulsars might have a lot to do with the emission from non thermal photon sources, especially the "filled" supernova remnants (cf. Caswell, 1979), and with some aspects of the γ-ray astronomy of the interstellar medium as well being fascinating objects in their own right. If the pair creation models are even vaguely on the right track (and I think they are), however, I doubt these schemes have much to do with cosmic ray nuclei.

I am indebted to C. Cesarsky and M. Cassé for informative conversations, and to C. Cesarsky for reminding me that IAU Symposium 94 is about cosmic rays, not pulsars. This work is supported in part by the US National Science Foundation and by the John Simon Guggenheim Memorial Foundation.

References

Akhiezer, A.I. and Polovin, R.V., 1956, Zh. Eksp. Teor. Fiz, $\underline{30}$, 915 (Soviet Phys., JETP, 3, 696).
Al'ber, Ya I., Krotova, Z.N., and Eidman, Y.Ya., 1975, Astrofizika, $\underline{11}$, 283 (Astrophysics, 11, 189).
Arons, J., 1979, Space Sci. Rev., $\underline{24}$, 437.
Arons, J., 1980a, to be submitted to Ap. J.
Arons, J., 1980b, ibid.
Arons, J., Kulsrud, R.M. and Ostriker, J.P., 1975, Ap. J., $\underline{198}$, 687.
Arons, J., Norman, C.A., and Max, C.E., 1977, Phys. Fluids, $\underline{20}$, 1302.
Arons, J., and Scharlemann, E.T., 1979, Ap. J., $\underline{231}$, 854.
Arons, J., and Smith, D.F., 1979, Ap. J., $\underline{229}$, 728.
Asseo, E., Kennel, C.F., and Pellat, R., 1978, Astr. and Ap., $\underline{65}$, 401.
Asseo, E., Llobet, X., and Schmidt, G., 1980, Phys. Rev., in press.
Axford, W.I., Leer, E., and Skadron, G., 1972, Proc. 15th Intl. Cosmic Ray. Conf., 11, 132.
Ayasli, S., and Ogelman, M., 1980, Ap. J., $\underline{237}$, 222.
Backer, D.C., 1976, Ap. J., $\underline{209}$, 895.
Barnard, J., and Arons, J., 1980, to be submitted to Ap. J.
Bell, A.R., 1978, M.N.R.A.S., $\underline{182}$, 142.
Bennett, K., Bignami, G.F., Boella, G., Buccheri, R., Hermsen, W., Kanbach, G., Lichti, G.G., Masnou, J.L., Mayer-Hasselwander, H.A., Paul, J.A., Scarsi, L., Swannenburg, B.N., Taylor, B.G., and Wills, R.D., 1977, Astr. and Ap., $\underline{61}$, 279.
Blandford, R.D., and Ostriker, J.P., 1978, Ap. J. (lett)., $\underline{221}$, 229.
Bodenheimer, P., and Ostriker, J.P., 1974, Ap. J., $\underline{191}$, 465.
Buffington, A., Orth, C.D., and Smoot, G.F., 1975, Ap. J., $\underline{199}$, 669.
Bussard, R.W., Ramaty, R., and Drachman, R.J., 1979, Ap. J., $\underline{228}$, 928.
Cassé, M., 1979, in Particle Acceleration Mechanisms in Astrophysics, ed. J. Arons, C.E. Max and C.F. McKee (New York : AIP), p.211.
Caswell, J.L., 1979, M.N.R.A.S., $\underline{181}$, 431.
Cheng, A.C., and Ruderman, M.A., 1977 a, Ap. J., $\underline{214}$, 598.
        1977 b, ibid., $\underline{216}$, 865.
        1980 ibid., in press.
Cheng, A.C., Ruderman, M.A., and Sutherland, P.G., 1976, Ap. J., $\underline{203}$, 209.
Cesarsky, C.J., 1980, Ann. Rev. Astron. and Ap., in press.
Cesarsky, C.J., and Volk, H.J., 1978, Astr. and Ap., $\underline{70}$, 367.
Cordes, J., 1979, Space Sci. Rev., $\underline{24}$, 567.
Cordes, J., 1980, Ap. J., $\underline{237}$, 216.
Deutsch, A.J., 1955, Ann. d'Ap., $\underline{18}$, 1.
Epstein, R., and Petrosian, V., 1974, Ap. J., $\underline{183}$, 611.
Fanselow, J.L., Hartman, R.C., Hildebrand, R.H., and Meyer, P., 1969, Ap. J., $\underline{158}$, 771.
Fawley, W.M., 1978, Ph.D. Dissertation, U. of California, Berkeley.
Fawley, W.M., Arons, J., and Scharlemann, E.T., 1979, Ap. J., $\underline{217}$, 227.
Flowers, E.G., Lee, J.F., Ruderman, M.A., Sutherland, P.G., Hillebrandt, W., and Muller, E., 1977, Ap. J., $\underline{215}$, 291.
Fujimura, F.S., and Kennel, C.F., 1980, Ap. J., $\underline{236}$, 245.

Gaffet, B., 1972, Ap. J., 216, 565.
Giacconi, R., 1979, Proc. 16th Intl. Cosmic Ray. Conf., p. 26-29.
Golden, R.A., Lacy, J.L., Stephens, S.A., and Daniel, R.R., 1979, Proc. 16th Intl., Cosmic Ray Conf., Paper OG-3.
Goldreich, P., and Julian, W.M., 1969, Ap. J., 157, 869.
Goldreich, P., Pacini, F., and Rees, M.J., 1972, Comm. Ap. and Space Sci., 3, 185.
Grindlay, J.E., Helinken, H.F., and Weekes, T.C., 1976, Ap. J., 209, 592.
Groth, E.J., 1975, Ap. J. (Suppl.), 29, 431.
Gunn, J.E., and Ostriker, J.P., 1969, Phys. Rev. Lett., 22, 729.
Gunn, J.E., 1970, Ap. J., 160, 979.
Heyvaerts, J., and Signore, M., 1980, Astr. and Ap., in press.
Jackson, E.A., 1976, Ap. J., 206, 831.
Jackson, E.A., 1980, ibid., 237, 198.
Jones, P.B., 1978, M.N.R.A.S. 184, 207.
Kanbach, G., Bennett, K., Bignami, G.F., Boella, G., Bonnardeau, M., Buccheri, R., D'Amico, N., Hermsen, W., Higdon, J.C., Lichty, G.G., Masnou, J.L., Mayer-Hassellwander, H.A., Paul, J.A., Scarsi, L., Swanenberg, B.N., Taylor, H.G., Wills, R.D., 1977, Proc. 12th ESLAB Symp., ESA SP-124, 21.
Kanbach, G., Bennett, K., Bignami, G.F., Buccheri, R., Canaveo, P., D'Amico, N., Hermsen, W., Lichti, G.G., Masnou, J.L., Mayer-Hasselwander, H.A., Paul, J.A., Sacco, B., Swanenburg, B., Wills, R.D., 1980, Astr. and Ap., in press.
Kennel, C.F., Schmidt, G., and Wilcox, T., 1973, Phys. Rev. Lett., 31, 1364.
Kennel, C.F., and Pellat, R., 1976, J. Plasma Phys., 15, 335.
Kennel, C.F., Fujimura, F.S., and Pellat, R., 1979, Space Sci. Rev., 24, 401.
Kulsrud, R.M., Ostriker, J.P., and Gunn, J.E., 1972, Phys. Rev. Lett., 28, 636.
Kundt, W., and Krotscheck, E., 1980, Astron. and Ap., 83, 1.
Leventhal, M., Mac Callum, C.J., and Stang, P.D., 1978, Ap. J., 225, L11.
Llobet, X., 1980, in preparation.
Machabelli, G.Z., and Usov, V.V., 1980, Ap. J., in press.
Manchester, R.N., 1978, Proc. Astron. Soc. Australia, 3, 200.
Manchester, R.N., and Lyne, A.G., 1977, M.N.R.A.S., 181, 761.
Manchester, R.N., and Taylor, J.M., 1972, Pulsars, W.M., Freeman Co. (San Francisco).
Mandrou, P., Vedrenne, G., and Masnou, J.L., 1980, preprint.
Max, C.E., 1973, Phys. Fluids, 16, 1272.
Max, C.E., and Perkins, F., 1971, Phys. Rev. Letters, 27, 1342.
Max, C.E., 1972, ibid., 29, 1731.
Massaro, E., and Salvati, M., 1979, Astr. and Ap., 71, 51.
Mestel, L., 1971, Nature Phys. Sci., 233, 149.
Mestel, L., Phillips, P., and Wang, Y.M., 1979, M.N.R.A.S., 188, 385.
Michel, F.C., 1969, Ap. J., 158, 727.
Michel, F.C., 1971, Comm. Ap. and Space. Sci., 3, 80.
Michel, F.C., 1980, Ap. and Sapce Sci., in press.

Novick, R., Weisskopf, M.C., Herthelsdorf, R., Linke, R., and
  Wolff, R.S., 1972, Ap. J. (Lett.), 174, L1.
Ogelman, H., Fichtel, C.E., Kniffen, D.A., Thompson, D.J., 1976,
  Ap. J., 209, 584.
Ostriker, J.P., and Gunn, J.E., 1969, Ap. J., 157, 1395.
Ostriker, J.P., 1971, ibid., 164, L95.
Pacini, F., 1967, Nature, 216, 567.
Rees, M.J., 1971, Nature, 229, 312.
Rees, M.J., and Gunn, J.E., 1974, M.N.R.A.S., 167, 1.
Ricker, G.R., Scheepmaker, A., Ryckman, S.G., Ballintine, J.E.,
  Doty, J.P., Downey, P.M., Lewin, W.H.G., 1975, Ap. J. (Lett.),
  197, L93.
Romeiras, F.J., 1979, J. Plasma Phys., 22, 201.
Ruderman, M.A., 1971, Phys. Rev. Lett., 27, 1306,
Ruderman, M.A., and Sutherland, P.G., 1975, Ap. J., 196, 51.
Scharlemann, E.T., Arons, J., and Fawley, W.M., 1978, Ap. J., 222, 297.
Scharlemann, E.T., 1979 in Particle Acceleration Mechanisms in Astro-
  physics, ed. J. Arons, C.E. Max and C.F. McKee (New York : AIP),
  p.373.
Shklovsky, I.S., 1970, Ap. J. (Lett.), 159, L77.
Smith, F.G., Disney, M.J., Hartley, K.F., Jones, D.H.P., King, D.J.,
  Wellgate, G.B., Manchester, R.N., Lyne, A.G., Goss, W.M.,
  Wallace, P.T., Peterson, B.A., Murdin, P.G., Danziger, I.J., 1978,
  M.N.R.A.S., 184, 39p.
Sturrock, P.A., 1971, Ap. J., 164, 529.
Sturrock, P.A., Petrosian, V., and Turk, J.S., 1975, Ap. J., 196, 73.
Sturrock, P.A., and Baker, K., 1979, Ap. J., 234, 612.
Tademaru, E., 1973, Ap. J., 183, 625.
Thompson, D.J., Fichtel, C.E., Kniffen, D.A., Lamb, R.C., Ogelman, H.B.,
  1926, Ap. J. (Lett.), 17, 173.
Weinberg, D.S.L., 1980, Ap. J., 235, 1078.

γ-RAY EMISSION FROM SLOW PULSARS

M. Morini and A. Treves
Istituto di Fisica dell'Università and
L.F.C.T.R., Milano, Italy

Among the COS-B Galactic γ-ray sources two have been identified with the young pulsars PSR0531+21 and PSR0833-45, while for most of the others there are no convincing counterparts (1). An obvious explanation is that at least part of them are also pulsars. Because of the rarity of pulsar formation, an important issue for testing this possibility is to discuss whether relatively old pulsars ($\sim 10^5$y) may be sources of γ-rays of the luminosity observed by COS-B (ϕ (100 MeV) 10^{-6}cm^{-2}s^{-1}). On this line it is of great interest the claim of observation of γ-rays with SAS II from PSR1747-46 (2) and with COS-B from PSR0740-28 and PSR1822-09, communicated at other conferences, but yet not confirmed (3). The three slow pulsars mentioned above were reported to have γ-ray luminosities of 10^{33}erg s^{-1}, close to their intrinsic energy loss $L_T = I\omega\dot\omega$. Their periods are about 0.1-1s and their distance $\sim$1 Kpc.

The scope of this communication is to calculate the expected γ-ray flux from such objects, neglecting the problem of the reliability of the observations. Our key hypothesis is that since the γ-ray luminosity is a substantial fraction of L_T it should be produced in the vicinity of the speed of light radius. This comes from the well known argument of simultaneous conservation of energy and angular momentum (4-5).

We refer to a model where a large fraction k of the total energy released $L_T = 2B_0^2 R^6 \omega^4/3c^3$ is in the form of a relativistic electron beam. Let k' be the fraction of the active polar cap ($R_c = R(\frac{\omega R}{c})^{\frac{1}{2}}$) occupied by the beam. Let r_0 be the distance out of which the pitch angle of relativistic particles departs from zero. At $r > r_0$ the electron distribution is taken isotropic within a cone of semiaperture Ψ. As typical values we take k=0,5; r_0=0.1 c/ω, k'=10^{-2}-10^{-3}, Ψ =1-10^{-1}rad, and a monochromatic electron spectrum with $\gamma_0=10^3$. These parameters are suggested by several pulsar models (6-7-8). One can verify that in these conditions synchrotron radiation does not contribute to the γ-ray luminosity. The contribution from Compton scattering derives predominantly from the scattering of the electrons with the synchrotron photons (Synchro-Compton), which move in the direction of the generating electrons. The calculation of the resulting spectrum is rather complicated, since the electron energy

	Small angle Compton Scattering		Isotropic Compton scattering	
$B_0 = 5 \times 10^{12}$ G, $R = 10^6$ cm	1822-09	1822-09	1822-09	0740-28
γ_0	1000	5000	3200	8000
k	0.5	0.5	0.5	0.1
k'	0.005	0.02	0.002	0.024
$r_0(c/\omega)$	0.12	0.19	0.09	0.36
Ψ (rad)	1.4	0.14	0.02	0.002
$m_V (A_V = 2 \text{mag/kpc})$	20.4	23.6	18.7	26.1
L_V (erg s^{-1})	2.1×10^{31}	1.3×10^{29}	1.7×10^{30}	2.7×10^{28}
ϕ (50-150MeV)(cm^{-2}s^{-1})	5.0×10^{-18}	2.2×10^{-22}	2.5×10^{-6}	3.5×10^{-8}
ϕ (150-500MeV)(cm^{-2}s^{-1})	1.1×10^{-19}	1.6×10^{-24}	1.9×10^{-6}	1.4×10^{-7}
L_γ(50-500MeV)(erg s^{-1})	3.4×10^{22}	1.7×10^{17}	8.8×10^{32}	5.4×10^{31}

varies because of radiative losses, the photon density depends on the synchrotron radiation, and varies with the radial distance. Moreover the scattering cross section depends strongly on the angle between electrons and photons (9), which in turn, is a function of the distance. The results are summarized in the Table. It is apparent that while the γ-ray fluxes are orders of magnitude below those of the three slow pulsars referred above, the luminosity in the visible is well above the limit of detectability, and nevertheless no optical counterpart was found thus far.

An important parameter in the calculations is the choice of the angle Ψ, which describes the anisotropy of electron and photon distributions. In the previous scheme the average electron-photon angle depends on Ψ. For small Ψ the synchro-Compton γ-ray flux diminish dramatically. As a limiting case one can consider small pitch angles, postulating that the electron-photon cross section is still that of isotropic distributions. The results are reported in the Table. This shows that, in order to account for $L_\gamma \simeq 10^{33}$ erg/s by synchro-Compton scattering at the speed of light radius, one should consider a situation which appears unrealistic, and still one finds an optical flux which is too large, requiring different beamings for optical and γ-ray photons.

References

1. Wills R.D. et al., 1980, Adv. in Space Explor., vol.7, Pergamon Press
2. Ögelman H., 1976, Ap.J. 209, 584
3. Lichti G.G., 1978, Communication at the IX Texas Symp., Munich
4. Cohen R.H., Treves A., 1972, Astr.Astrophys. 20, 305
5. Holloway N.J., 1977, M.N.R.A.S., 181, 9P
6. Ruderman M.A., Sutherland P.G., 1975, Ap.J., 196, 51
7. Arons J., Scharlemann E.T., 1979, Ap.J., 231, 854
8. Hardee P.E., 1979, Ap.J., 227, 958
9. Treves A., 1971, Nuovo Cimento, 4B, 88

DISTRIBUTION OF NEUTRINO FLUXES FROM PULSAR SHELLS

M. M. Shapiro and R. Silberberg
Laboratory for Cosmic Ray Physics
Naval Research Laboratory
Washington, D. C. 20375 U.S.A.

ABSTRACT

Young pulsars apparently have a distribution of initial power outputs $N(> P_0^{-\gamma})$, with $\frac{1}{2} < \gamma < 1$ and $P_0 \geq 10^{38}$ ergs/sec. Assuming that ultra-high-energy ($E \geq 10^{15}$ eV) cosmic-ray nuclei are accelerated at the central pulsar, a young, dense supernova shell can be a powerful source of high-energy neutrinos (Berezinsky, 1976). With an optical array placed in a volume of one km^3 at great ocean depths, as proposed for the DUMAND detector, it is likely that $\geq 10^3$ hadronic and electromagnetic cascades induced by neutrinos would be recorded for a stellar collapse within our Galaxy. Such supernovae occur about 8 times per century (Tammann, 1976). Neutrinos from young supernova shells in the Virgo supercluster would be marginally detectable via neutrinos with $N(> P_0) \propto P_0^{-\frac{1}{2}}$, but unobservable if $N(> P_0) \propto P_0^{-1}$.

- - - - - -

A young, dense supernova shell can be a powerful source of high-energy neutrinos (Berezinsky and Prilutsky 1976, and Berezinsky 1976). In this model, ultra-high energy ($E \geq 10^{15}$ eV) protons and other nuclei are accelerated at the central pulsar. The protons interact in the supernova shell and generate cascades of mesons, which in turn yield neutrinos upon decay.

One essential problem in the estimate of neutrino fluxes is the initial rate of energy loss P_0 of the pulsar. The observed radio luminosity function of pulsars follows a power law, and has a broad distribution of values. Probably P_0, too, has a broad range of values.

Consider the pulsar luminosity function based on all the observed Galactic pulsars. Taylor and Manchester (1977) find that the power output distribution among pulsars is $N(\geq P) \propto P^{-\gamma}$, where $\gamma \approx 1.0$. However, pulsars with large radio luminosities are relatively young; if these are subtracted out from the luminosity distribution, one gets a steeper ($\gamma > 1.0$) distribution for the old pulsars. Conversely, for young (or initial) pulsars, $\gamma < 1.0$. We shall assume that the frequency of supernovae with an initial power output greater than P_0 is $N(> P_0) \propto P_0^{-\frac{1}{2}}$ or P_0^{-1}, where P_0 ranges from 10^{38} to 10^{44} ergs/sec.

The efficiency of energy conversion into relativistic particles is high near a pulsar. Finzi and Wolf (1969) estimate that $\sim 1/3$ of the rotational energy loss of the Crab pulsar is converted into energy of relativistic particles and magnetic fields, observed through synchrotron radiation. We shall assume that the power input into relativistic protons is similar, i.e. 0.3 P_0.

Let us adopt the supernova frequencies expected by Tammann (1976), i.e. ~ 8 per century in our Galaxy (within a factor of two), and ~ 1800 per century in the Virgo supercluster at distances ≤ 20 Mpc. Table 1 shows the estimated frequency of supernovae per century as a function of the number of neutrino events for the exponents $\gamma_0 = 0.5$ and 1.0.

TABLE 1
Estimated Numbers of Neutrinos ($E \geq 4$ TeV) Detected in 10^{10} tons of water, in 4 months

	Initial rate of energy loss (ergs/sec)	Number of neutrino events	Frequency per century $N(\geq P_0) \propto$ $P_0^{-\frac{1}{2}}$	P_0^{-1}
Supernova in our Galaxy (at ~ 10 Kpc)	$10^{38} - 10^{39}$	10^3	~ 6	~ 7
	$10^{39} - 10^{40}$	10^4	~ 2	~ 1
	$10^{40} - 10^{41}$	10^5	~ 0.5	~ 0.1
Supernova in the Supercluster (at ~ 20 Mpc)	$10^{41} - 10^{42}$	0.1	40	1.3
	$10^{42} - 10^{43}$	1	13	0.1
	$10^{43} - 10^{44}$	10	4	0.01

We note from Table 1 that the high-energy neutrinos from supernovae in our Galaxy should be readily detectable, and with the assumptions stated earlier, the corresponding pulsars would have initial energy loss rates of 10^{38} to 10^{39} ergs/sec, i.e. they would be relatively low-powered pulsars. However, the neutrinos from the supernovae in the supercluster would be detectable with the DUMAND array only if $P_0 > 10^{42}$ ergs/sec (i.e. exceptionally high-powered pulsars), and if the power output distribution for young pulsars goes as $N(\geq P_0) \propto P_0^{-\frac{1}{2}}$.

REFERENCES:

Berezinsky, V.S.: 1976 DUMAND Summer Workshop, A. Roberts, ed., p. 229, Fermilab, Batavia, Ill.
Berezinsky, V. S. and Prilutsky, O. F.: 1976, Proc. Internat. Conf. (Neutrino-76), Aachen.
Finzi, A. and Wolf, R. A.: 1969, Ap. J. 155, L107.
Tammann, G. A.: 1976, Proc. 1976 DUMAND Summer Workshop, A. Roberts, ed., p. 137, Fermilab, Batavia, Ill.
Taylor, J. H. and Manchester, R. N.: 1977, Ap. J. 215, 885.

DISTRIBUTION OF NON-THERMAL EMISSION IN GALAXIES

R. Sancisi and P.C. van der Kruit
Kapteyn Astronomical Laboratory
Groningen University, The Netherlands

INTRODUCTION

The properties of the radio continuum emission from spiral galaxies have been reviewed by Van der Kruit and Allen (1976) and by Van der Kruit (1978). In more recent years the major developments in the understanding of the radio continuum properties and the underlying physical conditions of galaxies have come from a number of surveys of large samples of objects. Some of these surveys (e.g. Hummel, 1980a) have good sensitivity and sufficiently high angular resolution to allow for the first time a clear separation of central sources and disk emission and a study of the properties of these components in a large number of galaxies. As a consequence some results already found, suggested or only suspected in previous detailed investigations of a limited number of objects are put on a firmer basis or entirely new aspects are revealed.

We will summarize here mainly the results on the non-thermal radio emission from disks and halos of spiral galaxies and only make a few remarks on the relation with central sources in general and on the completely different behaviour of S0 and elliptical galaxies. The central radio sources are discussed in detail by Hummel (paper in this Symposium).

DISK AND HALO EMISSION

The radio continuum emission from spiral galaxies is mainly non-thermal (synchrotron radiation) and depends on the density of relativistic electrons and on the magnetic field strength. The main new results on the disk emission come from the survey of a large sample of galaxies at 1415 MHz by Hummel (1980a) with the Westerbork Synthesis Radiotelescope, and by the study of a number of nearby galaxies at high frequencies (up to 10.7 GHz) with the Effelsberg radiotelescope (Klein and Emerson, 1980). In these studies the emission from the disk and from the halo region are not separated. Some of Hummel's main results are:
1) The mean radio power of the disk emission is directly proportional to the mean optical luminosity.
2) The disk component gives in general the major contribution (more than

90 percent) to the total radio power. This contribution increases with morphological type.
3) The radio emissivity of the disk is not related to the radio power of the central source.
4) The radio emissivity per unit light of the disk does not depend on the presence of a bar, on the degree of development of the spiral arms (luminosity class), on the colour or on the presence of a companion. But it does depend on the morphological stage of the galaxy: it has a maximum for the intermediate (Sb-Sc) types and becomes significantly lower for the early (factor 2 to 3) and the late types (almost a factor 2).

These observational results lead Hummel to conclude that in general: i) central sources do not contribute a significant amount of relativistic electrons to the disk component, ii) density waves must play a very unimportant role in determining the radio-continuum emissivity, and iii) the main source of relativistic electrons does not belong to the young extreme population in the spiral arms, but rather to the old disk population, which provides most of the light in a galaxy. This was first suggested by Van der Kruit, Allen and Rots (1977) based on observations of NGC 6946.

The radial distribution - This has previously been discussed by Van der Kruit (1978). The most detailed information on the distribution in the disk of the non-thermal radio emission and of its spectral index comes from the high-resolution, multi-frequency observations of NGC 6946 (Van der Kruit et al., 1977), of M51 (Van der Kruit, 1977) and M31 (see paper by Beck and Klein in this Symposium). The studies of NGC 6946 and M51 indicate, in agreement also with Hummel's conclusions for a larger sample (see above), that the radial distribution of non-thermal emission correlates with the total stellar disk population rather than the extreme population I. It should be emphasized, however, that this does not necessarily prove that the sources of relativistic electrons are to be found mainly in the old disk population. But if the sources of cosmic ray electrons would belong mainly to the young population then there must be some way in which the mass density and total mass of the old disk population control the magnetic field strength and the propagation properties of the particles, and consequently also the distribution of volume emissivity and the total radio power.

The observed spectra of radio emission in these galaxies steepen with radius from the center to the other parts. It is not clear how much of this variation of spectral index is due to mixing with thermal emission in the inner parts, and how much is a real steepening of the non-thermal spectrum, to be explained by energy losses of the relativistic electrons and/or variation of the magnetic field strength. After some tentative correction for the thermal emission the spectrum of the non-thermal component still tends to steepen at the edges of the disk as in the case of M31 studied by Beck (see paper by Beck and Klein in this Symposium).

The thermal contribution - The question of the relative amount of thermal and non-thermal emission in a disk and of their separation is still a matter of debate. Observations at high radio frequencies (> 10 GHz), where the thermal contribution is expected to become dominant are important to settle this question. The recent observations of Klein and Emerson (1980) of a number of nearby normal spiral galaxies at 10.7 GHz indicate that the relative amount of thermal emission from galaxies at radio wavelengths is less than previous work (see Van der Kruit, 1978) had suggested. The galaxies studied show straight-line spectra (α = -0.70) over the frequency range at least from 0.4 to 10.7 GHz and none shows the flattening at the higher frequencies which would be expected in the case of substantial thermal emission. Klein and Emerson conclude that the thermal content of the integrated emission at 10.7 GHz must be less than 40 percent (and correspondingly less than 14 percent at 1.4 GHz) and that much higher frequencies would have to be observed before the thermal emission can be unambiguously separated from the non-thermal component and before it becomes dominant. Their conclusion depends critically, however, on the assumption of a straight spectrum of the non-thermal emission at high frequencies.

The integrated spectra - A striking result of this study by Klein and Emerson and of their compilation of the best available fluxes is that the spectra of these galaxies between 0.4 and 10.7 GHz are all power-law spectra and all possess a spectral index close to -0.70, with peak deviations of only ± 0.14. This is remarkably similar (same shape and value) to what is found in radio galaxies (see Moffet, 1975), possibly indicating that the same acceleration mechanism is at work.

Polarized emission - Some new results have been obtained recently on the polarized radio emission from the disk of galaxies, which give information on the general distribution and direction of the magnetic fields. The optical polarization of starlight has already been observed in the disks of M31, M51, M81 and some other spiral galaxies. The first detection of polarized radio emission was made by Segalovitz, Shane and De Bruyn (1976) at 21 and 6 cm in M51. Recently Beck, Berkhuysen and Wielebinski (1980) have mapped the polarized radio emission at λ 11.1 cm in M31. This emission is concentrated in a ring around the center of M31 in the same region of the optically brightest arms. The polarization is on average $\sim$ 15 percent and indicates a large-scale regularity of the magnetic field.

The z-distribution - The z-distribution of radio emission has been studied only in a small number of edge-on galaxies. The situation is not significantly changed since the review by Van der Kruit (1978). The two clearest cases with emission at high z-distances from the plane (up to 6 and 8 Kpc) at frequencies around 1415 MHz are NGC 891 (Allen, Baldwin and Sancisi, 1978) and NGC 4631 (Ekers and Sancisi, 1977). More recently Beck, Biermann, Emerson and Wielebinski (1979) have reported detection of emission at large z-distances also in NGC 253 at 8.7 GHz.

These studies have not gone much beyond the detection of a flattened halo (or thick disk) and the discovery that the non-thermal spectrum tends to become steeper with increasing z-distances from the galactic plane. The spectral index changes typically from values of -0.6 near the plane to values of $\sim$ -1.0, -1.5 at high z. But the shape of the spectral index dependance on the z-distance is certainly not determined accurately enough yet to allow a detailed comparison with theoretical models.

The relative amount of radio flux from the halo and from the disk is not well determined. There is no clear separation of these two components. In the case of NGC 891 a thin component confined to the plane and a broad component of steeper spectrum have been tentatively identified in the observations at 5 GHz (Allen et al., 1978). The broad component, which extends up to 4-6 Kpc from the plane and represents the thick disk or halo emission, contributes about 60 percent of the total emission at 5 GHz and becomes totally dominant at frequencies around 1.4 GHz or lower.

There are also other aspects not clear and questions which are still open:
1) how representative are these few objects for the existence of halos in galaxies in general and for their properties.
2) is the radio emissivity at high z related to the volume emissivity near the plane, as the cases known so far seem to suggest.
3) does the halo emissivity depend on the morphological type of the galaxy and is it related at all with the z-distribution of neutral gas. In NGC 891 (cf. Fig. 1 of Sancisi and Allen, 1979) and in NGC 4631 (compare Fig.1 of Weliachew et al., 1978 and Fig. 1 of Ekers and Sancisi, 1977) the distribution of HI and that of radio continuum emission at 21 cm appear quite different.

These results, in spite of all the uncertainties show that radio halos exist at least in some objects, and therefore support the hypothesis of a radio halo also around our Galaxy. As in NGC 891 and 4631 the halo extent and shape are likely to be dependent on the frequency at which the observations are done. This seems to be consistent with one of the most recent analyses of the radio data in our Galaxy (Webster, 1978).

INTERACTING GALAXIES

A large fraction of the spiral galaxies are members of multiple systems and some of these show peculiarities in their light distribution due to the gravitational interaction (Arp, 1966). There has been some discussion in the past on the effects of such interactions on the radio continuum properties of the galaxies involved. This is clearly relevant also for our Galaxy which is in interaction with the Magellanic Clouds. Stocke (1978) found excess radio emission from close pairs of galaxies, and Hummel (1980b) has shown that the central radio sources in galaxies with close neighbours are on average a factor 2 to 3 stronger as compared to those in isolated spirals, but the disk emissivity is about the same.

Detailed radio observations of some multiple systems with clear indications of interaction are revealing now an entirely new aspect, namely the existence of radio emission outside the optical images in the regions near or between the systems. Examples of the cases found so far are those of NGC 4038/39 (The Antennae) (Allen, Ekers, Burke and Miley, 1973; Ekers, 1980 private communication), and NGC 4438 (Kotanyi, Ekers and Van Gorkom, 1980 in preparation).

S0 AND E GALAXIES

It has already been emphasized (Ekers, 1978) that the continuum emission from S0 and elliptical galaxies is completely different from that of spiral galaxies. In the S0's no disk radio emission has been detected. This means that it must be at least a factor 4 weaker than in spirals in general and a factor 1.5 weaker than in early type spirals (Hummel, 1980a; Hummel and Kotanyi, 1980 in preparation).

The radio emission from ellipticals in no case resembles the distribution of light, contrary to spirals, and the range of the ratios of radio to optical luminosity is much larger than that for spirals. In most ellipticals the radio emission is substantially less than in spiral galaxies.

These properties must be related to the absence or the low content of gas, and hence the lower magnetic fields, in these types of galaxies.

ACKNOWLEDGEMENTS

We thank Beck and Klein and Emerson for providing information prior to publication, Ekers and Hummel for valuable discussions.

REFERENCES

Allen, R.J., Baldwin, J.E., Sancisi, R.: 1978, Astron. Astrophys. 62, 397
Allen, R.J., Ekers, R.D., Burke, B.F., Miley, G.K.: 1973, Nature 241, 260
Arp, H.: 1966, Atlas of Peculiar Galaxies (Calif. Inst. Tech., Pasadena)
Beck, R., Berkhuysen, E.M., Wielebinski, R.: 1980, Nature, 283, 272
Beck, R., Biermann, P., Emerson, D.T., Wielebinski, R.: 1979, Astron. Astrophys. 77, 25
Ekers, R.D.: 1978, I.A.U. Symp. 77 (Reidel, Dordrecht), p.49, Eds. Berkhuysen and Wielebinski
Ekers, R.D., Sancisi, R.: 1977, Astron. Astrophys. 54, 973
Hummel, E.: 1980a, Ph.D. Thesis, University of Groningen
Hummel, E.: 1980b, Astron. Astrophys. in press
Klein, U., Emerson, D.: 1980, Preprint
Kruit, P.C. van der: 1977, Astron. Astrophys. 59, 359
Kruit, P.C. van der: 1978, I.A.U. Symp. 77 (Reidel, Dordrecht), p.33, Eds. Berkhuysen and Wielebinski
Kruit, P.C. van der, Allen, R.J.: 1976, Ann. Rev. Astron. Astrophys. 14, 417

Kruit, P.C. van der, Allen, R.J., Rots, A.H.: 1977, Astron. Astrophys. 55, 421
Moffet, A.T.: 1975, in Stars and Stellar Systems, vol. IX, p.244, Eds. A. Sandage, M. Sandage and J. Kristian, University of Chicago Press
Sancisi, R., Allen, R.J.: 1979, Astron. Astrophys. 74, 73
Segalovitz, A., Shane, W.W., Bruyn, A.G. de: 1976, Nature, 264, 222
Stocke, J.T.: 1978, Astron. J. 83, 348
Webster, A.: 1978, Mon. Not. R. astr. Soc. 185, 507
Weliachew, L., Sancisi, R., Guèlin, M.: 1978, Astron. Astrophys. 65, 37

COSMIC RAYS AND GALACTIC RADIO NOISE

G.Sironi and G.De Amici
Laboratorio LFCTR/CNR and Istituto di Scienze Fisiche
dell'Universita' - Milano - Italy

Size and shape of the "confinement volume V_o", the region around our Galaxy where the c.r. particles live until they escape or loose completely their energy are essential parameters of models for the origin and propagation of the cosmic rays. Focussing the attention on the c.r. electrons, whose sources are definitely galactic and concentrated in the Disk, to confine V_o to the Disk or extend it to the Halo is full of consequences. For instance in the first case the sources can be everywhere inside V_o; by contrast in a Halo model the sources can reside only in some parts of V_o, the mechanisms of energy loss play different roles in different regions of the confinement volume and the particle lifetimes can be up to 100 times larger than in the Disk,(e.g. ref.1).

In the 50's Shklovskii, examining the distribution of the background radio emission, suggested that our Galaxy is surrounded by a Radio Halo produced by the synchrotron emission of c.r. electrons. Discussions on the Halo reality went on for a long time. In 1975 new evidence, based on analysis by the T-T plot method,(ref.2), of new systematic observations at VHF and UHF of the background emission was presented,(ref.3,4). Plotting against each other the sky temperatures measured at two frequencies allows to work out the frequency dependence of the anisotropic, i.e. galactic, component from the total,(galactic+extragalactic), background radiation. No precise knowledge of the zero level of the temperature scales is necessary. Between ~20 MHz and ~400 MHz the T-T plots of data obtained at various right ascensions and constant declination show typical V shapes suggesting that the frequency spectrum of the galactic radiation is steeper when looking away from the local spiral arm. Interstellar absorption is insufficient to explain this effect. Instrumental bias are also excluded because the same pattern can be obtained,(see fig.1), combining data obtained independently by different authors, using different equipments,(ref.5,6). An obvious interpretation (ref.7), is that the galactic radioemission is made of at least two components, a disk,(flat spectrum), and a halo,(steep spectrum), component.

Numerical models describing the emission of our Galaxy can be worked out following that line and agree with the observational data, (ref.3,4). In doing so it is important to remember that the radiation from particular regions of the Halo tend to appear isotropic and its contribution to the frequency spectrum deduced from the T-T plots has to be properly evaluated taking into account also absolute measurements. After 1975 radio haloes have been detected around external galaxies seen edge on, some of which similar to our Galaxy, (ref.8).

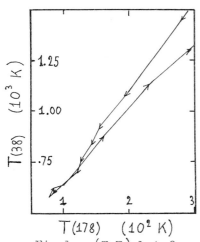

Fig.1 - (T-T) plot from 38 MHz, (ref.5), and 178 MHz (ref.6), observations smeared to 15°x15° for dec=+45 and 80° ≤ r.a. ≤ 290°

We believe therefore that the existence of the Galactic Radio Halo is sufficiently well established; physical parameters, shape and dimensions of that halo are however poorly defined for a number of reasons : i) low angular resolution of the observations; it allows to easily get rid of discrete sources contribution but reduce the number of statistically independent points and masks details; ii) loops and spurs, many degrees wide, have to be removed from the maps; iii) galactic absorption, below ∼20 MHz, and 3°K relict radiation, above ∼600 MHz, gradually efface the Halo signature.

Many models, some of which extremely different, can therefore be reconciled with the presently available radio data. New observations are necessary.

In particular we need : a) maps with angular resolution of few degrees, at various frequencies, of high galactic latitude regions of sky; aerial beam shapes and side lobes must be well known for subsequent convolution and comparison with other maps, (e.g. ref.6,9); b) accurate measurements of the absolute sky temperature toward the two minima of sky brightness and the Galactic Poles, at the same frequencies. A program of new observations covering the range 200 MHz - 2500 MHz is underway in our laboratory with ad hoc instruments and/or existing radiotelescopes, in collaboration with various groups.

References

1) Perola G.C., Scarsi L., Sironi G.:1968, Nuovo Cimento 53B , 459
2) Turtle A.J. et al.:1962 , Mon.Not.R.astr.Soc. 124 , 297
3) Gavazzi G., Sironi G.:1975 , Rivista Nuovo Cimento 5 , 155 and references therein
4) Webster A.S.:1975 , Mon.Not.R.astr.Soc. 171 , 243 and ref.therein
5) Turtle A.J., Baldwin J.E.:1962 , Mon.Not.R.astr.Soc. 124 , 459
6) Milogradov Turin J., Smith F.G.:1973 , Mon.Not.R.astr.Soc. 161 , 269
7) Webster A.S.:1972 , PhD Thesis , University of Cambridge
8) Allen R.J., Baldwin J.E., Sancisi R.:1978 , Astron.Astrophys. 62 , 397
9) Cane H.V.:1978 , Aust.J.Phys. 31 , 561

THE ALL-SKY 408 MHz SURVEY

C.G.T. Haslam, C.J. Salter and H. Stoffel
Max-Planck-Institut für Radioastronomie, Bonn, FRG

The 408 MHz All-sky Survey has been made from four radio continuum surveys observed between 1965 and 1978, using the Jodrell Bank MKI telescope (Haslam et al., 1970), the Effelsberg 100 metre telescope (Haslam et al., 1974) and the Parkes 64 metre telescope (Haslam et al., 1975). A detailed description of the survey data reduction and calibration methods, with preliminary astronomical results will soon be published (Haslam et al., 1980a) and a second paper will give an atlas of maps at the full survey resolution of 51' arc between half power points (Haslam et al., 1980b).

The original 408 MHz survey has a brightness sensitivity which is limited to some $1°K$ by the expected confusion at 408 MHz for a 51' arc beam. The temperature scale has been calibrated by smoothing and comparing the four component 408 MHz surveys in the regions where they overlap the absolutely calibrated 404 MHz survey of Pauliny-Toth and Shakeshaft (1962). Details of this comparison are given in Haslam et al. (1980a) and it is estimated that the absolute temperature scale is accurate to within 10% and the zero level known to within $3°K$.

With the advent of new satellite observations at X- and γ-ray wavelengths (e.g. Simpson, 1979; Mayer-Hasselwander et al., 1979), which have instrumental angular resolutions of around 3°, it is felt useful to present a radio continuum map of the whole sky, smoothed to 3° resolution, which has been made from the 408 MHz radio continuum data. This map will allow direct comparison of the large-scale Galactic structure as seen in the radio continuum, with the X- and γ-ray observations.

The map, smoothed to a gaussian beam with resolution between half power points of 3°, is presented in Figure 1. Cooperative work, using unfolding of the 408 MHz data to study the distribution of continuum emission within our Galaxy, has already begun (Kearsey et al., ibid). In addition, a first comparison with the 1420 MHz continuum survey of Reich (1978) and with the Parkes 5000 MHz survey (Haynes et al., 1978) to study spectral index variations at low and intermediate Galactic latitudes is already producing interesting results.

References

Haslam, C.G.T., Quigley, M.J.S., Salter, C.J.: 1970, Monthly Notices Roy. Astron. Soc. 147, 405.
Haslam, C.G.T., Wilson, W.E., Graham, D.A., Hunt, G.C.: 1974, Astron. Astrophys. Suppl. 13, 359.
Haslam, C.G.T., Wilson, W.E., Cooke, D.J., Cleary, M.N., Graham, D.A., Wielebinski, R., Day, G.A.: 1975, Proc. Astron. Soc. Australia 2, 331.
Haslam, C.G.T., Klein, U., Salter, C.J., Stoffel, H., Wilson, W.E., Cleary, M.N., Cooke, D.J., Thomasson, P.: 1980a, Astron. Astrophys. (to be submitted).
Haslam, C.G.T., Salter, C.J., Stoffel, H., Wilson, W.E.: 1980b, Astron. Astrophys. Suppl. (to be submitted).
Haynes, R.F., Caswell, J.L., Simons, L.W.: 1978, Australian J. Phys. Astrophys. Suppl. 45, 1.
Kearsey, S., Osborne, J.L., Phillipps, S., Haslam, C.G.T., Salter, C.J., Stoffel, H. (Ibid).
Mayer-Hasselwander, H.A., Bennett, K., Bignami, G.F., Buccheri, R., D'Amico, N., Hermsen, W., Kanbach, G., Lebrun, F., Lichti, G.G., Masnou, J.L., Paul, J.A., Pinkau, K., Scarsi, L., Swanenburg, B.N., Wills, R.D.: 1979 (to be published in Proceedings of 9th Texas Symposium in "Annals of the New York Academy of Sciences").
Pauliny-Toth, I.I.K., Shakeshaft, J.R.: 1962, Monthly Notices Roy. Astron. Soc. 124, 61.
Reich, W.: 1978, Astron. Astrophys. 64, 407.
Simpson, G.A.: 1979, NASA Technical Memorandum 80578.

THE LARGE-SCALE DISTRIBUTION OF SYNCHROTRON EMISSIVITY IN THE GALAXY

S. Kearsey, J.L. Osborne and S. Phillipps
Department of Physics, Durham University, U.K.
C.G.T. Haslam, C.J. Salter and H. Stoffel
Max-Planck-Institute for Radioastronomy, Bonn, F.R.G.

The all-sky radio continuum map at 408 MHz presented at this symposium by Haslam et al. can be interpreted in terms of the large-scale 3-dimensional distribution of synchrotron emissivity in the Galaxy when due allowance is made for the thermal emission. Its derivation from a 2-dimensional map must involve a number of assumptions so it is instructive to compare the results of alternative approaches (described in detail in forthcoming papers by the present authors). In both cases the variation of emissivity in the galactic plane is obtained from the observed intensity profile at $b=0°$ and then the z-variation is chosen to give the best fit to the complete map. The observed profile is shown in the figure with and without the contributions of catalogued supernova remnants and HII regions.

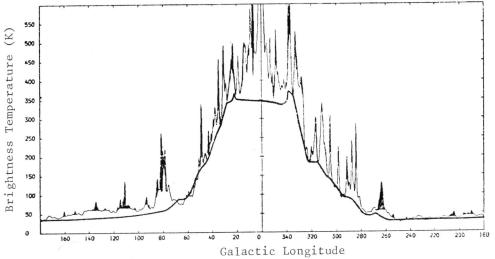

The profile along the galactic plane
Upper line: observations with contributions from identified discrete sources blacked in. Heavy line: the fit from the unfolding method with allowance for the thermal and extragalactic emission.

The first is an unfolding technique. The galactic plane is divided into logarithmic spiral sections and it is assumed that for each the form of the radial variation of emissivity, $\mathcal{E}(R)$, is the same. By an iterative method the azimuthal variation of emissivity from one section to the next is determined from the observed profile. The method, while similar to that of Kanbach and Beuermann(1979), differs from it in taking continuous spiral sections around the Galaxy rather than unfolding the two halves of the plane independently. The assumed pitch angle, p, was varied: $p=12°$ gave an unfolded emissivity with the sharpest features. Equally good fits are obtained with $\mathcal{E}(R) \propto R^{-1.9}$, $\propto \exp(-R/3.9 \text{kpc})$ or $\propto \exp(-(R/7.9\text{kpc})^2)$. The figure shows the fit for the exponential case. The unfolding applies only for $90°-p>l>270°-p$ and the fit has been made to the lower envelope of the observed profile. (In external galaxies the ridge lines of emissivity wander so that it is unrealistic to attempt a fit of the peaks to a smooth spiral structure). For the outer half of the plane the line results from an extrapolation extending to R=15 to 20kpc, dependent on the form of $\mathcal{E}(R)$, to fit the observed anticentre value. For $|l|<20°$ a fit to the profile cannot be obtained, indicating that the spiral structure is not present for $R \sim 3.6$ kpc. The heavy line indicates the emission for R>3.6 kpc only. A free parameter is the ratio of the strengths of the regular magnetic field along the spirals and the irregular, isotropic component. The fitting is not very sensitive to this provided that the ratio is $\lesssim 1$. The results place the sun in an interarm region with a local emissivity of 2.5 to 3 K kpc^{-1}. This would be given by equally strong regular and irregular fields of ~ 2.7 µG.

The second method is an extension of the work of Brindle et al. (1978) which used the earlier 150 MHz all-sky map. Here the basic structure of the Galaxy is taken to be known from a combination of the model of Georgelin and Georgelin (1976) within the solar circle and the HI arms without. The predictions of the density wave relate the synchrotron emissivity to it. Free parameters are the ratio of regular to irregular field and the underlying radial dependence of the emissivity. To reproduce the assymetric form of the longitude profile for $|l|<30°$ the spiral arms are required to emerge from an elliptical ring having axes of length 4 and 5 kpc. The emission from the $l \sim 80°$ region is provided by a local sub-arm which passes within 0.5 kpc of the sun in the anticentre direction. Apart from these non-spiral features the overall distribution of emission in the plane is similar to that from the unfolding technique. The sun is again in an interarm position with a similar local emissivity.

References

Brindle, C., French, D.K., and Osborne, J.L.: 1978 MNRAS 184, 283.
Georgelin, Y.M., and Georgelin, Y.P.: 1976 Astr. Astrophys. 49, 57.
Kanbach, G., and Beuermann, K.: 1979 Proc. 16th Int. Cosmic Ray Conf. 1, 75.

RADIO EMISSION FROM NEARBY GALAXIES AT HIGH FREQUENCIES

Rainer Beck and Ulrich Klein
Max-Planck-Institut für Radioastronomie, Bonn, FRG

High resolution observations of the radio continuum emission from nearby galaxies at several frequencies provide information about the cosmic ray electrons. Optimum results are expected by combining low-frequency synthesis observations with high-frequency single dish data, e.g. Westerbork 610 MHz / Effelsberg 10.7 GHz (1' resolution) or Cambridge 150 MHz / Effelsberg 2.7 GHz (4' resolution). Maps of 15 nearby spiral galaxies at 10.7 GHz have been made with the Effelsberg 100-m telescope (Klein and Emerson, 1980). Maps of M31 and M33 are available at 2.7 and 4.8 GHz (Beck, 1979; Berkhuijsen, 1978).

1. The spectra of the integrated emission from 15 nearby spiral galaxies between 10.7 GHz and 408 MHz (or lower) have a mean slope of -0.70 with a standard deviation of only 0.08. None of the spectra shows a positive curvature at high frequencies as would be expected in the case of substantial thermal emission; even at 10.7 GHz the non-thermal emission dominates. The thermal fraction appears to be always lower than 40%, corresponding to 14% at 1.4 GHz. For all galaxies observed, the spectral index α_n of the non-thermal emission lies in the range 0.7 to 0.85, corresponding to an electron energy spectral index of Γ = 2.4 to 2.7. This result suggests that the sources of cosmic ray electrons are similar in these spiral galaxies.

2. The radial variation of the thermal and non-thermal radio emission in the galaxies M31, M51 and NGC 6946 was determined by the variation of the spectral index, assuming α_n = 0.8 constant across the galaxy. The thermal radio emission was compared with the thermal Hα emission from HII regions. The two curves are proportional in the disc of each galaxy, indicating that the thermal emission is indeed responsible for the spectral index variation within the disc. Furthermore, any diffuse thermal emission must be similarly distributed as the HII regions.

The radial distribution of the non-thermal radio emission in M31 is neither proportional to the thermal radio emission nor to the blue light, but seems to be a curve in between these distributions. The scale length of the non-thermal emission beyond 10 kpc is 3.8±0.4 kpc, compared with 2.0±0.5 kpc for the thermal emission and 5.5±0.2 kpc for the blue light. It may therefore be concluded that the objects responsible for electron acceleration cannot solely belong to the old disc stellar population. However, these data do not exclude the possibility that the sources of cosmic ray electrons are to be found among the young stellar population. The distribution of the non-thermal emission in M51 and NGC 6946 is intermediate between the thermal emission and the blue light in both galaxies. For M33, Berkhuijsen (1978) obtained a similar result.

3. The variation of the spectral index in M31 with distance from the centre was published by Beck et al. (1980). The spectral index α_n was determined in rings of 1 kpc width after subtraction of the thermal emission. The range 0 - 4 kpc is dominated by the nuclear source with $\alpha_n \cong 0.75$. Between 5 and 12 kpc (where HII regions, OB stars, open star clusters, HI gas and X-ray sources are most frequent) α_n is nearly constant (0.88±0.01); beyond 12 kpc, α_n rises to 1.1. M51 and NGC 6946 show a similar steepening of α_n beyond the optical disc. The same behaviour was also found in the edge-on galaxies N891 (Allen et al., 1978), N253 (Beck et al., 1979), N3556 (de Bruyn and Hummel, 1979) and N4631 (Klein and Beck, in preparation). The present data are not accurate enough to distinguish between diffusion and convection models of electron propagation (e.g. Strong, 1978).

4. The possibility of a halo around M31 has been investigated at 408 and 842 MHz using data from the Effelsberg and Cambridge telescopes (Gräve et al., 1980). There is serious confusion from foreground radiation, but a small amount of excess radiation is found around M31. An upper limit of a possible halo emissivity can be set which is half of that deduced for our own Galaxy. Flattened 'halos' have been detected around NGC 4631 (Ekers and Sancisi, 1977) and NGC 891 (Allen et al., 1978). A weak radio 'halo' seems to be a common phenomenon among spiral galaxies and may be due to the diffusion of electrons out of the disc (e.g. Ginzburg and Ptuskin, 1976). Whereas the halos of NGC 891 and NGC 4631 could not be detected at 10.7 GHz, the NGC 253 halo is still visible at that frequency (Beck et al., 1979).

References

Allen, R.J., Baldwin, J.E., Sancisi, R.: 1978, Astron. Astrophys. 62, 397
Beck, R.: 1979, Ph.D. Thesis, University of Bonn
Beck, R., Biermann, P., Emerson, D.T., Wielebinski, R.: 1979, Astron. Astrophys. 77, 25
Beck, R., Berkhuijsen, E.M., Wielebinski, R.: 1980, Nature 283, 272
Berkhuijsen, E.M.: 1977, Astron. Astrophys. 57, 9
Berkhuijsen, E.M.: 1978, Structure and Properties of Nearby Galaxies (IAU Symposium No. 77), 149
Bruyn, A.G. de, Hummel, E.: 1979, Astron. Astrophys. 73, 196
Ekers, R.D., Sancisi, R.: 1977, Astron. Astrophys. 54, 973
Ginzburg, V.L., Ptuskin, V.S.: 1976, Rev. Mod. Phys. 48, 161
Gräve, R., Emerson, D.T., Wielebinski, R.: 1980, submitted to Astron. Astrophys.
Klein, U., Emerson, D.T.: 1980, submitted to Astron. Astrophys.
Pellet, A. et al.: 1978, Astron. Astrophys. Suppl. 31, 439
Strong, W.: 1978, Astron. Astrophys. 66, 205

HIGH-DENSITY, COOL REGIONS OF INTERSTELLAR MATTER IN THE GALAXY

W. B. Burton
Department of Astronomy, University of Minnesota

H. S. Liszt
National Radio Astronomy Observatory

The distribution of interstellar nucleons is dominated by gas at temperatures ranging from a few K to a few hundred K. At the warmer end of this temperature range, the gas is predominantly in the form of ubiquitously distributed atomic hydrogen. The colder gas is almost entirely molecular; it resides in compressed clumps, confined principally to the inner-galaxy, and is most effectively traced by observations of carbon monoxide. In this paper, we focus on some of the problems which currently hinder derivation of the morphology and total number of nucleons in the galaxy. For the atomic gas, these problems involve optical depth effects in HI profiles, the amount of cold HI residing in molecular clouds, and the form of the outer-galaxy rotation curve. For the molecular gas, the problems involve the uncertainties in the conversion from CO intensities to H_2 densities, including the possibility of composition gradients across the galaxy, the total number and typical size of molecular clouds, and the possibility that the molecular material in the region of the galactic nucleus is distributed differently from the material in the galaxy at large.

MORPHOLOGY OF COLD ATOMIC HYDROGEN

Atomic hydrogen coexists with molecules in dense clumps at temperatures of order 10 K and is diffused through much of the intercloud region at temperatures of about 6000 K. Between these temperature extremes there is a variety of structures whose physical characteristics are determined by the mechanical and radiation influences of their surroundings with which they exist in approximate pressure equilibrium (Heiles, 1980). The condition of pressure equilibrium allows study of the HI gas to proceed in terms of temperature hierarchies (see Baker's 1979 review), which are revealed in general more directly by the observations than are density hierarchies. The hottest gas is so diffuse that it contributes little to the overall density budget. The principal problems involved with estimates of this contribution are the very large thermal broadening which leads to blending in the spectral lines, the low intensities resulting from the low column densities, and the effects of stray radiation entering the antenna from other, more intense emission fields. Most HI nucleons reside in cooler material. The problems of blending and stray radiation persist to hinder the analysis; although the intensities are easily measurable, column densities are not because optical depth effects cannot be ignored.

The problem of line blending occurs generally for the entire range of structures in 21-cm profiles, except for a few isolated features found mostly at high latitudes so that descriptions of the galactic HI morphology refer to collective properties. The problem of stray radiation is one which has been generally neglected, but which is of importance for the HI case because of the properties of available antennas (see Kalberla, 1978) and because HI is distributed so widely. Relevant in the present context is Baker's (1976) emphasis of the importance of this problem to the determination of the outer boundary of galactic gas. The blending represents the combined influence of thermal broadening, turbulent and other small-size mass-motion broadening, and the accumulation of probably unrelated material from long lengths of path at overlapping velocities. The turbulent motions are evidently sufficient to prevent gravitational collapse of most HI structures. This line of sight blending is rendered especially important in a practial sense by the widespread distribution of atomic hydrogen. At low latitudes, no empty line of sight, and no region contributing negligible emission, has been found. At low latitudes emission features cannot be isolated for separate study. As a consequence most descriptions of the hierarchy of emission regions pertain to the local neighborhood. Even in the limited local volume, the evidence shows that this hierarchy is comprised of features which differ widely in their temperatures and intensities. The persistence of these differences in the presence of turbulence and systematic motions indicates that many HI features are transient.

Although few HI features at distances corresponding to galactic scales can be studied individually, as a group they are organized in a coherent layer whose scale height, kinematics, and gross optical depth properties can be studied rather directly. Regarding these and other properties of the galaxy, it is convenient to divide the layer into three zones. In the zone at galactic radii $4 \leq R \leq 10$ kpc, the layer parameters and the gross kinematics within it are known. HI cooler than a few hundred K is confined to a layer with a scale height ~120 pc and a representative velocity dispersion ~5 Km s^{-1}. The warmer, intercloud HI has a higher characteristic velocity dispersion (~8 km s^{-1}) and a correspondingly higher scale height. Deviations of the mean cool-HI layer from the galactic equator occur systematically with an amplitude ~50 pc, thus less than the thickness of the layer itself. The galactic kinematics in this layer follow directly from the HI data. There are perturbations to the basic galactic rotation which are probably related to spiral-arm structure, but for which the detailed production mechanisms are open to question. These perturbations amount to only a few percent of the amplitude of the basic rotation itself.

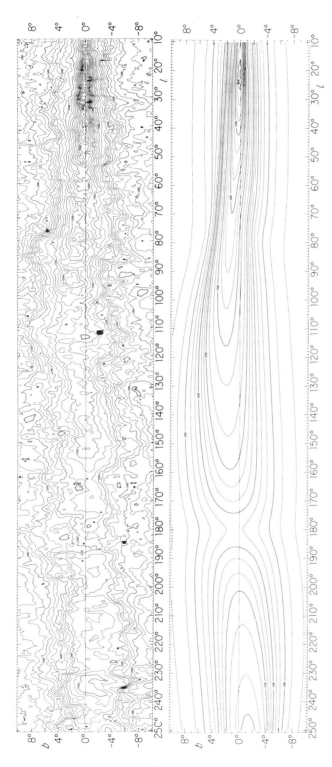

Figure 1. Arrangement on the plane of the sky of the total-velocity integrated HI emission. The upper panel contains the integrals $\int T(v)dv$ calculated from the observations of Weaver and Williams (1973). This integral gives the HI column density only if the gas is optically thin at all velocities. Although at high latitudes $\tau(v)<1$ is a reliable general approximation, at latitudes near the galactic equator, optical thinness pertains only for certain velocity segments of most profiles. The lower panel shows the total-velocity integrals calculated from synthetic data; it incorporates a description of the observed warp and flare of the outer-galaxy HI and accounts for the modulating effects of the kinematic parameter $\Delta v/\Delta r$ on the apparent column densities. Saturation is important at high $|v|$ in the first and second longitude quadrants and near $\ell \to 0°$ and $\ell \to 180°$. Although models based on a single component or on a "raisin pudding" gas distribution are too naive to have much relevance to the details of the distribution, they can serve in a pragmatically valid way to test morphological aspects of the gas distribution and radiative transfer through it.

In the outer-galactic zone, at R > 10 kpc, and in the innermost zone at R < 4 kpc, the layer has been less well described principally because the galactic kinematics are poorly known. At R < 4 kpc, the assumption of circular motion fails so that the rotation curve does not follow with much confidence. In the outer galaxy, there is no direct measure of distance corresponding to the geometrically produced maximum-velocity locus in the inner galaxy, so that indirect measures of the rotation curve must be used. Recent evidence of several sorts (e.g. Knapp, Tremaine, and Gunn, 1978; Blitz, 1979; Jackson, FitzGerald, and Moffett, 1979) shows that the rotation curve in our galaxy at R > 10 kpc is remarkably flat. This corresponds to the situation being observed in many spiral galaxies, and implies that the total mass distribution in the galaxy extends well beyond the horizon of presently detected constituents. The warp of the outer-galaxy gas layer away from the galactic equator and the flare of this layer to large thickness is well known, but not understood. Linear scales associated with early descriptions of the phenomenon (e.g. Baker and Burton, 1975) should be revised upward in view of the flat rotation-curve conclusions (see also Heiles, 1980).

MACROSCOPIC OPTICAL DEPTH OF THE HI LAYER

Determination of the total number of HI nucleons in the galaxy requires that the macroscopic optical depth characteristics of the gas be known. The HI column density $N_{HI} = 1.823 \times 10^{18} \int T_k \tau(v) dv$ cm^{-2} is only measurable if the gas along the particular line of sight is transparent at all velocities. If that is the case, the observed brightness temperature profile $T_B(v) = T_k\{1-\exp(-\tau(v))\}$ is $\sim T_k \tau(v)$. The volume density smoothed over a path of length Δr then follows directly from $\int T_B(v) dv / \Delta r$. In the limiting case of complete saturation, $T_B(v) \sim T_k$ and $\int T_B(v) dv/\Delta r = T_k |\Delta v/\Delta r|$, where Δv is the velocity extent of the portion of the profile considered. Thus the arrangement of the geometrical parameter $|\Delta v/\Delta r|$ is relevant for both the high and the low optical depth cases; a high optical depth situation will be revealed by length-corrected integrated profiles which vary in direct proportionality with this geometrical parameter.

The true HI macroscopic optical depth situation is confused: there are valid arguments for high overall optical depths as well as for low ones. The integrated intensities observed in the direction of the galactic anticenter are approximately the same as those observed from the much longer path through the galactic center. This fact was used by van de Hulst, Muller, and Oort (1954) to argue for high optical depths and a single harmonic mean temperature in these directions. Low optical depths, on the other hand, are indicated for general directions through the inner galaxy by the temperatures typically observed to be approximately twice as high at positive velocities (in the first longitude quadrant) as at negative velocities and by the ridge of emission at high velocities which is enhanced in accordance with the influence of the $|\Delta v/\Delta r|$ parameter (Burton, 1972). This discrepancy originates in the

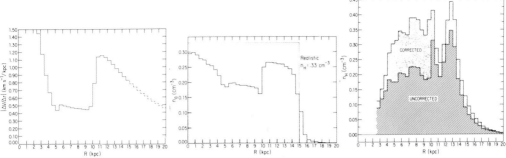

Figure 2. Influence of optical-depth effects on the space-averaged volume densities of HI gas near the galactic equator. The left panel shows the variation of the kinematic parameter $|\Delta v/\Delta r|$ resulting from galactic rotation observed from our vantage point within the disk. In the case of complete saturation, an abundance derived directly from the profile integrals would mimic this variation, because in that case $\int T_B(b) dv/\Delta r = T_k |\Delta v/\Delta r|$. Many basic properties of HI in the galaxy can be approximated with a one-component gas of constant volume density $n_{HI} = 0.33$ cm^{-3}; the middle panel demonstrates that the apparent morphology derived from such a model shows the influence of $|\Delta v/\Delta r|$, and suggests the optical-depth correction appropriate to the observed situation represented in the panel on the right.

double-valued velocity-distance relationship, which could only be of importance in this regard if the gas on the near side of the subcentral locus were sufficiently thin to allow the gas on the far side to contribute to the profiles.

Figure 2 and its caption describe the influence of the kinematic parameter. Many of the overall characteristics of HI observations in the galactic plane are simulated by model spectra corresponding to a ubiquitously distributed, one-component gas of $T_k = 135$ K and $n_{HI} = 0.33$ cm^{-3}. If the integrated model spectra are assumed to give column densities directly, as if the gas were thin, then the derived apparent volume density differs from the input constant density in the manner shown in the figure, mimicking the variations in the parameter $|\Delta v/\Delta r|$. The apparent density derived in the same straightforward way from observed integrated spectra also reflects this variation (except at R < 4 kpc where the decrease in HI gas density at b=0° is probably real). The controlled conditions inherent in the modelling procedure can be used to derive corrections that should be applied to the observed profile integrals to give reliable volume and column densities. The correction for partial saturation is greater at R < R_o because of the double-valued nature there of the velocity-distance relationship. At R > R_o, the form of $|\Delta v/\Delta r|$ and thus the details of the correction for partial saturation depend on the outer galaxy rotation curve, about which there is currently much discussion. It seems indicated, however, that the density of the atomic hydrogen gas remains roughly constant over the major part of the galactic disk from R ~ 4 kpc to at least R ~ 15 kpc. The numerical value of this density is about 0.4 cm^{-3},

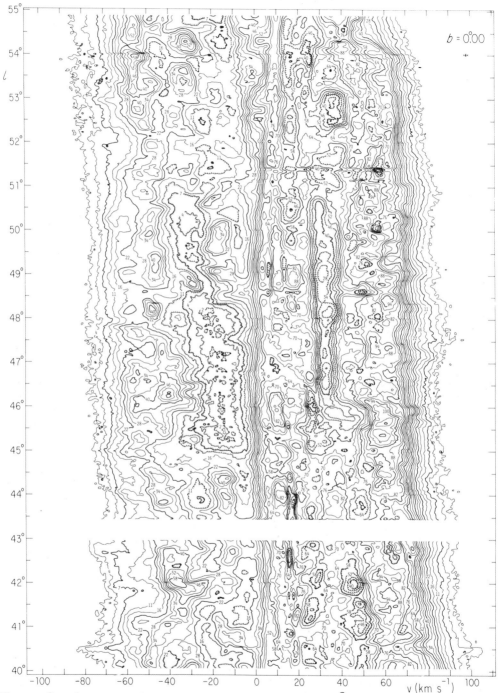

Figure 3. ℓ-v map of HI emission observed at b=0° at the high angular resolution, 3.2, afforded by the Arecibo telescope (Baker and Burton, 1979). The map is peppered with small-angular-size intensity minima identified with self-absorption in cold clouds.

although probably sufficient HI emission is concealed in cool, beam-diluted structures that this value should be treated as a lower limit. The rather constant density variation is in marked contrast to the distribution of total galactic mass density (contributed mostly by stars) which increases strongly toward the galactic center. The HI density decrease in the inner parts is a characteristic which our galaxy shares with other spirals.

COLD HI IN MOLECULAR CLOUDS

The evidence which reveals partial saturation in 21-cm spectra contains information also on the physical characteristics of the gas, but before this information can be interpreted it is necessary to decide if τ simply increases monotonically as the length of path at a particular velocity through a rather uniform gas increases (as in the model entering Figures 1 and 2), or if τ increases incrementally due to the collective behavior of cold, generally unresolved structures in the telescope beam (as is probably the case). A nonuniform distribution of temperature and density is expected on general grounds, and observational evidence for such a situation is well established. Thus Heeschen (1955) and Davies (1956) found low-latitude emission regions with kinetic temperatures about one-half the usually adopted value. Shuter and Verschuur (1964) and Clark (1965) showed that temperatures observed in 21-cm absorption spectra resemble those of cold emitting features. Because this cold gas is opaque, the higher temperatures generally observed require that a warmer, but thin gas be present also. Clark (1965) suggested a "raisin-pudding" model, in which cool opaque clouds are immersed in a warm transparent medium. Such a multicomponent situation has been given extensive theoretical justification (e.g. Field, Goldsmith, and Habing, 1969), although the details of the physical properties, and the distinctiveness of the different regions, are not yet accurately established by observation. General arguments and observations of local structures indicate that simple cloud/intercloud models are naive (see Heiles, 1980), although such models can validly serve useful purposes. Whatever the details of its distribution, the cold material must not be so prevalent that the spectra saturate at low temperature. Measurements of HI absorption against extragalactic continuum sources (e.g. Radhakrishnan et al., 1972) show high opacity HI to occur commonly; if this material were to dominate the emission spectra, the spectra would be quite saturated. This contadictory situation is avoided because of beam dilution of the opaque material.

Insight into the overall optical depth characteristics of the galactic gas layer is emerging from the identification of the common occurrence of residual cold atomic hydrogen in galactic molecular clouds. These clouds are generally sufficiently diluted in the telescope beam that emission profiles show large-scale optical thinness, although the clouds are individually opaque. The existence of residual atomic hydrogen in local, optically identified, dark dust clouds was demonstrated by Knapp (1974), and by others for individual local

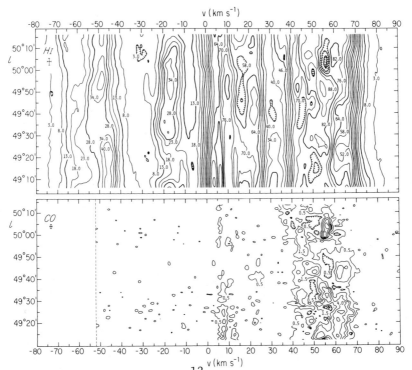

Figure 4. Comparison of HI and ^{12}CO emission observations made near $\ell=49°5$ with approximately similar resolution showing that molecular clouds have a residue of cold HI which is responsible for several aspects of the appearance of the HI emission (Liszt, Burton, and Bania, 1980). The cold HI in the molecular-cloud ensemble is severely diluted in the large beams of most surveys. Its effect is particularly important to derived HI overall densities and masses. The CO emission and HI absorption counterparts are apparently well-mixed, with comparable kinetic temperatures as implied by the CO data, $T_k \sim 10K$. The optical depths are ~ 0.4. The HI column densities are typically 2×10^{19} cm^{-2}.

---

features. It seems well-established now that residual cold HI occurs throughout the galaxy in the members of the molecular-cloud ensemble traced by observations of CO. Baker and Burton (1979) fully sampled all of the galactic equator accessible to the Arecibo telescope, and found that those observations (see Figure 3) are characterized by the striking appearance of a large number of narrow HI self-absorption features typically too small (~5') to have been resolved or even detected by the larger beams (~0°5) of earlier HI surveys. Burton, Liszt, and Baker (1978) showed that these HI features are strongly correlated with emission from carbon monoxide clouds. Figure 4 shows examples of this correlation. As a consequence of the correlation, HI emission observations can be used for investigation of the statistical and morphological characteristics of the cold, high density, star-forming regions in the galaxy.

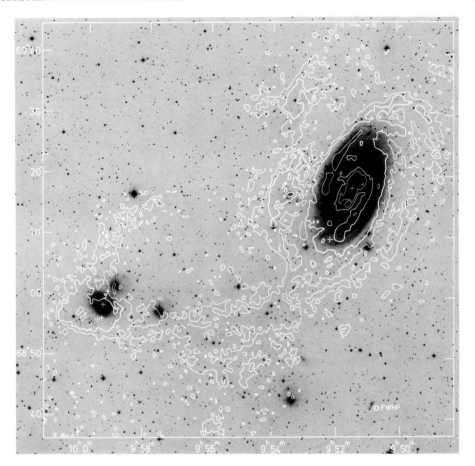

Figure 5. HI emission from M81 and its surroundings (van der Hulst, 1979) superimposed on a deep IIIa-J plate (de Ruiter et al., 1976). The HI distribution extends well beyond the luminous galaxy and the HI spiral arms are broader than the luminous ones; in these respects our galaxy is probably similar.

This correlation has important consequences. The number of HI nucleons in these clouds is underestimated in density derivations based on emission profiles, although probably only about 1% of the H nuclei residing in the clouds are in HI. More important are the modulations on the general HI emission intensities which result from the accumulated effects of opaque clouds which are individually non-overlapping in space and severely diluted in the cone of observation. The effects of this absorption from the galactic cloud ensemble seriously complicate interpretation of HI emission profiles (see Liszt, Burton, and Bania, 1980). They are most severe where $|\Delta v/\Delta r|$ is small; thus they can account for the frequent occurrence of maximum emission temperatures at velocities away from those corresponding to the subcentral locus,

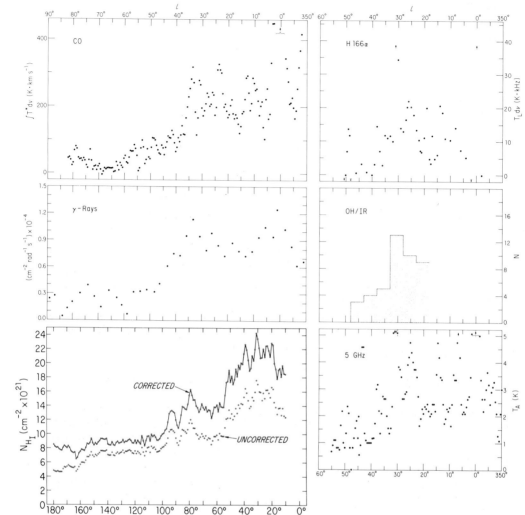

Figure 6. Comparison of the longitude distributions along the galactic equator of six galactic tracers (CO: Burton and Gordon, 1978; H166α: Lockman, 1976; γ-radiation above 100 MeV: Kniffen, Fichtel, and Thompson, 1977; 5 GHz continuum: Altenhoff et al., 1970; OH/IR: Johansson et al., 1977; HI: Burton, 1976). Excepting atomic hydrogen, all tracers accessible on transgalactic paths show a consistent morphological confinement to the inner galaxy.

where the maximum temperatures would generally lie for a thin gas. In the entirely plausible circumstance (not yet convincingly demonstrated for our own galaxy) that molecular clouds occur preferentially in spiral arms, the resulting modulation in HI emission profiles would probably be sufficient to cause the arms to appear as minima in these profiles. In the plots of $\int T_B(v) dv$ against ℓ in which peaks or steps have often been sought as evidence of arms, the eventual HI arms would

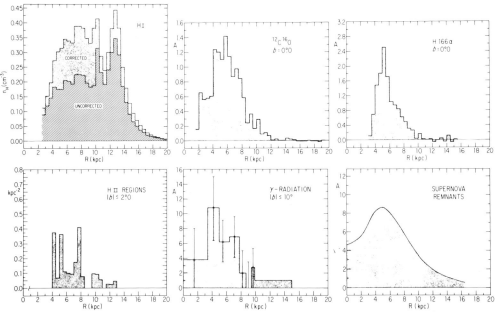

Figure 7. Comparison of the radial distributions of six constituents of the galactic disk (Burton and Gordon, 1978, and references there).

in such a case appear as minima.

MORPHOLOGY OF TRACERS OF THE COMPRESSED INTERSTELLAR MEDIUM

During the past decade it has become clear that among the observed constituents of the interstellar medium, atomic hydrogen is unique in its distribution. The galactic disk as defined by atomic hydrogen has a diameter at least twice as large as that defined by the ionized and molecular states of hydrogen, as well as by other molecules, supernova remnants, pulsars, γ-radiation, and synchrotron radiation. These other tracers refer either to high density regions of the interstellar medium or to consequences of active star formation. If our galaxy were viewed from an external perspective, the sun would be seen to lie near the outskirts of the optically luminous disk (see Figure 5).

The longitudinal distribution of those tracers accessible along transgalactic paths reveals the degree of confinement to the inner galaxy in a straightforward way (Figure 6). Conversion of the longitudinal distribution to the radial abundance distribution can be done using the kinematic information inherent in the spectral-line data or by using a geometrically-based unfolding process for tracers such as the radio continuum or γ-rays (Figure 7).

Information about the total amount of material requires, of course, measurements which extensively sample the relevant volume. Particularly important in this regard is the recent work by Cohen et al. (1980) and by Solomon, Sanders, and Scoville (1979) which has substantially aug-

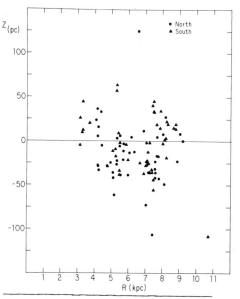

Figure 8. Distance from the galactic plane plotted against distance from the galactic center for luminous HII regions (Lockman, 1979). The HI and CO layers show similar systematic excursions from $b=0°$.

mented the CO data at latitudes away from $b=0°$. Available evidence for all of the inner-galaxy tracers shows that they are centered on the same mean layer, but that the centroid of this layer deviates systematically from $b=0°$ (see Figure 8). There is no consensus regarding the cause of these deviations. Layer-thickness measurements combined with the radial abundances yield projected surface densities (Figure 9).

H_2 DENSITIES FROM CO INTENSITIES

The most stable low-temperature form of the most abundant element in the interstellar medium is molecular hydrogen. It predominates over all other gaseous material in optically opaque, compressed regions where the molecule is shielded from photodissociation after formation on grain surfaces (Solomon and Wickramasinghe, 1969). This material is of course not represented in 21-cm observations of the hyperfine transition of atomic hydrogen. Having no dipole moment, H_2 has no observable transition in the radio or optical windows. Ultraviolet extinction due to interstellar dust limits observations of the H_2 Lyman absorption bands to the directions of reddened stars within a kiloparsec or so of the sun. The molecule second in abundance to H_2 is CO. Because the most important source of excitation of the CO rotational transitions involves collisions with H_2, observations of CO provide by implication much information on H_2. This information takes two important forms: pertaining to the distribution of H_2 and to cool, dense regions in general, the CO information can be interpreted without intervening assumptions; pertaining to the density of H_2 nucleons, the CO-based estimates involve critical uncertainties.

Emission lines from the $J=1\rightarrow0$ transition of the principal isotope ^{12}CO have a high optical depth and therefore provide the excitation

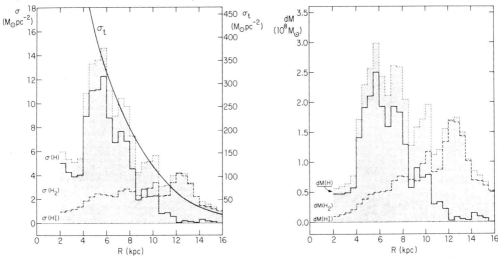

Figure 9. Radial distribution of projected surface densities and differential masses of atomic and molecular hydrogen (Gordon and Burton, 1976). Some of the uncertainties involved in the conversion of CO intensities to H_2 densities are mentioned in the text.

temperatures of molecular clouds, but not the column densities. Emission from the J=2→1 transition of the much less abundant isotope ^{13}CO is unsaturated, however, so that column densities follow directly from observed intensities. Because of the weaker intensities of the ^{13}CO spectra, the first estimates of H_2 densities were based principally on ^{13}CO data and on hypotheses, supported by limited data, regarding the abundance ratio $^{12}CO/^{13}CO$ (Scoville and Solomon, 1975; Gordon and Burton, 1976). The amount of ^{13}CO data available has now increased significantly (see Figure 10, and Solomon, Scoville, and Sanders, 1979). The ^{13}CO data show similar variations on a galactic scale as the ^{12}CO data and an approximately constant ratio of intensities. These results give pragmatic support to the use of the principal isotope for density-mapping purposes, and provide no evidence that the assumption (which is important to the density derivations) of constant isotopic abundance ratio across the galaxy is invalid.

Conversion of the CO intensities to H_2 densities involves several possible sources of uncertainty. Some of these potential uncertainties require for their resolution additional theoretical work on questions of molecule formation and dissociation, and on radiative transfer; others, in particular those regarding abundance ratios, require diverse observational work. Thus the numerical value of the abundance ratio $^{12}C/^{13}C = 40$ was suggested by Wannier et al. (1976), although some workers use the terrestrial value, 89. Additional sources of uncertainty concern the fraction of C which is bound in CO, and the abundance ratio C/H, as well as the degree of constancy of these quantities over the galaxy.

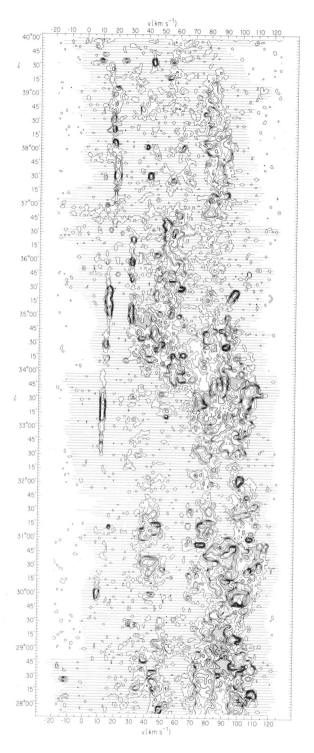

Figure 10. ℓ-v arrangement of emission from the ^{13}CO isotope in the galactic equator (Liszt, Burton, and Xiang, 1980). Emission from this isotope has lower opacity than emission from ^{12}CO, and therefore provides more direct information on the gas densities. Stochastic-ensemble models of this sort of observational material suggest that many of the complex features result from blending of physically unrelated clouds. Many of the small-scale features result from the sampling characteristics inherent in a single-latitude strip map. Straightforward measurement of the angular extent of the observed features might give incorrect cloud diameters, just as direct measurement of the velocity gradients might give misleading kinematic information.

There are other potential uncertainties which can be confronted through direct analyses of the CO survey observations themselves. Thus the question of optical depths of the ^{12}CO and ^{13}CO lines is a puzzling one in view of the proportionality between intensities observed and in view of the observed line shapes. Presumably the resolution of this puzzle is to be found in macroscopic turbulence within the clouds and in the manner in which the emission is sampled in the telescope beam. Interpretation of the sampling characteristics is also necessary before the role of shadowing can be understood. If shadowing of one opaque cloud by another at the same velocity on the same line of sight occurs commonly, the total number of emitters will be underestimated. This problem, and the related one concerning the volume-filling factor of the molecular clouds, require knowledge of the size and typical separation of the members of the galactic molecular cloud ensemble.

INTERPRETATION OF EMISSION FROM THE MOLECULAR CLOUD ENSEMBLE

The investigators agree concerning such morphological aspects of the molecular cloud distribution as its confinement to the annulus $4 \lesssim R \lesssim 8$ kpc and its z-dispersion of ~50 pc. Controversy remains, however, regarding basic intrinsic properties of the molecular clouds, their distribution within the cloud ensemble, and the manner in which the ensemble is perceived in the observations. Thus there has been an emphasis on "giant" molecular clouds (e.g. Solomon and Sanders, 1980; Szabo, Shuter, and McCutcheon, 1980) and on clouds with typical diameters $\gtrsim 5$ pc (e.g. Burton and Gordon, 1978). The random velocity component of the clouds has been held to be ~4 pc (Burton and Gordon, 1978) and 9 km s^{-1} (Stark and Blitz, 1978; Stark, 1979). Resolution of these disagreements is crucial to discussions of the total cold-gas mass, to questions of possible cloud-cloud accretion, to the role of clouds as sites of star formation, and to the statistical acceleration of stars through encounters with clouds. Some of the conclusions about which there is little consensus have been derived from the observational survey data in ways which are perhaps too straightforward for the peculiar task of studying (from an embedded perspective) an ensemble which, although composed of individually discrete, opaque clouds, is perceived as macroscopically transparent.

Recently Liszt and Burton (1980) have approached the general problem of interpreting the ^{12}CO survey data (of the sort shown in Figure 11) by simulating spectra from the galactic ensemble of clouds. The controlled conditions inherent in such modelling allow study of the consequences of the observing procedure, including those of telescope beam size and sampling interval, of the importance of such matters as cloud blending and shadowing, and of the influence of the galactic velocity field. Most aspects of the large-scale CO surveys can be simulated with stochastic-ensemble models in which cloud size and total number of clouds are the dominant parameters. Figure 12 shows an example of ^{12}CO emission synthesized for an ensemble of identical 18-pc clouds. Significant differences occur between the intrinsic

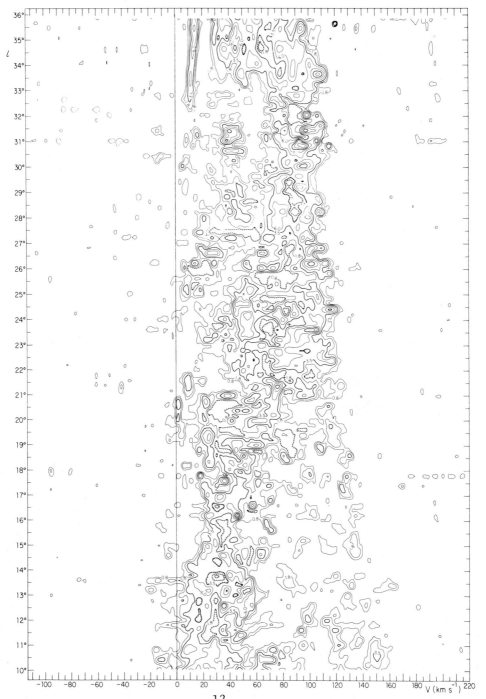

Figure 11. ℓ-v arrangement of ^{12}CO emission observed along the galactic equator (Gordon and Burton, 1976). This is the sort of material used to confront stochastic models of the molecular cloud ensemble.

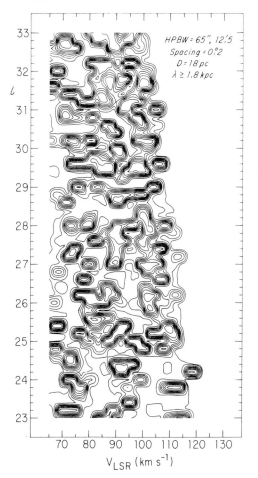

Figure 12. Synthesized ℓ-v map at b=0° of ^{12}CO emission from an ensemble of molecular clouds (Liszt and Burton, 1980b). The ensemble represented here contains 43200 identical clouds of 18-pc diameter, with a mean-free-path separation of 1.8 kpc at the peak of the molecular annulus. The spectral features which result from these identical clouds show, for several reasons, a great deal of inhomogeneity.

distribution of cloud diameters and the distribution of the sizes of features measured parallel to the galactic plane, principally because the scale height of the molecular distribution is not small compared to the typical cloud sizes and because of blending. Many of the large, complex features appearing in CO ℓ-v maps are due to blending of physically unrelated clouds. Straightforward measurement of the angular extent of these features might give too-large diameters; of their linewidths, too-large dispersions; of their intensity structure, false information on clumping; and of their velocity gradients, misleading kinematic information. The modelling results suggest to us that the typical cloud diameter in the inner galaxy is 15-20 pc. Neither very large ($D \gtrsim 40$ pc) nor very small ($D \lesssim 10$ pc) clouds contribute much of the aggregate cloud surface area. Larger clouds may contain an appreciable but not overwhelming fraction ($\lesssim 50\%$) of the total cloud volume. The one-dimensional cloud-cloud random velocity dispersion in the galactic plane is 3-4 km s^{-1}, too small to account straightforwardly for the observed z-dispersion of 50 pc. At galactocentric radii 5-6 kpc the volume filling factor of the molecular ensemble is

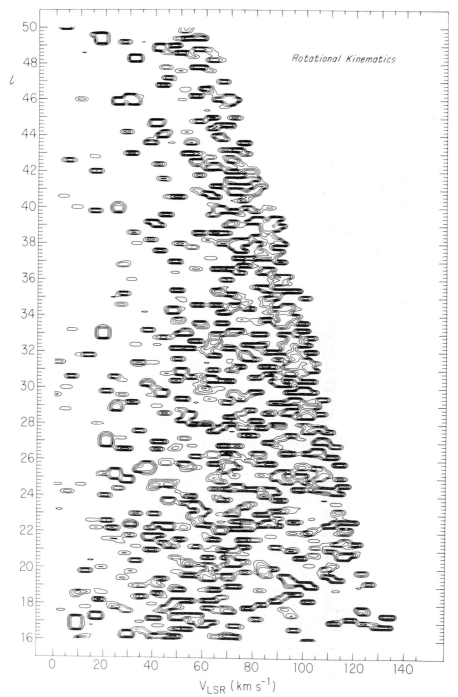

Figure 13. Synthesized map at b=0° of ^{12}CO emission from the molecular-cloud ensemble, calculated for a rotationally symmetric velocity and density distribution of 18-pc clouds (Liszt and Burton, 1980b).

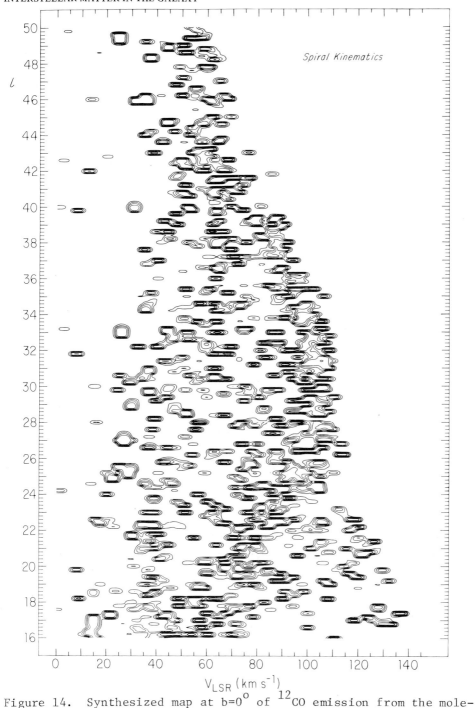

Figure 14. Synthesized map at b=0° of ^{12}CO emission from the molecular-cloud ensemble, calculated for a spiral-shock velocity field (Roberts and Burton, 1977), but with a rotationally symmetric cloud distribution (Liszt and Burton, 1980b).

~.007, the typical mean free path between clouds is 1.8 kpc, and the space-averaged H_2 number density is probably about 3 H_2 cm^{-3}. Molecular material almost certainly dominates the interstellar medium in the inner-galaxy.

Much of the motivation for carrying out the large-scale surveys of CO has focussed on the hope of mapping the spiral structure of our galaxy. This hope parallels the earlier situation regarding HI, but for the case of CO it was reinforced by a range of results reached by the late 1970's. Comparison of optical studies with aperture synthesis radio data showed that the narrow dust lanes in external galaxies portray spiral arms in a much more definite way than does the HI structure. Theoretical studies confirmed that the diffuse atomic gas would likely respond in a linear manner to the passage of a density wave, whereas the cold, dense molecular material would experience a non-linear response along a very narrow shock front. This narrow shock front might indeed precipitate the formation of the CO clouds. In addition, the early CO data showed that the molecular clouds occupy only about 1% of the total volume of the galactic layer, leading to a tendency to consider unimportant the velocity-crowding and blending responses to the kinematic parameter $|\Delta v/\Delta r|$ which dominate spectra from HI (which occupies essentially 100% of this volume in one state or other). It is particularly this tendency to which the ensemble modelling is relevant.

Figure 13 shows a large-scale simulated ^{12}CO ℓ-v map calculated for the case of a rotationally symmetric velocity and density distribution. The concentration of emission near the terminal-velocity locus indicates that the kinematic parameter is influential; otherwise there is in this map little of the structure which characterizes the Figure 11 observations and which might be viewed as true concentrations in space. Figure 14, however, does show such structure. The simulation responsible for the figure incorporates an azimuthally symmetric ensemble distribution, but the spiral velocity field predicted for a non-linear density wave model. A significant arm-interarm contrast occurs even though the input azimuthal density variation entails no such contrast. We conclude that the transformation from the galactic spatial and kinematic coordinates to the observed position-radial velocity maps involves for CO data the same sort of consequences that have frustrated galactic mapping efforts based on HI 21-cm data, despite the very small volume-filling factor of the ensemble of molecular clouds. Further numerical experiments show that the kinematic loops observed in the inner regions of the galaxy result primarily from perturbations of the galactic velocity field and only secondarily from localized density variations. Use of a pure-rotation velocity field to infer the spatial location of spiral arms in the molecular ensemble is probably inappropriate. Arm-interarm cloud abundance variations cannot be simply inferred from intensity-velocity profiles. Even if they are present, such arm-interarm contrasts do not necessarily constrain the lifetimes of molecular clouds to be small. Regarding evidence of spiral structure in the currently available CO surveys, we adopt an agnostic

INTERSTELLAR MATTER IN THE GALAXY

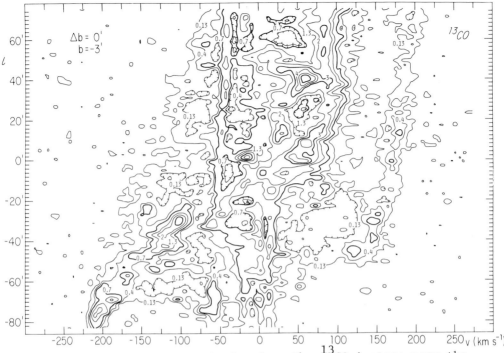

Figure 15. Distribution of emission from the ^{13}CO isotope near the galactic center. The molecular emission from the nuclear region is intense, and differs from the emission from the galaxy at large in its apparent ubiquity and its broad velocity dispersion.

position. In analogy to the HI structure, we believe the most direct evidence from the CO tracer for such structure to reside in the ordered perturbations to the variations of terminal velocities with longitude.

TWO SPECULATIONS RELEVANT TO THE TOTAL GALACTIC NUCLEON COUNT

<u>The possibility of very cold (3-5 K) generally undetected molecular gas.</u> The excitation temperature of the molecular clouds which are detected in the CO surveys and which contain no heat source from, for example, recently formed stars, is typically ~14 K. The constancy of this temperature implies that it is an intrinsic cloud property. The sensitivity of the current emission surveys would not provide detection of very cold gas at $T_k \lesssim 5$ K. If it exists, such gas would appear in absorption spectra. But, at millimeter wavelengths continuum-radiation sources are too weak to provide a suitable background. The CO line emission from the Sagittarius molecular complex represents a special case. This emission is sufficiently broad that it serves as a background, and, indeed, against the SGR complex deep, narrow dips in the CO emission spectra have been identified with absorption by very cold

foreground gas at T_k = 3.5 K (Liszt et al., 1977; Zuckerman and Kuiper, 1980). Linke, Stark, and Frerking (1980) show absorption evidence for very cold HCO^+ and HCN clouds in radiative equilibrium with the 3 K cosmic background radiation. It is necessary to remain open to the possibility that very cold regions are quite common throughout the galaxy, but that they are not adequately accounted for because of the lack of suitable background sources.

The possibility of smoothly distributed molecular gas in the innermost galaxy. Interpretations of CO surveys agree on the relative deficiency of CO in the region of the galaxy interior to R ~ 4 kpc. These interpretations involve techniques which, in essence, count discrete clouds. In the innermost galaxy, the shearing forces of rapid differential rotation are such that large, discrete clouds may not form (see Stark, 1979). Under these conditions there are plausible reasons to doubt that the ultraviolet shielding will be sufficient to allow extensive molecular formation. Nevertheless, the observations show all of the characteristics which would be expected for a ubiquitous distribution of molecular gas (Liszt and Burton, 1978). Thus they show smoothly varying intensities, with little of the patchy appearance characterizing emission from the galaxy at large, and an arrangement of intensities which follows in detail the predictions of a response to galactic kinematics and which would be expected only if the space were pervaded by the gas. Particularly important in this regard are the very large velocity widths of the observed molecular features (see Figure 15). Based on the assumption of a smooth molecular distribution, Liszt and Burton (1978) argue that the inner few kiloparsecs of the galaxy contains $M_{H_2} > 10^9 M_\odot$, and that the CO/HI ratio is anomalously large there. Linke, Stark, and Frerking (1980) support these high H_2 column densities using absorption data.

Acknowledgements: W.B.B. gratefully acknowledges support from the National Science Foundation through grant NSF/AST-7921812. The National Radio Astronomy Observatory is operated by Associated Universities, Inc., under contract with the National Science Foundation. The Arecibo Observatory is part of the National Astronomy and Ionosphere Center, which is operated by Cornell University under contract with the National Science Foundation.

REFERENCES

Altenhoff, W. J., Downes, D., Goad, L., Maxwell, A., and Rinehart, R.: 1970, Astr. and Astrophys. Suppl. 1, pp. 319-355.
Baker, P. L.: 1976, Astr. and Astrophys. 48, pp. 163-164.
Baker, P. L.: 1979, Proc. IAU Symp. 84, pp. 287-294.
Baker, P. L., and Burton, W. B.: 1975, Astrophys. J. 198, pp. 281-297.
Baker, P. L., and Burton, W. B.: 1979, Astr. and Astrophys. Suppl. 35, pp. 129-152.
Blitz, L.: 1979, Astrophys. J. Lett. 231, pp. L115-L119.
Blitz, L., and Shu, F.: 1980, Astrophys. J., in press.

Burton, W. B.: 1972, Astr. and Astrophys. 19, pp. 51-65.
Burton, W. B.: 1976, Ann. Rev. Astr. and Astrophys. 14, pp. 275-306.
Burton, W. B., and Gordon, M. A.: 1976, Astrophys. J. Lett. 207, pp. L89-L93.
Burton, W. B., and Gordon, M. A.: 1978, Astr. and Astrophys. 63, pp. 7-27.
Burton, W. B., Liszt, H. S., and Baker, P. L.: 1978, Astrophys. J. Lett. 219, pp. L67-L72.
Clark, B. G.: 1965, Astrophys. J. 142, pp. 1398-1422.
Cohen, R. S., Tomasevich, G. R., and Thaddeus, P.: 1979, Proc. IAU Symp. 84, pp. 53-56.
Cohen, R. S., Cong, H., Dame, T. M., and Thaddeus, P.: 1980, Astrophys. J. Lett., in press.
Davies, R. D.: 1956, Mon. Not. Roy. Astr. Soc. 116, pp. 443-452.
de Ruiter, H. R., Willis, A. G., and Arp, H. C.: 1977, Astr. and Astrophys. Suppl. 28, pp. 211-293.
Field, G. B., Goldsmith, D. W., and Habing, H. J.: 1969, Astrophys. J. 158, pp. 173-183.
Gordon, M. A., and Burton, W. B.: 1976, Astrophys. J. 208, pp. 346-353.
Heeschen, D. W.: 1955, Astrophys. J. 121, pp. 569-584.
Heiles, C.: 1975, Astr. and Astrophys. Suppl. 20, pp. 37-55.
Heiles, C.: 1980, Astrophys. J. 235, pp. 833-839.
Jackson, P. D., FitzGerald, M. P., and Moffat, A. F. J.: 1979, Proc. IAU Symp. 84, pp. 221-224.
Johansson, L. E. B., Andersson, C., Goss, W. M., and Winnberg, A.: 1977, Astr. and Astrophys. 54, pp. 323-334.
Kalberla, P. M. W.: 1978, Ph. D. Dissertation, University of Bonn.
Knapp, G. R., Tremaine, S. D., and Gunn, J. E.: 1978, Astr. J. 83, pp. 1585-1593.
Kniffen, D. A., Fichtel, C. E., and Thompson, D. J.: 1977, Astrophys. J. 215, pp. 765-774.
Linke, R. A., Stark, A. A., and Frerking, M. A.: 1980, Astrophys. J., in press.
Liszt, H. S., and Burton, W. B.: 1978, Astrophys. J. 226, pp. 790-816.
Liszt, H. S., and Burton, W. B.: 1980a, Astrophys. J. 236, pp. 779-797.
Liszt, H. S., and Burton, W. B.: 1980b, Astrophys. J., in press.
Liszt, H. S., Burton, W. B., and Bania, T. M.: 1980, Astrophys. J., submitted.
Liszt, H. S., Burton, W. B., and Xiang, D.-L.: 1980, Astrophys. J., in preparation.
Liszt, H. S., Burton, W. B., Sanders, R. H., and Scoville, N. Z.: 1977, Astrophys. J. 213, pp. 38-42.
Lockman, F. J.: 1976, Astrophys. J. 209, pp. 429-444.
Lockman, F. J.: 1979, Astrophys. J. 232, pp. 761-781.
Radhakrishnan, V., Murray, J. D., Lockhart, P., and Whittle, R. P. J.: 1972, Astrophys. J. Suppl. 24, pp. 15-47.
Roberts, W. W., and Burton, W. B.: 1977, in "Topics in Interstellar Matter," ed. H. van Woerden (Dordrecht: Reidel), pp. 195-205.
Scoville, N. Z., and Solomon, P. M.: 1975, Astrophys. J. Lett. 199, pp. L105-L109.
Scoville, N. Z., Solomon, P. B., and Sanders, D. B.: 1979, Proc. IAU

Symp. 84, pp. 277-283.
Shuter, W. L. H., and Verschuur, G. L.: 1964, Mon. Not. Roy. Astr. Soc. 127, pp. 387-404.
Solomon, P. M., and Sanders, D. B.: 1980, in "Giant Molecular Clouds in the Galaxy," eds. P. M. Solomon and M. Edmunds (London: Pergamon).
Solomon, P. M., and Wickramasinghe, N. C.: 1969, Astrophys. J. 158, pp. 449-460.
Solomon, P. M., Sanders, D. B., and Scoville, N. Z.: 1979, Proc. IAU Symp. 84, pp. 35-52.
Solomon, P. M., Scoville, N. Z., and Sanders, D. B.: 1979, Astrophys. J. Lett. 232, pp. L89-L93.
Spitzer, L., and Schwarzschild, M.: 1951, Astrophys. J. 114, pp. 385-397.
Spitzer, L., and Schwarzschild, M.: 1953, Astrophys. J. 118, pp. 106-112.
Stark, A. A.: 1979, Ph.D. Dissertation, Princeton University.
Stark, A. A., and Blitz, L.: 1978, Astrophys. J. Lett. 225, pp. L15-L19.
Szabo, A., Shuter, W. L. H., and McCutcheon, W.: 1980, Astrophys. J. 235, pp. 45-51.
van de Hulst, H. C., Muller, C. A., and Oort, J. H.: 1954, Bull. Astr. Inst. Netherlands 12, pp. 117-149.
van der Hulst, J. M.: 1979, Astr. and Astrophys. 75, pp. 97-111.
Wannier, P. G., Penzias, A. A., Linke, R. A., and Wilson, R. W.: 1976, Astrophys. J. 204, pp. 26-42.
Weaver, H., and Williams, D. R. W.: 1973, Astr. and Astrophys. Suppl. 8, pp. 1-503.
Zuckerman, B., and Kuiper, T. B. H.: 1980, Astrophys. J. 235, pp. 840-844.

COSMIC-RAY SELF-CONFINEMENT IN THE HOT PHASE OF THE INTERSTELLAR MEDIUM

Catherine J. Cesarsky
Section d'Astrophysique
Centre d'Etudes Nucléaires de Saclay, France

Russell M. Kulsrud
Plasma Physics Laboratory, Princeton, New Jersey

It is well known that, when cosmic rays stream along the field lines of the galactic magnetic field at a velocity which exceeds the Alfven velocity in the medium, they excite hydromagnetic waves of a wavelength comparable to their Larmor radius. These waves, in turn, scatter the cosmic rays, forcing them to reduce their bulk speed (Wentzel 1974, and references there-in). In a stationary situation, the bulk speed of the cosmic rays will depend on their scale height, and on the strength of the relevant damping mechanisms affecting the waves. Isotope observations imply that cosmic rays of energy $\lesssim$ 1 GeV are confined in the galaxy for a time $\sim 2.10^7$ yrs (Garcia-Munoz et al. 1977) ; a simple interpretation of the composition data in the energy range 1-100 GeV implies that the confinement time T decreases as the rigidity $\varepsilon = cp/eZ$ increases : $T \alpha \varepsilon^{-a}$, with a = 0.3 - 0.5. This energy dependence of the confinement time should hold at least up to $\varepsilon \simeq 3.10^6$ GV, to be consistent with the lack of structure in the cosmic ray spectrum in the range $10-3.10^6$ GV (see Cesarsky 1980 and references therein).

Until a few years ago, it was believed that the interstellar medium was mostly filled by a neutral gas, of density ~ 0.1 cm^{-3} and a temperature of several thousand degrees. Kulsrud and Cesarsky (1971) showed that, in such a medium, cosmic rays of energy $\gtrsim 100$ GeV are not confined at all, because the waves are damped very rapidly by the effect of the collisions between the neutral and the charged particles in the medium. The case of streaming in HII regions was considered by Wentzel (1974) and Skilling (1975), and did not lead either to a satisfactory solution.

At present, we think that a substantial fraction of the interstellar medium is filled with a hot ($T_6 \simeq 1$, where T_6 is the temperature in 10^6 °K) and diffuse "coronal gas" ($n_{-3} \simeq 1$, where n_{-3} is the density in 10^{-3} cm^{-3}). The strength of the magnetic field in such regions is unknown ; it is probably lower than the "normal" interstellar value, 2.5 µG, by a factor which may be in the range 3-30. If B_μ is the magnetic field strength in µ-gauss, the Alfven velocity is $v_A \simeq 66 \, B_\mu /(n_{-3})^{0.5}$ Km s^{-1}, while the thermal velocity is $v_i = 156 (T_6)^{0.5}$ Km s^{-1}. The main damping mechanisms affecting hydromagnetic waves in HI and HII regions are not operative in the coronal medium, where neutral particles are absent, and where, with $v_i > v_A$, Alfven waves cannot decay via two-wave interactions into sound waves. Still, waves propagating at an angle with the magnetic field are rapidly damped by collisionless processes while waves propagating along the field, but whose frequency is higher than the Larmor frequency of the thermal ions, are dissipated by Landau damping. However, all these "usual" damping mechanisms do not affect the Alfven waves propagating along the magnetic field, and resonating with cosmic rays ; thus the "coronal" phase of the interstellar medium appears to be a prime candidate as a region for cosmic ray self-confinement (Mc Cray and Snow 1979).

In fact, as pointed out by Lee and Völk (1973) and Kulsrud (1978), parallel propagating Alfven waves in hot plasma are also subjected to a non-linear damping mechanism, which is due to interactions of the thermal ions with beat waves produced by couples of Alfven waves. The non-linear damping rate of waves resonating with cosmic rays of Larmor frequency Ω is estimated by Kulsrud (1978) as :

$$\Gamma_{n\ell} \sim (\sqrt{\pi}/8)(v_i/c)(\delta B/B)^2 \Omega \quad (1)$$

where $(\delta B/B)^2$ represents the ratio of the energy density in resonant waves to that in the underlying magnetic field.

Setting the growth rate equal to the damping rate, we obtain the streaming velocity Δv of cosmic rays of rigidity ϵ with respect to a frame moving at the Alfven speed

$$\Delta v \simeq 10 \ \epsilon^{0.85} (T_6 n_{-3})^{0.25} L_1^{-0.5} \text{ Km s}^{-1} \quad (2)$$

where L_1 (Kpsec) is the scale height of cosmic rays along the magnetic flux tubes, and for a cosmic ray flux as in the solar neighborhood. Kulsrud (1978) observes that, when ($\delta B/B$) becomes large enough, the damping mechanism saturates, because thermal particles tend to get trapped by the beat wave, instead of continously exchanging energy with it. It turns out that the wave energy densities built up by the cosmic rays in the coronal medium are always high enough to ensure that this saturation effect sets in ; in that case, we obtain :

$$\Delta v \simeq 2.5 \ \epsilon^{0.8} T_6^{0.66} (B_\mu n_{-3})^{0.33} L_1^{-0.33} \text{ Km s}^{-1} \quad (3)$$

We expect the mean age T to be proportional to ($v_A + \Delta v$). Thus, formulae (2) or (3) give too steep a rigidity dependence of Δv to reconciled with the observations ; also, even in the saturated case (formula 3), the streaming speed attains a value as high as (c/3) at a rigidity $\epsilon \simeq 5 \times 10^5$ GV, while the observed anisotropy of cosmic rays in this range of ϵ does not exceed 0.1%. We conclude that, if cosmic ray propagation in the interstellar medium is governed by resonant wave-particle interactions in the coronal phase of the interstellar medium, the observations are easier to understand if the waves are due to an underlying interstellar spectrum rather than generated by the cosmic rays themselves (Cesarsky 1975, 1980).

REFERENCES

Cesarsky, C.J. 1975, Int. Cosmic Ray Conf., 14th, Munich, 12, 4166
Cesarsky, C.J. 1980, to appear in Vol. 18 of Ann.Rev.Astron.Astrophys.
Garcia-Muñoz, M., Mason, G.M., Simpson, J.A. 1977, Ap.J., 217, 859
Kulsrud, R.M. 1978, Astron. p. ded. to B. Stromgren, Copenhagen Univ.Obs.p.317.
Kulsrud, R.M., Cesarsky, C.J. 1971, Ap.J.Lett., 8, 189
Lee, M.A., Völk, H.J. 1973, Ap. Space Sc., 24, 31
Mc Cray, R., Snow, T.P., Ann.Rev.Astron.Astrophys., 17, 213
Skilling, J. 1975, M.N.R.A.S., 173, 255
Wentzel, D.G. 1974, Ann.Rev.Astron.Astrophys., 12, 71

RAYLEIGH TAYLOR INSTABILITIES IN THE INTERSTELLAR MEDIUM

Marc Lachièze-Rey
Section d'Astrophysique
Centre d'Etudes Nucléaires de Saclay, France

Parker (1966,1969) proposed that the interstellar medium could be subjected to instabilities, able to destroy its structure in a time shorter than the galactic evolution time. He emitted the idea that this phenomenon could lead to halo formation, escape of magnetic field and cosmic rays from the galaxy, via some sort of interstellar "bubbles", and also to cloud formation. He showed then that if the interstellar medium was well represented by a simple horizontal stratified equilibrium model, perturbations would effectively grow and destroy the structure.

The equilibrium state whose stability he studied was however very schematic. On the other hand, his work was discussed and criticized by various authors. We tried to have a new approach to the problem and to answer the questions which remained open.

In this perspective, we built a tool to test the stability of any 2 dimensionnal equilibrium configurations able to represent the interstellar medium (Asseo et al.1978,1980): following the idea of Zweibel and Kulsrud (1975), we adapted the energy principle used by Bernstein et al.(1958) to test hydromagnetic configurations. We developed it for the composite interstellar medium made of interstellar gas and cosmic rays gas, immersed in the galactic magnetic field and in the gravitational field of the galactic stars.

1) For horizontal configurations (where the magnetic field lines remain parallel to the galactic plane), we derived a global instability criterion (Lachièze-Rey et al.1980): it concerns all the types of perturbation which can apply to the configuration. This criterion shows us that all the configuration proposed by Badhwar and Stephens (1977) to depict the interstellar medium are unstable. Calculation of the growth rates ensures us that perturbations have enough time to grow.

2) For non horizontal configurations, we were not able to derive a necessary and sufficient instability criterion. In the case $\gamma = 1$ and $\Gamma = 0$ (γ and Γ are the polytropic indexes of the interstellar and cosmic ray gases respectively) we give however sufficient instability criteria, corresponding to special types of perturbation. They permit us to exhibit the destabilizing effects of the curvature of the magnetic field lines and of the cosmic rays (Asseo et al.1978,1980). We apply them to a family of possible equilibrium configurations, introduced by Mouschovias (1974). These configurations, related to horizontal states by a flux-conserving possible evolution, were proposed to be the result of the development of instabilities from horizontal states. We show that all are also unstable, including the one calculated numerically by Mouschovias. We give also the value of growth's rates.

3) We also studied the effect of hydromagnetic turbulence, as suggested by Zweibel and Kulsrud (1975), on the stability of horizontal configurations: the short wavelengths are completely stabilized, the longer ones (~ 100 pc) only partially (Lachièze-Rey et al.1980). We also discuss the values of γ and Γ and give some estimates of the influence of $\gamma \neq 1$ and $\Gamma \neq 0$ on the stability (Lachièze-Rey and Pellat 1980).

In conclusion, we showed that all the configurations already proposed to depict interstellar medium are unstable, and therefore not able to survive on the galactic time scale, although it is not clear if the unstabilities only generate small scale turbulence or lead to the formation of big scale structure, the escape of cosmic rays and magnetic field and the concentration of the gas near the plane.

REFERENCES

Asseo,E.,Cesarsky,C.J.,Lachièze-Rey,M.,and Pellat,R., 1978, Ap.J.Letters, $\underline{225}$, L21-25
Asseo,E.,Cesarsky,C.J.,Lachièze-Rey,M.,and Pellat,R., 1980, Ap.J. (1 May 1980)
Badhwar,G.D. and Stephens,S.A., 1977, Ap.J., $\underline{212}$, 494
Bernstein,I.B.,Frieman,E.A.,Kruskal,M.D.,and Kulsrud,R.M., 1958, Proc. Roy.Soc.London, $\underline{A244}$, 17
Lachièze-Rey,M.,Asseo,E.,Cesarsky,C.J.,and Pellat R., 1980, Ap.J. (15 May 1980)
Lachièze-Rey,M., Pellat,R., 1980 (in preparation)
Mouschovias,T.Ch., 1974, Ap.J., $\underline{192}$, 37
Parker,E.N., 1966, Ap.J., $\underline{145}$, 811
Parker,E.N., 1969, Space Sci.Rev., $\underline{9}$, 651
Zweibel,E.G., and Kulsrud,R.M., 1975, Ap.J., $\underline{201}$, 63

NONLINEAR LANDAU DAMPING OF ALFVEN WAVES AND THE PRODUCTION AND PROPAGATION OF COSMIC RAYS

R.J. Stoneham
Institute of Astronomy, The Observatories,
Madingley Road, Cambridge CB3 OHA, England

The existence of hydromagnetic waves (waves whose frequency ω is less than the ion gyrofrequency $\Omega_i = eB/m_i c$) in a collisionless magnetized plasma with β, the ratio of plasma pressure to magnetic pressure, much greater than unity is required in theories for Fermi acceleration of cosmic rays by converging scattering centres at a shock front (Axford et al. 1977, Bell 1978, Blandford and Ostriker 1978), in theories for the adiabatic cooling of cosmic rays due to trapping by plasma instabilities in an expanding supernova remnant (Kulsrud and Zweibel 1975, Schwartz and Skilling 1978) and in theories for resonant scattering of cosmic rays by hydromagnetic waves in the hot phase of the interstellar medium (Holman et al. 1979). Hydromagnetic waves may be damped by thermal ion cyclotron damping for wavenumbers $k \gtrsim \Omega_i/v_i$, where $v_i = (T_i/m_i)^{\frac{1}{2}}$ is the average thermal ion speed, and by linear Landau damping for non-zero angles of propagation with respect to the ambient magnetic field $\underline{B}$ (Foote and Kulsrud 1979). Damping by both these processes is strong in a high-β plasma where there are many particles travelling at the phase speed of the waves. Hydromagnetic waves propagating along $\underline{B}$ may be damped by nonlinear wave-particle interactions, the most important of which is thermal ion Landau damping of the beat wave of two Alfvén waves. This nonlinear process has the effect of transferring energy from the waves to the particles and can therefore be considered as a damping process for the waves.

When damping is weak, the dispersion relation for hydromagnetic waves propagating along $\underline{B}$ reduces to

$$(\omega^2 - k^2 v_A^2)^2 = \tfrac{1}{4} \omega^2 k^4 v_i^4 / \Omega_i^2 , \qquad (1)$$

where $v_A = B/(4\pi\rho)^{\frac{1}{2}} \simeq v_i/\beta^{\frac{1}{2}}$ is the Alfvén velocity and terms of order $1/\beta$ have been neglected. For long wavelength waves (i.e. for $k \ll \Omega_i/\beta v_A$), the two positive-frequency solutions of (1) have phase velocities of order v_A, while for short wavelength waves ($k \gtrsim \Omega_i/\beta v_A$) the phase velocity of the right-hand (left-hand) circularly polarized wave mode is significantly greater than (less than) v_A. For $k \gg \Omega_i/\beta v_A$, the two positive-frequency solutions of (1) are

$$\omega_R = k^2 v_i^2 / 2\Omega_i \quad \text{and} \quad \omega_L = 2\Omega_i / \beta. \tag{2}$$

For weak damping, nonlinear Landau damping of long wavelength hydromagnetic waves propagating in the same direction along $\underset{\sim}{B}$ in a high-β plasma may be described by the damping coefficients (Lee and Völk 1973)

$$|\gamma_1| = \tfrac{1}{4}\pi^{\tfrac{1}{2}} \frac{|\underset{\sim}{B}_2|^2}{B^2} \beta^{\tfrac{1}{2}} \omega_1 \quad \text{and} \quad |\gamma_2| = \tfrac{1}{4}\pi^{\tfrac{1}{2}} \frac{|\underset{\sim}{B}_1|^2}{B^2} \beta^{\tfrac{1}{2}} \omega_2, \tag{3}$$

where $\underset{\sim}{B}_1$ and $\underset{\sim}{B}_2$ are the wave magnetic fields. The particles and the lower-frequency wave both gain energy from the higher-frequency wave when the two waves have the same sense of polarization. Both waves are damped when the two waves have the opposite sense of polarization. The damping coefficients for nonlinear Landau damping of long wavelength hydromagnetic waves propagating in opposite directions along $\underset{\sim}{B}$ are more complicated functions of plasma and wave parameters. For certain values of the ratio of frequencies the damping may be much greater than for waves propagating in the same direction (Lee and Völk 1973). Resonant wave-wave interactions (e.g. Sagdeev and Galeev 1969) are possible between the strongly dispersive short wavelength hydromagnetic waves. Short wavelength waves propagating along $\underset{\sim}{B}$ will be coupled to the strongly damped hydromagnetic waves propagating at nonzero angles to $\underset{\sim}{B}$ and will therefore be damped by this process as well as by nonlinear Landau damping.

Damping of all hydromagnetic waves in a high-β plasma is likely to be strong. This needs to be taken into account in theories in which the existence of hydromagnetic waves in a hot collisionless magnetized plasma is postulated.

References

Axford, W.I., Leer, E., and Skadron, G.: 1977, Paper presented at the 15th Int. Cosmic Ray Conf., Plovdiv.
Bell, A.R.: 1978, Mon. Not. R. astr. Soc., 182, 147; 182, 443.
Blandford, R.D., and Ostriker, J.P.: 1978, Ap. J., 221, L29.
Foote, E.A., and Kulsrud, R.M.: 1979, Ap. J., 223, 302.
Holman, G.D., Ionson, J.A., and Scott, J.S.: 1979, Ap. J., 228, 576.
Kulsrud, R.M., and Zweibel, E.G.: 1975, Proc. Int. Cosmic Ray Conf., Munich, 2, 465.
Lee, M.A., and Völk, H.J.: 1973, Ap. Space Sci., 24, 31.
Sagdeev, R.Z., and Galeev, A.A.: 1969, *Nonlinear Plasma Theory*, Benjamin, New York.
Schwartz, S.J., and Skilling, J.: 1978, Astron. Astrophys., 70, 607.

COSMIC RAY ANTIPROTONS 5-12 GEV

T. K. Gaisser[*], A. J. Owens, and Gary Steigman[*]
Bartol Research Foundation of The Franklin Institute,
University of Delaware, Newark, DE 19711

Secondary antiprotons are a potentially interesting probe of cosmic ray propagation because their production cross section is strongly energy-dependent, increasing by more than two orders of magnitude between 10 and 1000 GeV/c. This is quite unlike the case for fragmentation cross sections of complex nuclei, which are virtually constant with energy. Moreover, the $\bar{p}$ flux depends primarily on the environment seen by protons which need not be identical to that probed by other nuclei.

Stimulated by the recent report by Golden et al. (1979) of a measurement of the flux of cosmic ray antiprotons, we have reevaluated calculations of the flux to be expected assuming that antiprotons are due to collisions of cosmic rays with interstellar matter. We find that the reported result, $\bar{p}/p \sim (5.2 \pm 1.5) \times 10^{-4}$, is 3-4 times greater than would be expected if cosmic ray protons sample the same distribution of interstellar matter seen by nuclei with $Z > 1$. This conclusion differs from that reached by Golden et al. because they used an earlier calculation by Badhwar et al. (1975) which gave a $\bar{p}/p$ ratio significantly higher than the original estimate of Gaisser and Maurer (1973).

To track down the source of this difference, we have repeated the calculation using the parametrization of the data on $\bar{p}$ production given by Badhwar et al. They had pointed out that data on $\bar{p}$ production at FNAL momenta (100-400 GeV/c) were not available to Gaisser and Maurer and that consequently those authors had to make a significant interpolation between low energy data and data from ISR at equivalent laboratory momenta of 1000-2000 GeV/c. This is not, however, the source of the difference between the two calculations: with the parametrization of Badhwar et al. we find a result only $\sim$25% greater than that of Gaisser and Maurer in the range of $\bar{p}$ momenta 5-12 GeV/c. On the basis of independent calculations, Szabelski et al. (1980) and Mauger and Golden (private communication) have also reached the conclusion that the $\bar{p}/p$ ratio is $1-2 \times 10^{-4}$ in the momentum range covered by the experiment of Golden et al. This

[*]Work supported in part by the U. S. Department of Energy under contract AS02-78ER05007.

result is for a mean path length of 5 gm/cm² of equivalent hydrogen and includes a contribution of about 30% for production of $\bar{p}$ by nuclei with Z > 1.

It is interesting to speculate on possible explanations if the $\bar{p}$ flux is indeed significantly larger than conventional expectations. We note that, even though the median primary energy responsible for a given $\bar{p}$ is about 10 $E_{\bar{p}}$, the path length probed is that at the energy of the observed $\bar{p}$. This is because the source spectrum of $\bar{p}$ depends on the primary spectrum, which is assumed to be that measured at Earth. The observed $\bar{p}$ spectrum then depends on the mean confinement time of the $\bar{p}$; i.e., on the path length at $E_{\bar{p}}$. Secondly, we note that if some of the matter seen by the primaries is traversed before they are fully accelerated, this would reduce the $\bar{p}$ flux relative to the flux of secondary complex nuclei because of the difference in behavior of the two types of cross section as a function of energy.

Steigman (1977) pointed out that a closed galaxy model, such as that of Peters and Westergaard (1977), could give a considerably increased $\bar{p}$ flux. This is because a significant fraction of the primary proton flux is "old", whereas the flux of primary nuclei with Z >1 and E/n $\lesssim$ 100 GeV is essentially all from the "young", local component and therefore properly treated in a leaky box model. A simple estimate suggests $\bar{p}/p \sim 1.5-3$ times that expected in the standard leaky box model. We are currently making a quantitative estimate of $\bar{p}$ production in this model.

References

Badhwar, G. D., et al.: 1975, Astro. and Space Sci., 37, 283.
Gaisser, T. K. and Maurer, R. H.: 1973, Phys. Rev. Letters 30, 1264.
Golden, P. L. et al.: 1979, Phys. Rev. Letters 43, 1196.
Peters, B. and Westergaard, N. J.: 1977, Astro. and Space Science 48, 21.
Steigman, Gary: 1977, Ap. J. 217, L131.
Szabelski, J., Wdowczyk, J. and Wolfendale, A. W.: 1980 (this volume).

ANTI-PROTONS IN THE PRIMARY COSMIC RADIATION

J. Szabelski and J. Wdowczyk,
Institute of Nuclear Research, Uniwersytecka 5, Lodz,
Poland, and
A.W. Wolfendale,
Department of Physics, Durham University, U.K.

A knowledge of the flux of antiprotons is of value in examining the manner in which cosmic rays propagate, assuming, as is conventional, that the antiprotons arise from interactions in the I.S.M. Golden et al. (1979) have recently measured the $\bar{p}/p$ ratio in the region of 10 GeV and, although confirmatory measurements are needed, it is instructive to compare with expectation. In a recent paper (Szabelski et al., 1980) we presented results of a new calculation of the expected $\bar{p}/p$ ratio using the standard (equilibrium) 'leaky box model'.

The Figure shows a comparison of the measured ratio (including a measurement of Bogomolov et al., 1979) with our own, and other predictions (see caption for key); the predictions having been made

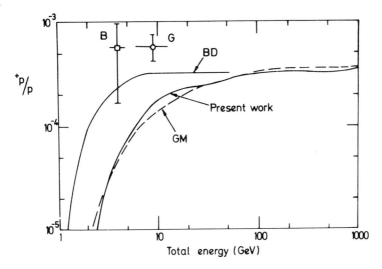

Comparison of observed and expected $\bar{p}/p$ ratios.
B, Bogomolov et al. (1979); G, Golden et al. (1979); BD, Badhwar et al. (1975); GM, Gaisser and Maurer (1973).

for a mean grammage traversed by cosmic rays of 5 gcm^{-2} (hydrogen) as expected from analysis of mass composition data. It is immediately apparent that not only is there a large discrepancy between the measured ratio and our own predictions but that the different predictions are disparate.

The source of the difference in predictions is not known, only a little is due to known input differences; using the same input cosmic ray spectrum the ratio of the GM value to ours for $\bar{p}$ of all energies is 0.74 and the ratio of the BD value to ours is 6.1. The closeness of GM to our value adds confidence.

A number of possibilities arise to account for the difference between the measured $\bar{p}/p$ ratio and our predictions. The measured ratio may be too high; however, we have analysed the sea level data with the same instrument for the μ^+/μ^- ratio and find good agreement with the Durham (and other) measurements of this quantity, thus giving confidence. Another possibility is that there is a contribution from 'genuine' extragalactic $\bar{p}$ from anti-galaxies; this cannot be discounted but seems unlikely for a number of reasons. The most likely explanation, in our view, is that the 'leaky-box' model of propagation is not accurate.

Our choice of propagation effects as the explanation stems in part from our earlier contention (Giler et al., 1977) that the grammage derived from an analysis of e^+ was not the same as that expected from mass composition studies. Specifically we found, using the leaky box model, $\lambda(e^+)$ increasing from $\simeq 2.5$ gcm^{-2} at 2GeV to ~ 8 gcm^{-2} at 50 GeV compared with λ (nuclei) falling from $\simeq 6$ gcm^{-2} to $\simeq 3$ gcm^{-2} between the same limits. Now we find $\lambda(\bar{p}) \simeq (18\pm5)$ gcm^{-2} at $\simeq 10$ GeV. There are several ways of modifying the propagation characteristics to achieve these values; an attractive model is one where the lifetime distribution is not the usual exponential but has a much longer tail, energy losses for the positrons would reduce their flux (and thus reduce $\lambda(e^+)$) even at quite low energies. The difference between $\lambda(\bar{p})$ and λ(nuclei) could arise from protons (and α-particles) and heavier nuclei having sources differently distributed in the Galaxy.

References

Badhwar, G.D., et al.: 1975, Ap. Space. Sci., 37, 283.
Bogomolov, E.A., et al.: 1979, Proc. Int. Cosmic Ray Conf., Kyoto, 1,330.
Gaisser, T.K., and Maurer, R.H.: 1973, Phys. Rev. Lett., 30, 1264.
Giler, M., Wdowczyk, J., and Wolfendale, A.W.: 1977, J. Phys. A, 10, 843.
Golden, R.L., et al.: 1979, Phys. Rev. Lett. 43, 1196.
Szabelski, J., Wdowczyk, J., and Wolfendale, A.W.: 1980, Nature (in the press).

THE X-RAY SKY

J. Trümper

Max-Planck-Institut für Physik und Astrophysik
Institut für Extraterrestrische Physik
D-8046 Garching, W.-Germany

ABSTRACT   Today X-ray astronomy encompasses almost all astronomical objects, from nearby stars to the most distant quasars. We review a few selected recent results obtained by X-ray astronomy on galactic objects like normal stars, supernova remnants, neutron stars and the black hole candidate Cyg X-1.

1. INTRODUCTION

Many of us are old enough to remember the Cosmic Ray Conferences in the mid sixties when X-ray astronomy could be reviewed in half an hour. Today this has become completely impossible - the X-ray sky comprises about a thousand sources representing almost all astronomical objects between nearby normal stars and the most distant quasars at the edge of the known universe. Therefore, I have to make a selection, and I have chosen a few topics of galactic X-ray astronomy which perhaps may be relevant to Cosmic Ray Origin. In particular, I will deal with normal stars, the chemical composition of supernova remnants, neutron stars and their magnetic fields and the question of stellar black holes.

As in any other branch of astronomy there are three major observational avenues in X-ray astronomy. Historically, studies of source variability have been most important and they will continue to play a major role because chaotic, periodic and quasi-periodic variability is characteristic of many sources. The large area detectors used for such studies also provide coarse spectral resolution. So far, high resolution spectroscopy with dispersive instruments and polarimetry have been used as powerful diagnostic tools only for a limited sample of bright objects. Recently, the power and beauty of imaging has become evident to a large community. The Einstein telescope with its much finer angular resolution and its one-thousand times greater sensitivity for point sources - compared with previous instruments - renders the first deep look at the X-ray sky.

2. NORMAL STARS

One of the first surprises of the Einstein mission was the detection of OB associations as clusters of bright X-ray stars (Harnden et al., 1979). In the meantime a first survey of the sky has been completed which contains ~ 140 objects of almost all spectral types and luminosity classes (Vaiana et al., 1980). Figure 1 summarizes the ratios of X-ray to optical fluxes measured for a number of main sequence stars. The distribution shows a rather large spread and a dramatic increase towards K stars, which radiate roughly one tenth of their energy in X-rays. For the late type stars a correlation between stellar X-ray flux and rotational velocities has been noted (Vaiana et al., 1980). Since our sun seems to be rather underluminous in X-rays, previous predictions for stellar coronal X-ray fluxes have been too pessimistic.

For giants and supergiants of spectral types between O and K the ratio of X-ray to optical luminosities is similar to that of main sequence stars. However, in the case of late type giants only upper limits of the X-ray fluxes could be detected indicating a drastic decrease of f_x/f_o for cool giants. This is a rather dramatic effect indeed. For α Sco, a M1 supergiant, the mean X-ray flux at the stellar surface is more than a hundred times lower compared with that of solar coronal holes (Vaiana et al., 1980).

Clearly all these new results have a great impact on our notion of stellar coronae, their heating and their confinement, as well as on the physics of stellar winds and their interaction with the interstellar medium. They also can be relevant for the origin of cosmic rays, at least those of rather low energies.

3. CHEMICAL COMPOSITION OF YOUNG SUPERNOVA REMNANTS

Supernova remnants are bright X-ray sources and belong to the classical objects in X-ray astronomy. In general, the X-ray emission of supernova remnants comes from a mixture of ejected stellar and shocked interstellar material. In sufficiently young supernova remnants the contribution of the ejected component may not be negligible, opening the possibility to observe freshly synthesized material. Figure 2 shows the Einstein observatory high resolution imager picture of Cas A, the youngest known galactic supernova remnant (Murray et al., 1979). The brightest parts of this shell source are at a temperature of $\sim 10^7$ K. The faint halo which can be seen just outside the bright emission region may be produced by the shock wave moving ahead of the expanding debris.

The chemical composition of the glowing material can be deduced from the spectra obtained with the Einstein solid state spectrometer as shown in Fig. 3 (Becker et al., 1979 a). The data suggests a strong enhancement of elements between silicon and calcium but nearly normal iron abundance, compared with the solar composition. Similar results

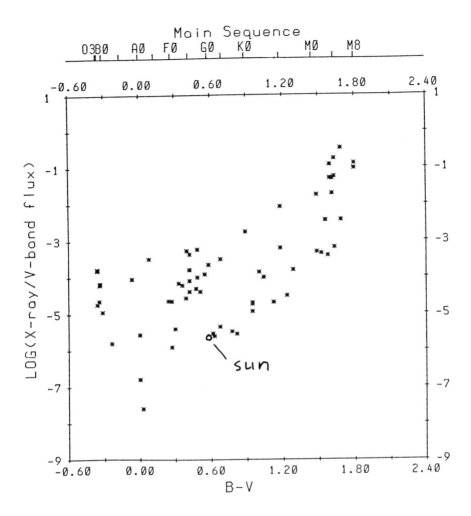

Figure 1. The ratio of soft X-ray to V-band fluxes for main sequence stars detected in the Einstein stellar survey (from Vaiana et al., 1980; the solar point has been added by the present author).

have been obtained for the Kepler and Tycho supernova remnants (Becker et al., 1979 b; Becker et al., 1980).

4. NEUTRON STARS

X-ray emitting neutron stars have been probably found in four different astrophysical situations:
(1) The Crab pulsar remains a unique object. Obviously its pulsed X-ray emission is part of the non-thermal radiation emitted synchronously

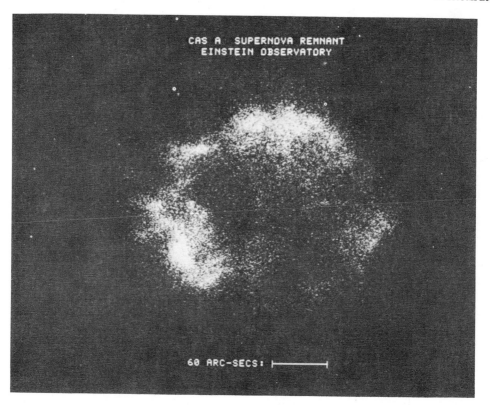

Figure 2. X-ray image of the Cas A supernova remnant, obtained with the Einstein high resolution imager (from Murray et al., 1979).

at all frequencies between radio waves and gamma rays.

(2) The large sensitivity of the Einstein observatory has made it possible to look for thermal emission from the surface of isolated neutron stars. An extensive survey of radio pulsars and supernova remnants has led to a number of upper limits on the surface temperature (Helfand et al., 1980). Of course the Crab pulsar as the youngest of the known pulsars is of particular interest. A limit on the temperature of 3×10^6 K had been obtained by lunar occultations (Toor et al., 1977) which may be improved by the final analysis of the data obtained with the Einstein high resolution imager. Another candidate is the Vela pulsar. Einstein observations have revealed the presence of an extended ($\sim$ 1 arcmin) diffuse source around the Vela pulsar which may be a Crab nebula type object. At the center a point like feature has been found which could be attributed to thermal emission of the neutron star. The corresponding black body temperature is then $\sim 1.5 \times 10^6$ K (Harnden et al., 1979).

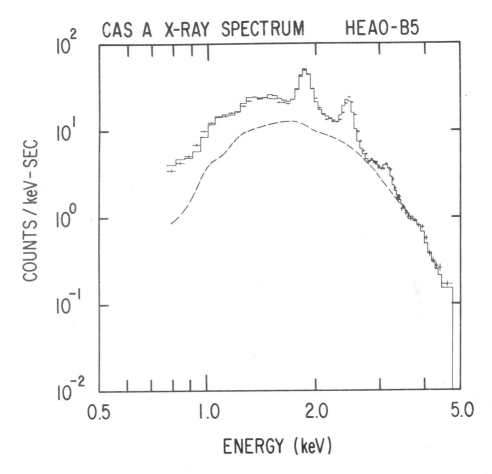

Figure 3. X-Ray spectrum of the Cas A supernova remnant, obtained with the Einstein solid state spectrometer. The dominant line features are due to emission from helium-like silicon and sulfur ions (from Becker et al., 1979 a).

Recently a third case has been reported by Tuohy et al. (1980), who find a point source in the supernova remnant RCW 103. If it is a neutron star, its black body temperature is $\sim 2.1 \times 10^2$ K. Figure 4 shows a compilation of these data and several other upper limits in comparison with theoretical cooling curves of neutron stars (after Tsuruta, 1979). For most cases the data are consistent with the predictions of standard cooling models. The upper limits for Tycho SNR and SN 1006 suggest that these objects either do not contain a neutron star or cooling is much faster, e.g. due to neutrino emission of a pion condensate.

(3) About 25 X-ray bursters are known which form a galactic bulge popu-

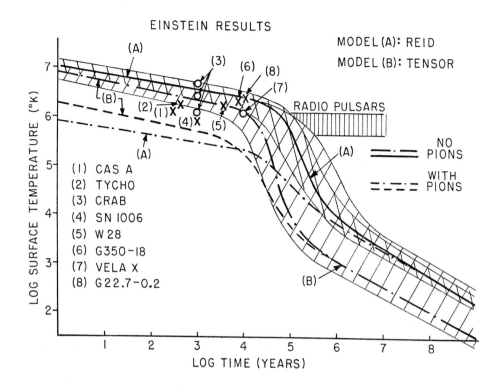

Figure 4. Comparison between surface temperatures predicted from theoretical work and preliminary results from the Einstein observatory (Tsuruta, 1979). The shaded stripes represent surface temperatures expected for a range of masses, magnetic fields and neutron star models, without effects of a pion condensate on cooling. The latter effects lead to the dashed curves.

lation, and a few of these objects are found in globular clusters (e.g. Lewin and Clark, 1979). There is good evidence that bursters represent weakly or non-magnetized neutron stars in binary systems with low mass companions ($\leq 0.3\ M_\odot$). The bursts are probably a result of thermonuclear explosions of accreted material (He-explosions) piled up at the neutron star surface as first suggested by Maraschi and Cavaliere (1977), and worked out in more detail by Joss (1978). The burst spectra are of black body type showing a temperature decrease during the bursts. The 'black body radii' of the objects are of the order ~ 10 km, consistent with neutron star dimensions (Hoffmann et al., 1977; von Paradijs, 1978).

(4) The classical pulsating X-ray sources like Cen X-3 and Her X-1 have provided a wealth of information on neutron stars. There are now

some 18 objects known with periods ranging from 0.7 to 835 seconds (cf the review of Rappaport (1979) and recent results by Lamb et al. (1980) and Kelley et al. (1980)). For most of them the binary nature has been well established and the companion masses range from $\sim 2\ M_\odot$ to $\sim 30\ M_\odot$. They are powered by gravitational energy release of the accreting matter and the occurence of X-ray pulsations suggests an efficient magnetic channelling ($B \gtrsim 10^{11}$ G) of plasma flow onto the neutron star's polar caps. The X-radiating spot must have rather small dimensions (~ 1 km).

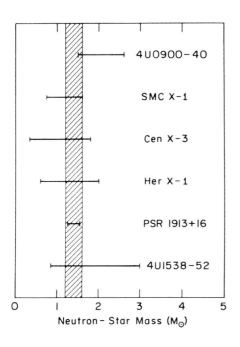

Pulse timing measurements have been used to determine the orbital parameters of several of these systems and to derive neutron star masses (Rappaport, 1979). Figure 5 shows that all observations are consistent with a typical neutron star mass of $1.3 - 1.4\ M_\odot$, in good agreement with theoretical expectations. In contrast to radio pulsars, pulsating X-ray sources generally show a spin-up which is obviously a result of momentum transfer to the neutron star by the accreting matter. In addition, erratic period changes have been observed in several sources, notably in Her X-1 and Vela X-1, which may be due to fluctuations in the accretion rate or to intrinsic torques (cf Boynton et al., 1979).

Figure 5. Empirical knowledge of neutron star masses (from Rappaport and Joss, 1979). PSR 1913+16 is the binary radio pulsar. The other masses are derived from observations of X-ray binaries.

5. MAGNETIC FIELDS OF NEUTRON STARS

Already before the discovery of radio pulsars it had been suggested that neutron stars possess very strong magnetic fields ($B \sim 10^{12}$ G) due to magnetic flux conservation during the gravitational collapse (Woltjer, 1964; Pacini, 1967).

This order of magnitude estimate has been confirmed by observations of radio pulsars: The spin-down of these objects - interpreted as a consequence of magnetic braking - leads to magnetic moments of the order of 10^{30} Gauss cm^3, corresponding to surface field strengths of $\sim 10^{12}$ Gauss. A further estimate can be obtained from the spin-up of pulsating X-ray sources which is due to the transfer of angular momentum onto the neutron star by the accreting matter. The expected period change $\dot{P}$ depends on the mass accretion rate $\dot{M}$ (derived from the X-ray luminosity L), the magnetic moment of the neutron star μ, its moment of inertia I, and the details of the interaction between the stellar wind/accretion disk and the rotating magnetosphere (Pringle and Rees, 1972; Lamb et al., 1973; Gosh and Lamb, 1979; Anzer and Börner, 1980). Under a wide range of conditions one expects a dependence

$$\dot{P} \sim \mu^{2/7} \, R^{6/7} \, \dot{M}^{-3/7} \, I^{-1} \, (PL^{3/7})^2$$

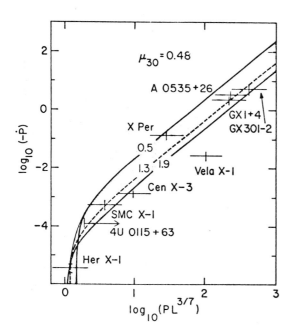

Figure 6. Relation between spin-up rate $\dot{P}$ and X-ray luminosity L predicted by disk accretion theory. Superimposed are the data of nine pulsating X-ray sources (from Lamb, 1979).

Figure 6 shows that the pulsating X-ray sources as a class indeed behave as expected for standard magnetic neutron stars with a magnetic moment of $\sim 10^{30}$ G cm^3, in good agreement with the other estimates. Of course, these cannot be more accurate than to an order of magnitude, since the very details of the braking and spin-up mechanisms as well as the radii and moments of inertia of neutron stars are poorly known.

A rather direct determination of neutron star magnetic fields may be obtained by cyclotron line spectroscopy of pulsating X-ray sources. Figure 7 depicts the hard X-ray spectrum of the Her X-1 1.24 sec pulses which shows strong structures above 40 keV (Trümper et al., 1977, 1978, 1979). It can be seen easily that for abundance reasons atomic or nuclear lines cannot be responsible for these features.
On the other hand, electrons

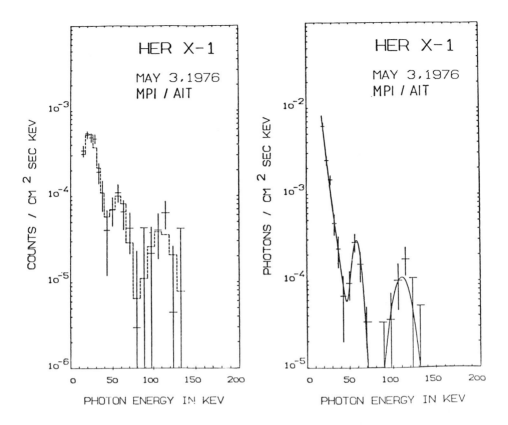

Figure 7. Hard X-ray spectrum of Hercules X-1 obtained during the pulse phase of the 1.24 sec pulsations. The left diagram shows raw count rate spectra. The other diagram shows the deconvoluted spectrum, assuming spectral lines at 58 keV and 110 keV (from Trümper, 1979).

are the most abundant species in the radiating polar hot spot and their (transverse) energies are quantized in units of the cyclotron frequency

$$\hbar\omega_B = \hbar \frac{eB}{mc}.$$

The corresponding Landau transitions may show up in the spectrum as emission or absorption lines depending on the details of the density and temperature distributions in the outer layers of the emitting plasma (Trümper, 1979). For Her X-1 one finds surface magnetic field strengths of $\sim 5.3 \times 10^{12}$ G (emission line) or $\sim 4 \times 10^{12}$ G (absorption line), which need to be corrected for the unknown gravitational red shift (15-30 %) of the neutron star.

It should be noted that the detection of cyclotron resonances requires that the magnetic field is sufficiently homogeneous over the emission region. In the present case $\Delta B/B \lesssim .3$ which means that for a dipole configuration the radial extent of the emission region is less than a tenth of the neutron star radius, i.e. $\lesssim 1$ km in agreement with 'black body' estimates of the radiating surface (Trümper et al., 1978).

While the existence of the feature at ~ 58 keV has been confirmed by later balloon (Voges et al., 1979) and HEAO-1 (Gruber et al., 1978) observations, the second harmonic line has not been seen again. This may be explained by smaller source luminosities (and temperatures) in the later observations. Recently, similar features have been found in 4U 0115+63 (P = 3.6 s; B $\sim 2 \times 10^{12}$ G, Wheaton et al., 1979) and possibly in 4U 1626-67 (P = 7.7 s ; B $\sim 2 \times 10^{12}$ G, Pravdo et al., 1979).

6. STELLAR BLACK HOLES

The question of the existence of stellar black holes is one of the most important issues of modern astronomy, but despite of many attempts there has been embarrassingly little real progress in this context. The present evidence still rests mainly on the few compact sources with strong variability and apparently large masses among which Cyg X-1 is the clearest case (cf Eardly et al., 1978). However, the mass argument is not irrefutable and rapid variability is also seen in the case of neutron stars. The hopes to establish the black hole nature of globular cluster X-ray sources have not materialized so far. The deviations of X-ray source positions from the cluster centers obtained by Einstein high resolution measurements lead to masses of the sources of $M_x \geq 2 M_\odot$ (2 σ). Since M_x includes the mass of a likely present low mass companion the result is consistent with normal neutron star binaries (Grindlay, 1979).

In the case of Cyg X-1, also spectroscopic data have been used to support the black hole evidence. Figure 8 shows the hard X-ray spectrum measured with high precision by Voges et al. (1980). It can be fitted very well by the spectrum expected from Comptonization of soft photons in a hot plasma cloud which is at a temperature of $kT \sim 27$ keV with a Thomson optical thickness of $\tau = 5$ (Sunyaev and Trümper, 1979). Such comptonizing plasma clouds are an integral part of models describing disk accretion onto black holes. While the outer disk is cool (Pringle and Rees, 1972), it becomes unstable and hot in the vicinity of the black hole. In this hot inner part of the disk free-free radiation processes are rather inefficient and Comptonization of soft photons in the hot plasma becomes the main cooling process leading to hard X-ray spectra as observed in Cyg X-1. Interpreting the observed spectrum in terms of this model it is possible to obtain some important parameters of the disk. Apart from the temperature ($kT \sim 27$ keV) and luminosity ($L_x = 10^{37}$ erg/sec) which are directly measured one can derive the radial inward velocity of matter in the disk ($v \sim 10^{-2}$ c) and the turbulence parameter ($\alpha \sim 1$) (Sunyaev and Trümper, 1980). Of course, this

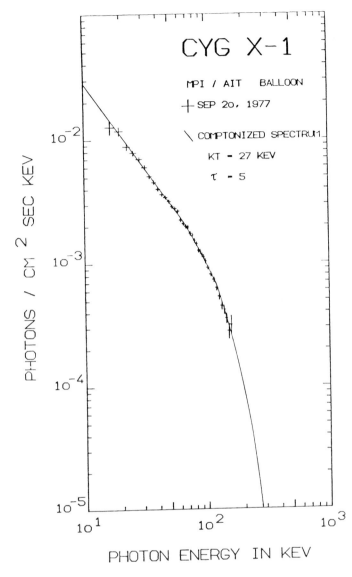

Figure 8. Hard X-ray spectrum of Cyg X-1, observed by the MPI/AIT balloon experiment. The solid line represents a photon spectrum expected from Comptonization of soft photons in an electron cloud with a temperature of 2.5×10^8 K and a Thomson optical thickness $\tau = 5$ (from Sunyaev and Trümper, 1979).

does not proof the black hole nature of Cyg X-1, but is one more piece of indirect evidence.

7. FINAL REMARKS

Although many surprising discoveries and a wealth of quantitative information has been obtained by galactic X-ray astronomy, we are still at the beginning of the development of this field. As usual in new and rapidly developing fields, there are now more questions than before. We can be sure, that further progress will come from Einstein observations. We can also hope that the upcoming, powerful X-ray missions now in preparation will answer many of the present questions. They may also help to solve one of the most important problems of astrophysics: the question of the Origin of Cosmic Rays.

Anzer, U. and Börner, G.: 1980, Astron. Astrophys. 83, 133.
Becker, R.H. et al.: 1979, Astrophys. J. (Lett) 234, L73
Becker, R.H. et al.: 1979, to be published in Astrophys. J.
Becker, R.H., Boldt, E.A., Holt, S.S., Serlemitsos, P.J. and White, N.E.: 1980, to be published in Astrophys. J. (Lett).
Boynton, P. and Deeter, J.: 1979. Workshop on Compact Galactic X-Ray Sources, ed.: Lamb, F. and Pines, D., 168.
Eardly, D.M., Lightman, A.P., Shakura, N.I., Shapiro, S.L. and Sunyaev, R.A.: 1978, Comments on Astrophysics.
Ghosh, P. and Lamb, F.K.: 1979, Astrophys. J., 234, 296.
Grindlay, J.E.: 1979, CFA, preprint 1249.
Gruber et al.: 1978, BAAS, 10, 433.
Harnden, F.R., Jr., et al.: 1979, Astrophys. J. (Lett), 234, L51.
Helfand, D.J., Chanan, G.A. and Novick, R.: 1980, Nature 283, 337.
Hoffmann, J.A., Lewin, W.H.G. and Doty, J.: 1977, M.N.R.A.S., 179, 57.
Joss, P.C.: 1978, Astrophys. J. (Lett) 225, L123.
Kelley, R.L., Apparao, K.M.V., Doxsey, R.E., Jernigan, J.G., Naranan, S. and Rappaport, S.: 1980, preprint.
Lamb, F.K., Pethick, C.J. and Pines, D.: 1973, Astrophys. J. 184, 271.
Lamb, F.K.: 1979, Workshop on Compact Galactic X-Ray Sources, ed.: Lamb, F. and Pines, D., 143.
Lamb, R.C., Markert, T.H., Hartmann, R.C., Thomson, D.J. and Bignami, G.F.: 1980, preprint.
Lewin, W.H.G. and Clark, G.W.: 1980, Ann. New York Acad. Sci. 336, 451.
Marashi, L. and Cavaliere, A.: 1977, Highlights of Astronomy, Dordrecht: Reidel, vol. 4, part 1, 127.
Murray, S.S., Fabbiano, G., Fabian, A.C., Epstein, A. and Giacconi, R.: 1979, Astrophys. J. (Lett) 234, L69.
Pacini, F.: 1967, Nature 216, 567.
Pravdo, S.H. et al.: 1979, Astrophys. J. 231, 912.
Pringle, J.E. and Rees, M.J.: 1972, Astron. Astrophys. 21, 1.
Rappaport, S.: 1979, NATO Adv. Study Institute of Galactic X-Ray Sources, Cape Sounion, Greece.
Rappaport, S. and Joss, P.: 1979, preprint.
Sunyaev, R.A. and Trümper, J.: 1979, Nature 279, 506.
Sunyaev, R.A. and Trümper, J.: 1980, to be published.
Tuohy, J. and Garmire, G.: 1980, submitted to Astrophys. J. (Lett).
Toor, A. and Seward, F.D.: 1977, Astrophys. J. 216, 560.
Trümper, J., Pietsch, W., Reppin, C., Sacco, B., Kendziorra, E. and Staubert, R.: 1977, Ann. New York Acad. Sci. 302, 538.
Trümper, J., Pietsch, W., Reppin, C., Voges, W., Staubert, R. and Kendziorra, E.: 1978, Astrophys. J. (Lett) 219, L105.
Trümper, J.: 1979, IAU/Cospar Symp. on X-Ray Astronomy, Innsbruck, June.
Tsuruta, S.: 1979, preprint.
Vaiana, G.S. et al.: 1980, submitted to Astrophys. J.
Voges, W., Pietsch, W., Reppin, C., Trümper, J., E. Kendziorra and Staubert, R.: 1979, IAU/Cospar Symp. on X-Ray Astron., Innbruck, June.
Voges, W. et al.: 1980, to be published.
von Paradijs, J.: 1978, Nature 274, 650.
Wheaton, W.A. et al.: 1979, Nature 282, 240.
Woltjer, L.: 1964, Astrophys. J. 140, 1309.

QUASAR CONTRIBUTION TO THE X-RAY BACKGROUND

Ajit Kembhavi and A.C. Fabian
Institute of Astronomy,
Madingley Road,
Cambridge CB3 0HA, England

Recent observations by the Einstein Observatory have shown that a majority of known quasars are powerful X-ray emitters. The 107 objects observed as of Feb. 1980 (Zamorani et al. 1980) have X-ray, optical and radio luminosities scattered over a wide range. Until a large enough X-ray selected sample of quasars becomes available, it is necessary to study statistical correlations in the available sample, so that some insight into X-ray production may be obtained, and the contribution of quasars to the X-ray background estimated.

In an attempt to find correlations, Zamorani et al. (1980) have divided the sample into radio-loud and radio-quiet. A quasar in their scheme is defined to be radio loud if it shows positive radio emission and if $\alpha_{RO} > 0.35$, where $\alpha_{RO} = -\log(L_{OP}/L_R)/\log(\nu_{OP}/\nu_R)$. Such a definition of radio loudness however introduces a selection effect: suppose there is a correlation between X-ray and optical luminosities so that $L_X = kL_{OP} + b$, with k and b approximately constant. Then in selecting members of the radio loud class, all quasars below the line $L_R = (a/k)(L_X - b)$ in the $L_R - L_X$ plane are rejected, where the constant a depends upon the value of α_{RO} used to distinguish between loud and quiet. This distorts any correlation present. The selection effect does not materially affect results for the present sample. It may be avoided by defining as radio-loud all quasars with radio flux greater than some limiting value (see Smith and Wright 1979). Zamorani et al. (1980) also report a correlation between α_{RO} and $\alpha_{OX} = -\log(L_X/L_O)/\log(\nu_X/\nu_O)$. Such a correlation could be spurious, because $\log(L_R/L_O)$ and $\log(L_X/L_O)$ will show a correlation even if L_R and L_X are not correlated, and L_O is chosen at random from a fixed range.

As a preliminary step in a detailed analysis of the X-ray data we have estimated the contribution of quasars to the X-ray background by convolving the observed distribution of α_{OX} with the source counts of optically selected quasars given by Braccesi et al. (1980). The contribution to the α_{OX} distribution for the 28 sources with 3σ upper limits on L_X(2 keV) is obtained by assuming that the distribution of

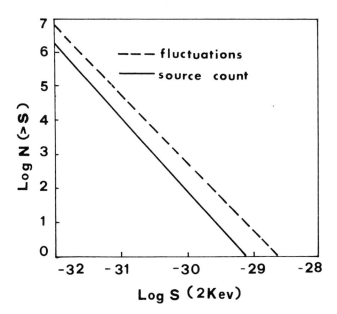

L_X is a single-sided Gaussian. The sample is split up into radio-loud and radioquiet groups, and each group is divided into subgroups with redshift $Z \leq 1$ and $Z > 1$. The subgroups are all averaged over after assuming that 10% of all quasars are radio-loud and that 75% of all quasars are at $Z > 1$. We find that $\log N (> S) = -63.0 - 2.16 \log S_X$. The slope is the same as that of the Braccesi counts, and it can be estimated that at the limiting flux of the Einstein deep survey ($\sim 5.2 \times 10^{-32}$ erg sec^{-1} cm^{-2} Hz^{-1} at 2 keV), there should be ~ 11 source (sq. deg.)$^{-1}$ and that the contribution to the background should be $\sim 10\%$.

It has been pointed out by Fabian and Rees (1978) that evolved sources contribute fluctuations of amplitude $\leq 1\%$ in the Uhuru 5° x 5° field of view. We find that for the present sample this implies $\log N(S) \leq -2 \log S - 57.27$. The X-ray source counts are close to but consistent with this fluctuation limit. We are investigating how strong the constraints set by the fluctuations are, and the dependence of these on the average quasar spectral index (see also Cavaliere et al. 1980).

References

Braccesi, A., Zitelli, V., Bonoli, F., and Formiggini, L.: 1980, Astron. Astrophys., 85, pp 80-92.
Cavaliere, A., Danese, L., De Zotti, G., and Franceschini, A.: 1980, Preprint.
Fabian, A.C., and Rees, M.J.: 1978, MNRAS, 185, pp 109-122.
Smith, M.C., and Wright, A.E.: 1979, Anglo-Australian Observatory, Preprint.
Zamorani, G. et al.: 1980, Preprint.

DYNAMICAL BEHAVIOUR OF GASEOUS HALO IN A DISK GALAXY

S. Ikeuchi and A. Habe
Department of Physics, Hokkaido University,
Sapporo 060, Japan

Assuming that the gas in the halo of a disk galaxy is supplied from the disk as a hot gas, we have studied its dynamical and thermal behaviour by means of a time dependent, two-dimensional hydrodynamic code.

We suppose the following boundary conditions at the disk. (i) The hot gas with the temperature T_d and the density n_d is uniform at r=4 - 12 kpc in the disk and it is time-independent. (ii) This hot gas rotates with the stellar disk in the same velocity. (iii) This hot gas can escape freely from the disk to the halo. These conditions will be verified if the filling factor of hot gas is so large as f=0.5-0.8, as proposed by McKee and Ostriker (1977).

These models have already been examined by Bregman (1979, 1980). In the present paper, we have studied the gas motion in the halo for wider ranges of gas temperature and its density at the disk than those studied by Bregman (1979, 1980). At the same time, we have clarified the observability of various types of gaseous haloes and discuss the roles of gaseous halo on the evolution of galaxies.

Depending upon the values of T_d and n_d, the resultant gaseous halo can be classified into three types, i. e., (a) a wind type, (b) a bound type and (c) a cooled type. This is shown in the Figure.

In the wind-type halo, the gas expands with the velocity higher than the escape velocity, sometime with the supersonic velocity, and the radiative cooling hardly affects the gas motion. This wind-type halo is realized when $T_d \gtrsim (2-4) \times 10^6$ K and $n_d \lesssim 10^{-2}$ cm^{-3}. The gas temperature in the halo exceeds $\sim 1 \times 10^6$ K, and its density below $z \sim 4$ kpc is larger than 10^{-4} cm^{-3}. The isointensity curves at E_x=0.28-0.53 keV become parallel to the disk plane, and the region $r \lesssim 10$ kpc and $z \lesssim 2$ kpc is brighter than 10^{-7} erg cm^{-2} s^{-1}. The total X-ray luminosity at E_x=0.4-4 keV is 10^{37-39} erg s^{-1}.

The cooled type of gaseous halo is formed if $T_d \lesssim 10^6$ K or $n_d \gtrsim 10^{-2}$

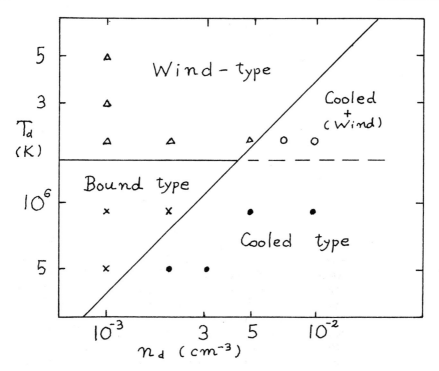

cm^{-3}. Two distinct flow patterns are realized, depending upon T_d and n_d. One is the case when $T_d \lesssim 10^6$ K and the radiative cooling is efficient within $t \lesssim 10^7$ y. Once the gas expands to the halo, it collapses to the disk due to rapid cooling as a whole. Because of centrifugal force, the gas is trasferred to the outer region of the disk and falls with the velocity 30-70 km s^{-1}. On the other hand, when $T_d \gtrsim (2-4) \times 10^6$ K but $n_d \gtrsim 10^{-2}$ cm^{-3}, the gas initially expands extensively and a part of the gas escapes from the galaxy as a wind. However, the radiative cooling is efficient for the disk gas within $r \lesssim 8$ kpc and it falls to the outer region of the disk after rapid expansion. Since the falling gas and the outflowing gas encounter at the region r= 6-12 kpc and z=1-4 kpc, the compressed and then cooled gas extends there. These gas components may be observed as high velocity clouds because the falling velocity would attain $\gtrsim 100$ km s^{-1}.

If these characteristics of dynamical and thermal behaviours are observationally confirmed, the properties of hot gas in the disk would be clarified.

References

Bregman, J. N. : 1979, Ap. J., 229, 514.
Bregman, J. N. : 1980, Ap. J., 236, 577.
McKee, C. F., and Ostriker, J. P. : 1977, Ap. J., 211, 148.

THE FLUCTUATIONS OF THE COSMIC X-RAY BACKGROUND AS A SENSITIVE TOOL TO THE UNIVERSAL SOURCE DISTRIBUTION

P. Giommi[*] and G.F. Bignami
Istituto di Fisica Cosmica - CNR - Milano
[*]Istituto di Fisica Università - Milano

Recent experimental results (Giacconi et al,79, Tananbaum et al 79) ascribe an increasingly important role to the contribution of discrete sources to the low-energy (few Kev) cosmic X-ray background (CXB). While the astrophysical nature of the objects involved is not yet clear, distant and powerful emitters like QSO play probably an important role (e.g. Setti and Woltjer 1979, Field 1980). For them, often the number-flux curve (LogN-LogS) provides useful hints on such properties as space distribution and/or evolution. For the case of the X-ray sources, moreover, a definite relation exists between their LogN-LogS and the granularity of the sky emission as described by the fluctuations of the X-ray background (Cavaliere and Setti, 1976).
This relation refers to the case of a LogN-LogS graph characterized by a unique, continuous slope in the flux region of interest, the value of the slope bearing a precise relation to the distribution and/or evolution of the sources supposed to generate the fluctuations. It is appropriate now to consider the more general case of the impact on the percentage fluctuations of a LogN-LogS graph with one (or more) change(s) in slope, allowing for a greater variety of contributing populations and/or evolution models. Such assumption is suggested by (a) the recent Einstein Observatory data (see e.g. Maccacaro et al, 1979) (b) by the comparison of the Giacconi et al (1979) point at $\sim 3 \times 10^{-3}$ UEFU (UFU 1 - 3 Kev) with the 1 UFU point of Forman et al 1978), especially when considering that the latter point might include unidentified sources of galactic nature; (c) expected, or at least not unreasonable on theoretical grounds, from the recent blue object survey by Braccesi et al, 1980, Bonoli et al,1980 as well as from the well known QSO-QSS distribution data (see e.g. Setti and Zamorani 1978).
Starting then from the results of Cavaliere et al, 1973, and of Cavaliere et al, 1979, assuming the differential LogN-LogS graph to steepen for weak fluxes from the "euclidean" -2.5 value to a slope $-\beta$ ($\beta > 2.5$) the expression relating the fluctuations to the numbers and fluxes of sources will change into:

$$\delta S^2 = \Omega h / S_L^{\beta-3} \left\{ 1/(3-\beta) \left[(S_L/S_1)^{\beta-3} - 1 \right] + 2(S_L/S_1)^{\beta-3} \left[(S_1/S_U)^{1/2} - 1 \right] \right\}$$

where Ω is the field of view considered, h is the differential LogN-LogS normalization constant, S_U is the flux corresponding to the brightest unresolved source, S_L to the faintest source contributing to the CXB flux and S_1 is the flux value when the slope changes from -2.5 to -β

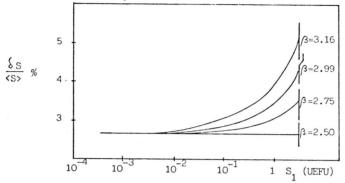

Fig. 1 shows the expected percentage fluctuations as a function of the S_1 flux value (in UEFU), for different β values. Note that for β = 2.5 the expected value (horizontal line) is < 3% for a field of view of 100 sq. deg. It is seen that the change in slope of the LogN - LogS is reflected quantitatively on the percentage fluctuations, and that, at least in principle, an accurate measurement of the latter could determine β. Obviously, fig. 1 is computed for the case of the totality of the 1-3 Kev CXB (1.9 10^{-8}ergs/cm^{2}sec sr.)being due to point sources. Since a significant fraction of it, however, could well be due to truly diffuse processes (see e.g. Field, 1980, Boldt, 1980), it is important to note that the present results still hold true because this is equivalent to truncate the LogN-LogS graph on the side of the very weak source which contribute very little to the fluctuations.

The conclusions of the present work are to some extent academic, if compared to the capabilities of the present generation X-ray detectors of measuring both very faint sources and low-level fluctuations of the CXB. It is hoped, however, that the nearly contemporary advent of very much improved, long-lived X-ray facilities (e.g. AXAF) and of the ST, will make it possible to investigate more completely on the various cosmological populations contributing to the (optical and) CXB.

References:
Boldt, E.A., 1980 NASA T.M. n° 80659
Bonoli, F. et al. 1980 preprint
Braccesi A. et al. 1980, A&A, 85, 80.
Cavaliere, A et al, 1973 Ap; J. 182, 405
Cavaliere A. and Setti G., 1976 A&A 46, 81
Cavaliere A. et al, 1979 A&A, 79, 169
Field, G.B., 1980 Ap. J. in press
Forman, W. et al. 1978 Ap. J. suppl. 38, 357
Giacconi R. et al 1979 Ap. J. 234, L1
Maccacaro T. et al. 1979 B.A.A.S. 11, 771
Setti G. and Zamorani G. 1978 A&A 66, 249
Setti G. and Woltjer L. 1979 A&A 76, L1
Tananbaum H. et al., 1979, Ap. J. 234, L9.

THE GAMMA-RAY SKY

L. Scarsi
Istituto di Fisica dell'Università, Palermo, Italy.

R. Buccheri, G. Gerardi, B. Sacco
Istituto di Fisica Cosmica e Tecnologie Relative CNR-Palermo

1. INTRODUCTION.

When we talk of Gamma-Ray Astronomy we refer in general to celestial photons covering the energy domain from a fraction of MeV up to the highest detected energies at hundreds or thousands of GeV. The six to seven decades altogether spanning the overall energy range constitute a very wide domain to be compared with those of the optical (less than one decade) or of the neighbouring X-Ray Astronomy (about two decades). The observational technique varies widely with changing energy; detector performances and characteristic parameters are not, in general, comparable. All of this has inevitably introduced, for the Gamma-Ray field, an additional subdivision in at least three sub-sectors:

Low and Medium Energy   :   0.1 to 20-30 MeV.
Intermediate and High Energy: 30 MeV to 5-10 GeV.
Very High Energy   :   the region beyond.

We will devote this review essentially to the High Energy sector; the choice stems from the fact that for these energies the observational situation has reached a rather advanced status with the Satellite SAS-2 and especially with COS-B which is still operational as of today. For the other two sectors the situation is moving slowly. The potential interest, specifically on the low energy side, is enormous; it is enough to recall the search for nuclear lines and the fact of looking for a connecting bridge between the typical X and Gamma-Ray Skies which look so different. On the other hand life is hard because of technical difficulties connected mainly with problems of background: HEAO 1, for example, has been rather disappointing with respect to expectation. On the "Very High Energy" side the problems derive mainly from the extremely weak signals to look for.

2. THE 30-5000 MeV ENERGY BAND

The Gamma-Ray Sky is best known in this energy range for which the exploration started some 20 years ago with a sizeable effort in balloon and satellite experiments, following a "call for attention" from theorists going as far back as 1952 (1-3).

2.1. Experiments before COS-B

Balloons, although limited by the interference of the atmospheric background and by reduced exposure time, succeeded in discovering the gamma-ray emission above 50 MeV from the Crab Nebula and its 33msec pulsar, the Vela Pulsar, the Galactic Center and the Cygnus regions, together with the diffuse emission at all galactic latitudes. As an example see the following references (4-8).

The satellite programme has listed a series of major missions with two of them (SAS-2 and COS-B) involving the full spacecraft capacity. We recall here the main chapters and major advancements; at the end the full story, as far as we know today, is that told by COS-B, the last experiment in the row.

Explorer XI, launched in 1961, carried on board a Gamma-ray experiment of the Group of MIT (9) and represented also the first space venture in High Energy Astrophysics; in its lifetime the experiment collected 31 gamma rays, but it failed in producing the evidence of their cosmic origin.

A clear evidence of galactic gamma-ray emission at $E_\gamma > 100$ MeV was first given by OSO-3 in 1968, again with an experiment of the Group of MIT (10); the gamma-ray telescope detected, within its rough angular resolution, a "line source" associated with the galactic disc, with a broad maximum in the inner Galaxy ($300° < \ell^{II} < 60°$), superimposed on a diffuse flux interesting all galactic latitudes. No localized sources were found and this can be explained, as we understand today, as due to the observational limits of the experiment. The interpretation put forward by Kraushaar et al. was of a galactic diffuse gamma-ray emission produced as a consequence ($\pi°$ decay) of the interaction of cosmic ray particles with the diffuse interstellar matter; the high latitude emission was attributed to an extragalactic origin. The model, which fitted the observational data within a factor much better that an order of magnitude (quite satisfactory at the time, taking into account the level of incertitude for the parameters involved), was on the "economy" side, calling in only the most conventional factors and reducing to the minimum the role of any new assumption.

The third milestone has been SAS-2 of the NASA-Goddard Space Flight Center Group. The satellite operated between November 1972 and June 1973 and the experiment collected in its lifetime about 8000 gamma-ray

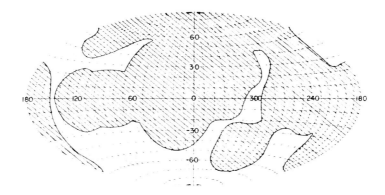

Fig. 1. Region of the celestial sphere viewed by SAS-2 (shaded area in the figure).

events above ~ 35 MeV attributed to cosmic emission (11,12); the exposure covered 60% of the celestial sphere (Fig. 1).
The Goddard Group has squeezed out of the data all possible information and it has published recently the "very final" picture of the Gamma-ray sky "as seen" by SAS-2. Ref. 13 contains in form of tables the details of the analyzed gamma-ray data from the entire SAS-2 data base. Conclusions on the "Galactic plane gamma-radiation" can be found in ref. 14, 15 and 16.
Most if not all of the SAS-2 reported results are overrided by the data of COS-B which can count on much higher statistics collected during its five years operation, a capability of measuring gamma-ray energies up to several GeV, timing accuracy down to a fraction of millisecond in absolute time, systematic and repeated observations of the various regions of the sky. It is worthwhile however to present the SAS-2 results and conclusions with some detail for two main reasons:
a) They represent a coherent picture as given by a single experiment.
b) For some time to go the COS-B data will not be available both because the experiment is still operating and because of the inherent, and by now well known viscosity of the Caravane Collaboration in releasing data (even more in interpreting them).

SAS-2: Localized galactic sources. The SAS-2 experimenters report the observation of four localized sources along the galactic plane revealed by a strong spatial enhancement coherent with the presence of a point-like emitting object (14). The first two are the well known Crab Pulsar PSR0531+21 and the Vela Pulsar PSR 0833-45. The third, γ 195+5 is confirmed by COS-B as CG 195+4; the 59 sec periodicity suggested by SAS-2

(17) and by COS-B in the first observation of the source (18) has not been any more confirmed by COS-B in successive observations. Because of the poor statistical significance of the data when a positive result was claimed, it seems clear that the presence of the periodicity should be considered doubtful and most probably the apparent result of random fluctuations. The fourth spatial enhancement has been identified (19) as due to Cygnus X-3, the identification being supported by the observation of the 4.8h periodicity characteristic at other wavelengths. COS-B does not confirm the existence of the 4.8h periodicity at the position of Cyg X-3 or for the nearest source found, CG 78+1 (20). These discrepancies put serious dubts on the source identification claimed by SAS-2. Also listed are two Radio Pulsars tentatively identified as gamma-ray sources (21): PSR 1818-04 and PSR 1747-46. The first one is not confirmed by COS-B; for the second the value of the period derivative $\dot{p}$ has been recently remeasured (22) and the value found (1.3×10^{-15} ss^{-1} instead of 70×10^{-15} ss^{-1}) would make the gamma flux claimed more than 100 times the rotational energy loss. We should consider therefore the effect observed for the two Radio Pulsars as most probably due to statistical random fluctuations.

In considering the distribution of gamma-rays in galactic longitude (binned in $\Delta \ell^{II} = 2.5°$) for $-10° \leq b^{II} \leq 10°$, Hartman et al. (14) report intensity peaks near the longitudes: 312°, 332°, 342° and 37°, which they consider associated with tangential directions to galactic spiral arm features.

SAS-2: Galactic plane gamma radiation. We report here directly the conclusions given by the SAS-2 Experimenter Group in their final paper(14).
- The large scale distribution of the gamma-ray flux (E $\gtrsim$ 35 MeV) from the Galactic Disc has several similarities to other tracers of galactic structure. The radiation is primarily confined to a thin disc, with off sets from $b^{II} = 0°$ similar to the "hat brim" feature revealed by radio-frequency measurements (+2°± 0.5° for $90° < \ell^{II} < 175°$; -2° ± 0.5° for $205° < \ell^{II} < 250°$).
- The distribution in galactic latitude of the gamma-ray flux at different galactic longitudes suggests a local component for the broad distribution observed in directions away from the inner Galaxy and a predominant contribution from distant features ($\geq$ 3 Kpc) for the narrower distribution observed toward the inner Galaxy.
- Enhancements are seen in the gamma-radiation in the galactic center and regions deduced from 21 cm measurements to be associated with spiral arms.
- Excluding the four strong localized sources, the energy spectrum of the gamma radiation seems to show no significant differences along the galactic plane. The overall spectral index (1.70 ± 0.14) is con-

sistent with the value (1.5 ± 0.3) deduced for the galactic component at high latitudes ($|b^{II}| > 10°$).

By observing that:

the uniformity of the galactic gamma-ray energy spectrum, the smooth decrease in intensity as a function of galactic latitude and the absence of (SAS-2 observed) galactic sources at high galactic latitudes, all argue in favour of a diffuse origin for the bulk of the galactic gamma-radiation rather than a collection of localized sources

and on the assumption that:

cosmic-ray density be not uniform through the Galaxy but the density in the plane be correlated with the matter density on the scale of galactic arms,

the SAS-2 group favours the conclusion that the majority of the observed gamma radiation above 35 MeV coming from the galactic plane seems best explained in terms of diffuse emission with a cosmic-ray interaction origin, the net contribution of localized sources, however, being very uncertain primarily because of the limited angular resolution of the telescope on board SAS-2.

SAS-2: Diffuse gamma-radiation away from the galactic plane. The data from SAS-2 for the region of the sky corresponding to $10° < |b^{II}| < 90°$

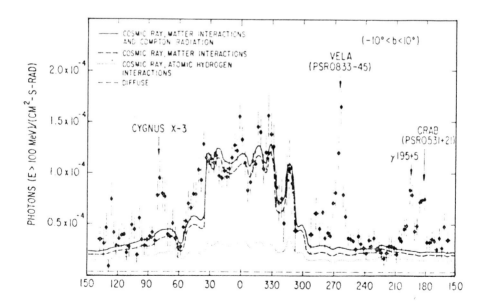

Fig. 2 (ref. 14). Gamma-ray flux (E ≥ 100 MeV) as a function of galactic longitude. The continuous lines show the comparison with the gamma-ray intensity predicted by the model of Kniffen et al.(23).

have confirmed the existence of a diffuse high galactic latitude gamma-radiation first reported by OSO-3. By considering the latitude, longitude and energy dependence of the gamma-ray flux revealed, the Goddard Group reaches the following conclusions (24-26):

- The diffuse gamma-radiation ($|b^{II}| > 10°$; $E \gtrsim 35$ MeV) consists of two components:

a) One considered to be a "galactic" component. It has the intensity strongly dependent on galactic latitude, joining smoothly to the intense regions of the plane; it is well correlated with the atomic hydrogen column density as deduced from the 21 cm measurements and the galactic synchrotron emission. The following are characteristic values:
spectral index (assuming a power law): 1.5 ± 0.3
intensity for galactic latitudes near the pole:
\> 35 MeV : $(1.5 \pm 0.4) \times 10^{-5}$ photons/cm^{2}s sr.
\>100 MeV : $(0.9 \pm 0.2) \times 10^{-5}$ " "
intensity for a typical region with a galactic latitude of about 15°:
\> 35 MeV : $(6.9 \pm 1.7) \times 10^{-5}$ photons/cm^{2}s sr
\>100 MeV : $(4.0 \pm 1.0) \times 10^{-5}$ " "

b) One "isotropic", at least on a coarse scale, with galactic latitude. The following are the characteristic values:
spectral index (assuming a power law) : 2.7 $(+0.4,-0.3)$
intensity:
\> 35 MeV $(5.7 \pm 1.3) \times 10^{-5}$ photons/cm^{2}s sr
\>100 MeV $(1.0 \pm 0.4) \times 10^{-5}$ " "
When extrapolated to 10 MeV with the values quoted above the power law joins smoothly to the "diffuse" isotropic intensity measured at low energies.

- No evidence is found of a cosmic ray halo surrounding the Galaxy in the shape of a sphere or oblate spheroid with galactic dimensions.

<u>SAS-2: High galactic latitude localized sources. Extragalactic sources</u> (27). SAS-2 does not report evidence for the existence of localized gamma-ray sources at high galactic latitudes. The experimenters list a set of upper limits for possible gamma-ray emission from active galaxies including those known to be X-ray emitters. The upper limits regard two energy regions: 35-100 MeV, and greater than 100 MeV. The values are spread over a rather wide range following differences in the diffuse emission level, the instrument exposure and the number of gamma-rays actually observed within the source region. The list includes a sample of Seyfert and N Galaxies, BL Lacertae objects, quasars, sharp emission line galaxies. The 95% confidence level limits for 35 -100 MeV expressed in keV/keV cm^2 s range from 1 to 10 ($\times 10^{-6}$); the cor

responding values for E >100 Mev (expressed in photons/cm^2 s) range from 0.5 to 5 (x 10^{-6}).

2.2. The "COS-B Gamma-Ray sky"

Until the second half of the years 1980's, i.e. until the G.R.O. mission (the NASA Gamma Ray Observatory) will start to produce some data after 1985, the gamma-ray sky in the 70-5000 MeV band will be essentially that of COS-B, presumably with some adjustement for specific targets coming from dedicated balloons, or less probably Spacelab experiments. Important results concerning the discrete sources are also expected from the Franco-Russian satellite Gamma 1.

Before showing the COS-B data it is worthwhile introducing some basic information on the satellite instrumentation, the mission profile, the data analysis procedures and the data publishing policy of the Caravane Collaboration. This with the purpose of an objective as possible presentation of the observational data, trying to explicitate the steps carried out by the experimenters towards the interpretation and presentation of the results.

2.2.1. The COS-B Experiment

The general characteristics are described in ref. 28. The gamma-ray telescope is based on a magnetic-core wire-matrix Spark Chamber triggered by a three element counter telescope. Beneath the telescope is an energy calorimeter consisting of a CsI scintillator which absorbs the secondary particles produced by the incident photons. Fig. 3 shows the effective sensitive area, angular resolution and energy resolution of the COS-B gamma-ray detector for axially incident gamma-rays satisfying the selection criteria applied in the analysis. The angular resolution is given as the FWHM of the spatial angular resolution or point-spread function, which best describes the experiment capability to resolve two neighbouring point sources. It decreases asymptotically to about 2° at high energies. The parameters described degrade with increasing angle of incidence and the sensitive area falls to zero at an angle of about 30°. Most of the results reported by the Caravane Collaboration are derived from measurements within 20° of the axis pointing direction. Timing of gamma ray events is available with a final accuracy better than a fraction of millisecond in either the satellite or the solar barycentre reference frame.

The COS-B experiment was designed, constructed and tested under the responsibility of a collaboration of research laboratories known as the "Caravane Collaboration", whose members are listed in Table 1. The definition of the observation programme and the analysis of the data are also collaborative activities.

Data are analyzed independently in the various Laboratories of the Collaboration, but they are jointly discussed in meetings and working ses-

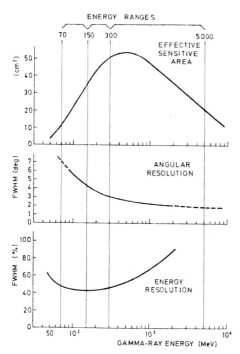

Fig. 3. Effective sensitive area, angular resoltution and energy resolution of the COS-B detector.

sions of a "Data Reduction Group" composed of representatives of each member Laboratory. Results are released for publication only when a unanimous agreement is reached on the data treatement, elaboration and interpretation. This policy, while guaranteeing a maximum of objectiveness, has on the other hand the drawback of slowing down considerably the release of results and of reducing the "Caravane Collaboration" published papers to bare straightforward presentation of observational data with practically no comments and model interpretation reduced to an absolute minimum.

Launched on August 9, 1975 COS-B has operated since then with practically unchanged performance; the instrument present status and the onboard consumable goods (gas for spark chamber flushing and gas for attitude control system) would permit an operational life at least until the end of 1981. At the moment, for budgetary reason, the COS-B operational life is assured only to september 1980. The orbital ele-

Table 1

THE CARAVANE COLLABORATION

Max-Plank-Institut für Physik and Astrophysik. Institut für Extraterrestrische Physik. Garching-bei-München.

Service d'Electronique Physique. Centre d'Etudes Nucleaires de Saclay.

Cosmic-Ray Working Group. Huygens Laboratory. Leiden.

Laboratorio di Fisica Cosmica e Tecnologie Relative CNR. Milano.

Laboratorio di Fisica Cosmica e Tecnologie Relative CNR. Istituto di Fisica Università di Palermo.

Space Science Department of ESA. ESTEC. Noordwijk.

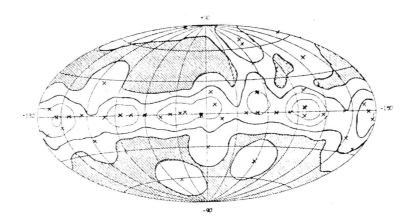

Fig. 4. Relative sky coverage between 17 August 1975 and 20 April 1980. The unshaded area indicates the region of the sky observed within 20° of the pointing direction (galactic coordinates). The contour intervals are chosen to indicate the number of times a given direction has been observed. Crosses indicate the pointing directions.

ments at launch were: height of perigee = 340 Km; height of apogee = = 99000 Km; inclination = 90.2 degree; drifted now (as for May 1980) respectively to the values: 13.000 Km; 86.000 Km; 98°. The eccentric orbit was chosen for scientific reasons (satellite for most of the time outside the radiation belts) and for technical reasons (for most of the operational part of the orbit satellite in sight of an ESA ground station).

The mission profile has been based on pointing to sky targets for periods of four to five weeks. Presently is "on" the observation period Nb.52 with the Cygnus region ($\ell^{II}=80°$; $b^{II}=0°$) as a target.

Fig. 4 shows the relative sky coverage between launch and April 1980. Two thirds of the observation time have been devoted to a systematic scanning of the region of the galactic disc, with repeated or overlapping observations (pointing direction $|b|<10°$). The remaining third has been addressed to high galactic latitudes, but with no claim to a systematic coverage. The presence of a non-negligible gamma-ray instrumental background induced by interaction of the Cosmic Radiation (non modulated by the geomagnetic field) precludes the evaluation of a possible diffuse isotropic component. This background has been reduced significantly with the progressing of the solar cycle.

In the following paragraphs the main results obtained by COS-B will be published. In doing that, we will specify, case by case, the "status" of the result reported: if fully and finally agreed by the

COS-B Data Reduction Group, it will be reported on behalf of the Caravane Collaboration; if not yet at this stage, and therefore susceptible of possibile modifications, the announcement engages only the responsability of the authors of this paper.

2.2.2. The Large Scale Galactic Gamma-ray Emission

The available data from COS-B constitute a complete high energy ($\gtrsim 70$ MeV) gamma-ray survey of the Galactic disc. A first release of the Sky-map of the Milky Way has been done by the Caravane Collaboration in December 1978 at the 9th Texas Symposium on Relativistic Astrophysics (29); it referred to the satellite operation until February 1978 and covered the whole range of galactic longitudes. The data base was constituted by 64.000 gamma-ray events collected during ~ 700 days corresponding to the 20 low $|b^{II}|$ out of 28 observation periods, lasting ~ 1 month each. An updated version covering the operation period up to the end of 1979 and enlarging the data base to about 100.000 events is presently under elaboration and it will be completed within the next few months. Details of the analytical procedure can be found in (29). The spark-chamber data are first analysed by an automatic pattern-recognition programme to select likely gamma-ray events and subsequently visually scanned in order to remove background and improve the direction determination. The events are then grouped in three energy bands (70-150 MeV; 150-300 MeV and 300-5000 MeV; the band boundaries are chosen to provide comparable statistical accuracy for each group of events. Following the reconstructed direction of arrival, the events are assigned to $0.5° \times 0.5°$ bins; the resulting map is treated with a smoothing procedure to obtain the best representation compatible with the angular resolution of the gamma-ray telescope. The smoothing procedure chosen should not suppress the sharpest physically possible peaks, without on the other hand reproducing peaks significantly narrower than the point-spread function and therefore evidently produced by statistical fluctuations.

Fig. 5 refers to the results obtained in the first analysis run (August 1975-February 1978) already published (29); the figure shows the longitude profiles in the three energy ranges. As it can be seen, more fine structures are visible going from low energies to high energies in relation to the varying angular resolution of the telescope. Latitude profiles have also been derived by these data and have been used to determine the intrinsic thickness of the emissivity taking into account the instrument point-spread function. Fig. 6 shows an example together with the unfolded angular thickness of the emitting region along the galactic disc.

In Fig. 5 and Fig. 7 the intensity is given in "on axis" counts/s sr,

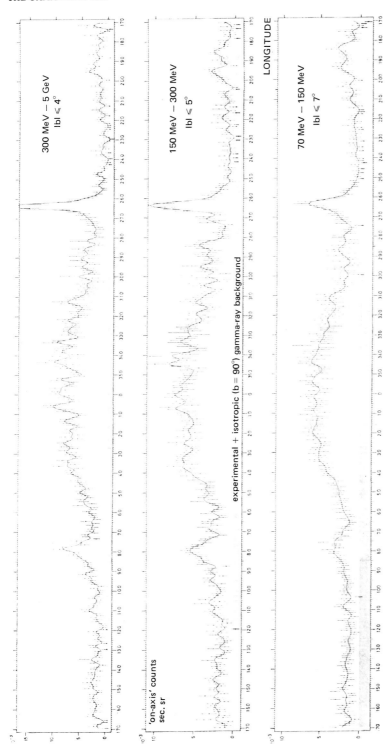

Fig.5. Longitude profiles of "on-axis" counts in three energy ranges. The full line indicates the fitted surface. The background is indicated by the shaded area.

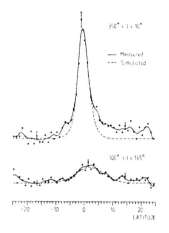

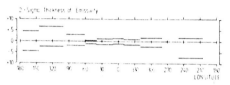

Fig. 6. Examples of latitude distributions. Lower figure gives the thickness of the gaussian distributions describing the unfolded intrinsic emissivity at varying galactic longitudes.

which represent the number of gamma-ray events which would have been recorded in the bin chosen if the experiment had been pointing there observing it "on axis". This presentation of the spatial distribution of the photon arrival directions has the advantage over the "photon flux" maps of not being dominated by the low energy component which is measured with least accuracy in angular resolution.

The definition of "photon flux" F_k from a given sky bin k and the corresponding "on axis counts" are given in Table 2.

Fig. 7 shows a map of the galactic disc region, always seen by COS-B, but derived on a data base 1.5 time larger (corresponding to a comparatively longer observation time) than that used in ref. 29.

This sky-map (courtesy of Dr. H. Mayer-Hasselwander) is still preliminary, but it confirms substantially the picture given in ref. 29, adding some more detail.

Table 2

Definition of "photon flux" and of "on-axis counts" from a sky bin k of angular size ω_k from which N_k gamma-ray events were recorded. E_i is the energy of the event; θ_i is the incidence angle relative to the axis of the experiment, T is the observation time; $\varepsilon(E)$ is the experiment's effective area for axial incidence; $\eta(\theta_i, E_i)$ describes the decrease of effective sensitive area with incidence angle.

$$F_k = \frac{1}{T\omega_k} \sum_{i=1,N_k} \frac{1}{\varepsilon(E_i)\,\eta(\theta_i, E_i)} \quad \text{photons cm}^{-2}\,\text{s}^{-1}\,\text{sr}^{-1}$$

$$G_k = \frac{1}{T\omega_k} \sum_{i=1,N_k} \frac{1}{\eta(\theta_i, E_i)} \quad \text{counts s}^{-1}\,\text{sr}^{-1}$$

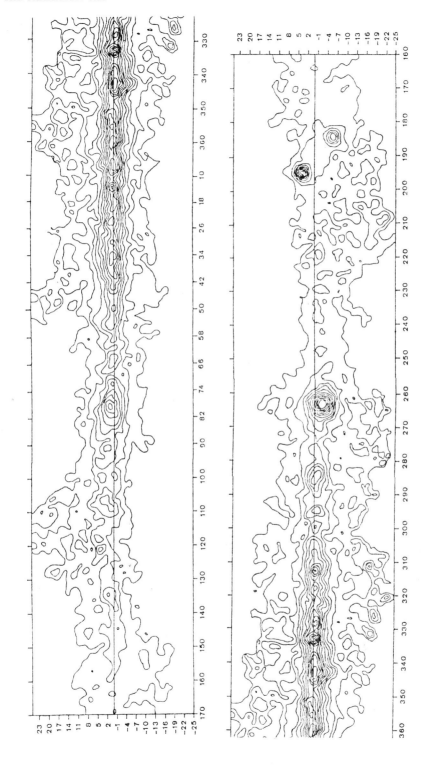

Fig. 7. Structure of the Galactic gamma-ray emission as measured by COS-B. Sky map of the Galactic Plane in the Energy Range 70 – 5000 MeV. The isointensity contour intervals represent "on--axis counts" s^{-1} sr^{-1} and are separated by 4×10^{-3}.

The observational data presented in the sky-map and the longitude, latitude profiles shown above give a projected picture of the Galaxy with no direct information of distances for the emitting regions (apart from those independently existing for the identified discrete sources).
A few basic points can be made:

a) The sky map taken along the galactic disc, even at a first rough inspection, shows the existence of several enhancements (localised excess counts) corresponding to the expected appearance for a point source; Vela and Crab, between others, are clear examples. By a systematic investigation with a cross-correlation procedure of the data with the point-spread function of the instrument (see paragraph 2.2.3.) numerous point-like sources have been evidenced. SAS-2 with an angular resolution comparable, if not better, to COS-B failed to recognize the importance of discrete sources most probably because of the limited statistics of the data base available for ana lysis. The discrete sources, following COS-B, can easily account for up to 50% of the total galactic emission; the pronounced granularity remaining in the maps after subtraction of the resolved sources suggests that possibly the 50% should be considered as a lower limit. The capability of detecting localized sources is conditioned by the limited angular resolution of the experiment against the background of the unresolved and diffuse galactic emission.

b) Strong and Worral (30) from the SAS-2 data, assuming a cylindrical symmetry, have estimated the absolute galactic luminosity above about 50 MeV: $L_\gamma \sim 5 \times 10^{38}$ erg/s. This estimate is coherent with the COS-B data.

c) Taking latitude profiles as a function of longitude at high energies (>300 MeV) for which the angular resolution is better, the radiation appears to be confined to a thin disc. The peak of the excess has a small but significant latitude displacement from $b^{II}=0°$ corresponding to the "hat-brim" effect seen in the 21 cm line. The existence of two components in the disc emission first revealed by SAS-2 appears to be clearly established. A broad component appears at all galactic longitudes, with superimposed a narrow one for directions corresponding to the inner Galaxy ($55° < \ell^{II} < 285°$). The two components can be related to features at different distances in the Galaxy. By assuming a constant thickness of ~ 200 pc for the gamma-ray disc, the broader component can correlate well with local regions at $\lesssim 1$ Kpc, while the narrow component appears to have its origin at a 3 to 6 Kpc distance and shows a strong variation with galactic longitude.

d) The wide component for the disc emission shows a broad excess at positive latitudes in the central region of the Galaxy and at negative latitudes in the anticentre. Fichtel et al. (11) noticing a similar behaviour in a qualitative analysis of the SAS-2 data suggested this feature to be an indication of gamma-ray emission from the Gould Belt of dust, gas and young stars at distances varying up to 450 pc from the Sun. The gamma ray emission would derive from cosmic ray (protons and electrons) interactions with the interstellar clouds in the Belt (31-33). The discrete gamma-ray source of the COS-B catalogue CG353+16 is positionally compatible with the ρ-Oph dark cloud complex and its emission could be attributed to cosmic ray interactions in the cloud (34). COS-B also reports the observation of high-energy gamma-ray emission from the Orion cloud complex (37). In general, without making a detailed correlation with individual clouds and assuming a mass distribution for the Gould Belt as given in ref. 35 and 36, with a production rate of gamma-rays of 2-3 x x 10^{-25} photons ($\geq$ 100 MeV) sec^{-1}, it is possible to justify approximatively the gamma-ray flux observed.

e) In considering the narrow galactic component, its longitude distribution can be well correlated with many different galactic tracers and it has been interpreted by several authors as implying the presence of an enhanced flux of cosmic rays in the inner Galaxy. On the other hand the presence of an important contribution from discrete sources with an unknown integrated effect of the unresolved ones, together with the arbitrary way in which several assumption can be made, makes the picture rather confuse.

2.2.3. Localised Gamma-ray Sources

One of the outstanding results of the COS-B mission is the compilation of a catalogue of discrete gamma-ray celestial sources. This has produced a significant shift from the previous "way of thinking" of the galactic emission as being substantially due only to diffuse type processes. A first Catalogue of 13 objects in a survey covering 50% of the galactic plane has been published in 1977 (38), showing their predominantly galactic nature. A new list of 29 has been successively released by the Caravane Collaboration at the Erice Europhysics Conference on Gamma-ray Astronomy (1979) and at the COSPAR Meeting at Bangalore (40). The second COS-B Catalogue (2CG), which will be published shortly by the Caravane Collaboration, has been included by W. Hermsen in his Ph.D. Thesis at the University of Leiden (41). Although evidently not yet representative of the final results of the COS-B mission (a more advanced edition is currently under preparation), the 2CG Catalogue represents most of the data acquired during the first three years of COS-B operation and it is the result of a systematic analysis covering the entire galactic plane. The data used were obtained in 32 separate observation periods corresponding each to about one moth poin-

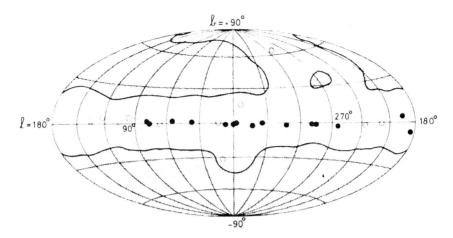

Fig. 8. Region of the sky searched for gamma-ray sources (unshaded) and sources detected above 100 MeV by spatial analysis. The closed circles denote sources with measured flux $\geq 1.3 \times 10^{-6}$ photons/ /cm^2 s. Open circles denote sources below this threshold.

ting to a given target (Fig. 8).
For each observation a useful field of view inside 20° from the pointing axis was retained. For overlapping regions the data were merged; in these cases information on possible time variability was lost.
A gamma-ray source is defined as a significant excess above the local background which has a spatial distribution consistent with the instrument intrinsic point-spread function. This function corresponds to the distribution of the photon arrival directions expected from a point source in the sky as determined by preflight calibration of the telescope at accelerators and confirmed by the actual flight data for the strong source PSR 0833-45. It must be stressed that a truly point-like object and a feature extending to about 1° are not distinguished; for this reason the meaning of a "Gamma-ray discrete source" is in principle different from the current meaning associated to a source in an X--ray catalogue.

Explication of Notes to the 2CG Catalogue
1. Expressed in Units of Standard deviations of local background (included contribution from the excess counts).
2. Expressed in MeV.
3. In degrees.
4. In units of 10^{-6} (photons/cm^2·s), for $E \geq 100$ MeV.
5. Intensity ($E \geq 300$ MeV)/intensity ($E \geq 100$ MeV); both intensities are calculated assuming an Energy spectrum E^{-2}.

The 2CG Catalogue of Gamma-Ray sources (ref. 41)

(For note expl. see foll. page)

Source name	Observation periods.	(1) Signif- icance	(2) Energy thresh.	(3) Position ℓ b	(3) Error radius.	(4) Flux	(5) Spectral parameters	Identification. Comments
2CG006-00	2,18,25	10.2	300	6.7 -0.5	1.0	2.4	0.39±0.08	
2CG010-31	30	5.7	100	10.5 -31.5	1.5	1.2	-	
2CG013+00	2,8,18,25	5.3	300	13.7 0.6	1.0	1.0	0.68±0.14	
2CG036+01	9,25,26	4.9	300	36.5 1.5	1.0	1.9	0.27±0.07	
2CG054+01	9,25,26	5.3	100	54.2 1.7	1.0	1.3	0.20±0.09	
2CG065+00	4,9,22,26	5.5	100	65.7 0.0	0.8	1.2	0.24±0.09	
2CG075+00	4,16,22,26,36	5.8	100	75.0 0.0	1.0	1.3	-	could be an extended feature
2CG078+01	4,16,22,26,36	11.9	100	78.0 1.5	1.0	2.5	-	
2CG095+04	4,16,22	4.9	150	95.5 4.2	1.5	1.1	-	
2CG121+04	11,16,28	4.9	100	121.0 4.0	1.0	1.0	0.43±0.12	
2CG135+01	11,16,28	4.9	100	135.0 1.5	1.0	1.0	0.31±0.10	GT0236+610?
2CG184-05	1,14,17,29	20.6	100	184.5 -5.8	0.4	3.7	0.18±0.04	PSR0531+21
2CG195+04	1,14,29	27.1	100	195.1 4.5	0.4	4.8	0.33±0.04	
2CG218-00	14,18,21	6.2	100	218.5 -0.5	1.3	1.0	0.20±0.08	
2CG235-01	19,21	5.0	150	235.5 -1.0	1.5	1.0	-	
2CG263-02	3,5,12,21	35.7	100	263.6 -2.5	0.3	13.2	0.36±0.02	PSR0833-45
2CG284-00	5	6.5	100	284.3 -0.5	1.0	2.7	-	could be an extended feature
2CG288-00	5	4.8	100	288.3 -0.7	1.3	1.6	-	
2CG289+64	10,32	6.5	100	289.3 64.6	0.8	0.6	0.15±0.07	3C 273
2CG311-01	5,7	5.6	150	311.5 -1.3	1.0	2.1	-	
2CG333+01	7,13,24	5.4	300	333.5 1.0	1.0	3.8	-	
2CG342-02	2,7,13,18,24	8.9	300	342.9 -2.5	1.0	2.0	0.36±0.09	
2CG353+16	2,13,18,24	5.1	100	353.3 16.0	1.5	1.1	0.24±0.09	ρ-Oph dark cloud?
2CG356+00	13	5.3	300	356.5 0.3	1.0	2.6	0.46±0.12	may be variable
2CG359-00	2,18,24	6.3	300	359.5 -0.7	2.0	1.8	-	

The 2CG Catalogue

- As for the first source catalogue, the source name is expressed by the symbol CG followed by the truncated value in degrees of the galactic longitude and of the galactic latitude.
- Because of the complex structure of the galactic gamma-ray emission and its variation with ℓ and b, it is difficult to maintain a sufficiently uniform source visibility throughout the Galaxy. To improve uniformity only events with energy greater than 100 MeV have been accepted, their direction of arrival being sorted in $0.5° \times 0.5°$ bins; moreover the exposure in regions of high sky background (e.g. Galactic Center region) has been increased.
- Statistical significance of the associated count excess at the level of at least 4.75 σ has been required; this procedure together with the fact that sources which have been at least twice in the field of view have been seen in each observation, assures that the number of spurious detections is insignificant. The one exception of non repeated observation (2CG356+00) is a possible example of variability.
- Some of the sources listed in the first CG or in lists previously published are not present in the 2CG Catalogue; the reason can be either because of the more extended data set the source has been better analysed and it has revealed itself as an extended feature, or because it has too soft an energy spectrum to show up above threshold at 100 MeV.
- The direction ℓ and b listed corresponds to the peak direction of the correlated spread function. The error radius (90% confidence level) has been estimated from simulations.
- To derive an integral flux above 100 MeV, a power law E^{-2} has been adopted for the differential energy spectral shape; this is certainly not correct in most cases, but sufficient for a 30% approximation.
- The spectral parameter indicated in the Catalogue list gives a rough indication of the hardness of the spectrum above 100 MeV; the spread in the value for the different sources indicates significant differences between the spectra.
- The 2CG Catalogue, although representative of the general galactic picture, is certainly not complete; this fact has to be taken in mind in making extrapolations and in deriving conclusions.

3. GAMMA-RAY SOURCE IDENTIFICATION

I - The Crab and Vela Pulsars

Sure identification of CG sources is established only for 2CG263-02 and 2CG185-04 corresponding to PRS0833-45 (the Vela Pulsar) and to PRS0531+21 (the Crab Pulsar) respectively; the timing signature gives a proof of the identification beyond any doubt. The two sources represent the most intense and the third most intense source in the gamma-

-ray sky. They are extensively described in the literature; for COS-B data see ref. 42-44.

Table 3 shows the relevant parameters for the two sources:

Table 3

CG Source	PSR	Distance	Age	$L_\gamma^{(+)}$ erg/s	$\dot{E}$ erg/s	$L_\gamma/\dot{E}$
2CG263-02	PSR0833-45	500 pc	10^4 yr	4.0×10^{34}	7.0×10^{36}	5.7×10^{-3}
2CG185-04	PSR0531+21	2000 pc	10^3 yr	1.9×10^{35}	7.8×10^{38}	2.5×10^{-4}

(+) A beaming factor 0.184 is included. $L_\gamma = 0.184 \cdot \text{Flux} \cdot E_\gamma \cdot 4\pi d^2$.

II - The Quasar 3C273

The location at high galactic latitude, in a region of low and rather uniform background, makes possible the identification of the source 2CG289+64 with the Quasar 3C273 on the basis of a positional coinciden ce. The source has been seen in two observation periods (39-45) and the probability of a chance coincidence of the gamma-ray source with 3C273 is estimated at 10^{-3} (45) (Fig. 9). In Fig. 10 the energy spectrum obtained by COS-B is shown together with the data obtained by HEAO-1/A2 and A4.

III - The Source 2CG135+01

Fig. 11 shows the error cicle obtained by COS-B with the cross-correla-

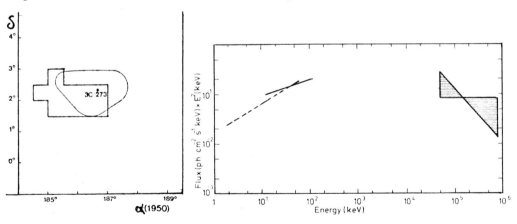

Fig. 9. COS-B observation of 2CG289+64. Thin line: 90% con fidence level contour with the likelihood method. Thick line: 6 contour with cross corr. method.

Fig.10. High energy spectrum of 3C273; dotted line: HEAO1/A2; thick line: HEAO1/A4; shaded area: COS-B.

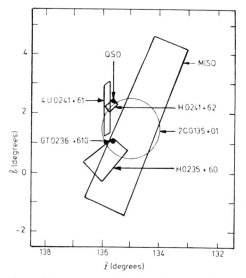
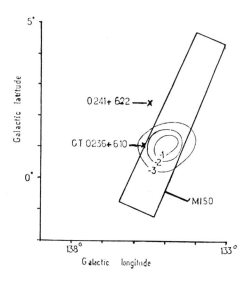

Fig. 11. Error circle for 2CG135+
+01 obtained with the crosscorre-
lation method in the COS-B data.
Indicated are also the position
and error boxes of objects identi
fied at other w.l.

Fig. 12. Contours of constant likelihood for the location of a point
source of 150 to 5000 MeV in the
region of QS00241+622 and the radiostar GT0236+610.

tion method for the source 2CG135+01, together with the positions of
the QSO 0241+622 and of the variable radio star GT0236+610. Fig. 12
shows the same data with the indications deriving from the likelihood
method which tests the hypothesis that the presence of the one or the
other source explain the COS-B observation best (46). This procedure
gives evidence for the identification of the gamma-ray source with the
radiostar rather than the QSO. The optical counterpart of GT0236+610
has been reported as an 11th magnitude OB$^+$ star (LSI+61°303) at 2.3Kpc
distance (47,48). A probable association with the source is represented
by the low energy gamma-ray source reported by the MiSo Group (49) at
energy above 120 KeV; recently (41) the Eistein Observatory has identified an X-ray source in the 1-3 KeV energy range, compatible with the
position of LSI+61°303.

IV - 2CG Source-Cloud Complex associations

Association between a gamma-ray source and a cloud complex, always because of positional coincidence, has been suggested for some elements
of the Catalogue of COS-B.

a) <u>2CG353+16 and the ρ-Oph dark cloud complex</u>
The source lies in a region of comparatively low confusion due to back-

ground. The ρ-Oph dark cloud complex, which is at a distance of $\sim 160 pc$, was suggested to be detectable as a gamma-ray source above 50 MeV several years ago by Black and Fazio (50), the gamma ray flux deriving from the interaction of the cosmic radiation with the matter of the cloud.
The cloud is expected to be a "passive" actor, with no indipendent compact gamma-ray emitter. To account for the gamma-emission detected by COS-B it is however necessary to speculate on the possibility of a cosmic ray flux locally higher than that at the Sun, or otherwise to some other mechanism at work to supply the missing intensity (51); on the other hand everything appears normal to other authors (52).

b) The Carina Region

W. Hermsen (41) reports the coincidence in position between 2CG284-00 and the small cluster Wd1 and of 2CG288-00 with NGC3372. Montmerle et al. (53) also suggest that the source at $\ell=288°$ and $b=0°$ should be identified with the Carina complex; in this case the source should be at ~ 2.7 Kpc and its gamma-ray luminosity, above 100 MeV, $\sim 2.10^{35}$ erg s^{-1}. Because of the nature of the region considered, it is difficult to identify straigthforward and simple mechanisms for the gamma-ray source. It is interesting the model proposed in (53) involving the interaction of locally injected cosmic rays with the matter of the molecular cloud.

c) The Orion Cloud complex

To complete the list of gamma-ray localized emission possibly associated with molecular cloud complexes, we report the finding by COS-B of strong emission above 100 MeV from two centroids roughly located at $\ell==209°$, $b=-20°$ and $\ell=206°$, $b=-16°$, coinciding with the dark clouds L1630 and L1641 (54). The observation represents the first extended, and yet localized "source" of gamma-rays, coinciding with a known astronomical object. For its peculiarity of not being compatible with the COS-B point spread function corresponding to a point-like source, the Orion gamma source will not be listed in the COS-B Catalogue.

3.1. The localized sources as a "class" of objects

What preceeds, shows that only a small fraction of the gamma-ray sources of the Cos-B Catalogue can be associated with an astronomical object known at other wavelengths. In reality this is possible with certainty with only the two Pulsars (Crab and Vela), while for the other few cases the identification, if real, is not unambiguously sustained by the positional coincidence.
Searches for counterparts have been attempted far various classes of astronomical entities, but up to now with little success, mainly because of the large error box associated with the CG sources (typically 3 to 4 square degrees, with few examples down to half a square degree). In these conditions, if no timing signature is present, only a "class"

association on a purely statistical basis is possible, with the caution that in any case the significance of the result is essentially speculative. We recall here, as examples, the search for optical counterparts by Van den Bergh (55), who reached the conclusion that statistically a relevant fraction of the CG sources is associated with young SNR or with compact objects related to these remnants. Montmerle (56) identifies a possible association, for about half of the CG sources, with "SNOBs" (=Supernova + OB association). A search for very high energy (100 - 1000 GeV) gamma-ray emission from a sample of the 100 MeV sources has been made by Helmken and Weekes (57), with only upper limits as result.

The Caravane Collaboration itself is presently carrying on a programme for the systematic search in the error box of selected sources, of counterparts in the X-ray KeV region with the Satellite Einstein and for the presence of not yet discovered Radio Pulsars in a "very deep" survey covering the period interval from about 10 msec to the order of seconds; this part of the programme is made in collaboration with R.N. Manchester at the Parks and Tidbinbilla Radiotelescopes and with R. Isaacman and D. Ferguson at Arecibo.

3.2. "2CG Source" average characteristics

A qualitative inspection of Fig. 8 shows immediately the galactic nature of the great majority of the sources of the 2CG catalogue. In fact, leaving out the three high latitude entries (2CG289+64, supposedly extragalactic and associated with the QSO 3C273, 2CG353+16 with the proposed association to the nearby ρ-Oph dark cloud and the still unidentified 2CG010-31), the remainig 22 objects appear rather uniformly aligned along the galactic equator, region which has been extensively and uniformly covered by COS-B for the data concerning the construction of the catalogue.

The (b,ℓ) distribution of the sources shows the well known "hat brim" effect typical of the relevant galactic tracers (Table 4).

We can divide the galactic longitude in two regions: a first one corresponding essentially to the outer galaxy ($60° < \ell < 300°$) and the other

Table 4

ℓ	180°-90°	90°-30°	30°-330°	330°-270°	270°-180°
b	+3°.2	+0°.9	-0°.3	-0°.8	-1°.1
No. of sources	3	5	6	3	5

corresponding to the inner galaxy, inside the solar circle ($60° > \ell > 300°$).

3.2.1. The outer Galaxy

13 of the low latitude 2CG sources belong to this region. The LogN-LogS distribution (Fig. 13) is in reasonable agreement with the assumption of a thin disk uniformly filled with sources ($N \sim S^{-1}$). The absence of a flattening down to the limit of detectability at $\sim 10^{-6}$ ph/cm^2 s (above 100 MeV) favors the assumption of an unbiased collection of data down to the visibility limit: this is reasonable taking into account the comparable level of the continuum background all over this region.
By considering the source spread around the galactic equatorial plane we obtain:

$|b| = 2.1°$ around $b=0°$

$|b| = 1.4°$ around the slanting plane defined by the "hat brim" figure.

For a population with a scale height of 50 pc (we take this value as a reference not as an absolute indication) we would have for the outer galaxy sources a characteristic distance of $d \sim 1.4$ Kpc with reference to the equatorial plane $b=0°$, turning to $d \sim 2$ Kpc if the "hat brim" figure is taken as reference. Considering the average flux of $2.8 \, 10^{-6}$ ph/cm^2 s above 100 MeV and an average photon energy $\overline{E} = 250$ MeV, we derive a characteristic luminosity $L_\gamma = 2.5 \, 10^{35}$ ergs/s (ref. $b=0°$) or $L_\gamma = 5.3 \, 10^{35}$ ergs/s (ref. "hat brim").

Table 5 shows a comparison between the characteristic values deduced and those of the sources identified in the sample (2CG184-05 and 2CG263-02) or for which an identification is suggested (2CG288-00). No beaming factor is introduced.

Table 5

	characteristic values $b=0°$	"hat brim"	2CG184-05	2CG263-02	2CG288-00
d(pc)	1400	2000	2000	500	2700
L (ergs/s) ($E_\gamma > 100$ MeV)	2.5×10^{35}	5.3×10^{35}	1×10^{36}	2×10^{35}	2×10^{35}

3.2.2. The inner Galaxy

The region contains 9 sources of the 2CG catalogue. The average $|b|$ for this sample is $1°$ with no significant difference between the reference to $b=0°$ or to the "hat brim" figure of the galactic equatorial plane.
The characteristic distance is ~ 3 Kpc, always on the assumption of a scale height of 50 pc above the plane. The characteristic luminosity

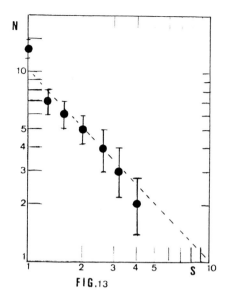

FIG.13

LogN-LogS distribution for the 2CG sources in the outer galaxy ($60° < \ell < 300°$). S is in units of 10^{-6} ph/cm^2 s.

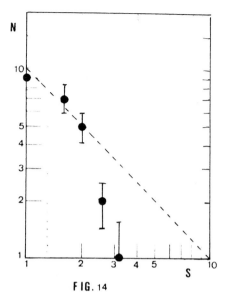

FIG. 14

LogN-LogS distribution for the 2CG sources in the inner galaxy ($60° > \ell > 300°$). S is in units of 10^{-6} ph/cm^2 s.

deduced from the average flux turns out to be $L_\gamma = 9.10^{35}$ ergs/s. This value seems to be higher than that deduced for the outer galaxy; on the other hand the higher background level for the central region has apparently raised the minimum detectable flux value for a discrete source to $1.3 \; 10^{-6}$ ph/cm^2 s (for $E_\gamma > 100$ MeV) instead of $1 \; 10^{-6}$ ph/cm^2 s for which, in the outer region, the sample appears to be unbiased. When this is taken into account and the comparison is made for similar conditions (namely sources with flux higher than $1.3 \; 10^{-6}$ ph/cm^2 s), an identical value for the luminosity is obtained in the case of dispersion around the "hat brim" figure.

Considering the distribution in ℓ close to the galactic centre, within about 30°, Hermsen (41) derives a maximum distance of ~ 7 Kpc for the individual entries of the 2CG catalogue for the inner galaxy.

Fig. 14 shows the LogN-LogS graph for the inner galaxy sources; the distribution seems difficult to be reconciled with a uniform density in the disk and best fitted by a law $S^{-\alpha}$ with α much higher than 1. On the other hand we must take into account the very limited statistics we are relying upon; in fact it can be seen that all the discrepancy from the S^{-1} law would disappear should one or two additional sources be present with $S > 5 \; 10^{-6}$ ph/cm^2 s.

For considerations on the LogN-LogS graph for the COS-B galactic gamma-ray sources see also (58).

3.3. Nature of the galactic gamma-ray sources: discussion

The population of gamma-ray sources in the inner and outer galaxy appears to be the same with a luminosity $L_\gamma = 10^{35} - 10^{36}$ erg/s (for $E_\gamma > 100$ Mev). The dynamic range of COS-B for the visibility (somewhat less than a decade around the Crab source value) limits the maximum distance to less than 7 Kpc towards the galactic centre while presumably all of the outer galaxy is reachable.
The scale height is compatible with 50 pc, which is typical of various galactic populations.
Table 6 lists the average characteristics of the 2CG sources in a similar way as presented by Hermsen in his thesis (41).

Table 6

angular size	$(1° - 2°)$
intensity (>100 MeV)	$(1 - 5) \, 10^{-6}$ ph/cm^2 s
energy flux (>100 MeV)	$(0.4 - 2) \, 10^{-12}$ W/m^2
energy spectrum	diverse; average intensity ratio consistent with E^{-2} spectrum
time variability	not excluded
characteristic distance	2 Kpc (outer Galaxy)
	3 Kpc (inner Galaxy)
X-ray luminosity (+)	$< 0.1 \; L_\gamma$
Radio luminosity (+)	$<< L_\gamma$

(+) By integrating over the conventional energy interval

A hint on the nature of the 2CG sources is given by the few identified cases:
a) Compact objects, e.g. pulsars, as suggested by the identification of PSR 0531+21 and PSR 0833-45. Although no other association of the sources with radio pulsars has been found up to now, the attribution of an important role to this class of objects is reasonable: the energetics involved and the scale height for young pulsars is coherent; the available radio pulsar surveys cannot be considered sufficient for an exhaustive comparison and it is necessary to wait for the results of the specific search which is being carried out at Arecibo and Parkes in the 2CG error boxes.
b) Molecular cloud complexes are reasonable candidates. Few interesting possible associations have been suggested.
c) A variety of degenerate objects going from neutron stars to black

holes can be called in. To these models the only limit is the lack of fantasy.

3.4. Contribution of discrete sources to the total galactic emission

With the increased angular resolution and the larger data base going from OSO-3 to SAS-2 to COS-B, the picture of the gamma-ray sky has acquired numerous details. The total luminosity has been suggested to be $5 \cdot 10^{38}$ ergs/s.

The original OSO-3 picture, dominated by a diffuse smooth emission region along the galactic plane, has acquired some structure with SAS-2, while COS-B has definitely revealed a highly structured panorama. The localized sources, absent for OSO-3 and a minority component for SAS-2, are now dominating the picture given by COS-B, no matter what they represent, compact objects or features with angular dimension of 1° or so. A quantitative estimate of their percentual contribution to the total galactic emission has been attempted by several authors: Bignami et al. (59), Salvati et al. (60) and Hermsen (41) give a lower limit of 40-50% but the contribution can be as high as to account for almost all the galactic emission. A somewhat lower value, around 30%, is quoted by Wolfendale at this Symposium.

The remaining galactic flux (besides that due to the localized sources) can be accounted for by the interaction of the cosmic radiation with the diffuse interstellar matter or the matter present in large concentrated structures ($\phi_\gamma = 2\text{-}3 \cdot 10^{-25}$ ph/s H_{atom}, for $E > 100$ MeV) and by low luminosity sources ($10^{33} - 10^{34}$ erg /s), unable to show up individually but adding together to simulate a diffuse emission. Salvati et al. (60) suggest for this component the radio pulsars with age greater than 10^5 years accounting for $\sim 25\%$ of the total galactic emission above 100 MeV, the remaining 75% being attributed for 25% to the truly diffuse processes involving the cosmic rays and the I.M. and for 50% to the discrete high luminosity sources of the category listed in the 2CG catalogue.

3.5. The extragalactic sky

The data analyzed up to now by the COS-B experimenters refer to the region shown in Fig. 15 (61). With the exception of the identification of 2CG289+64 with the QSO 3C273, no positive observation of extra-galactic localized sources has been made. Pollock et al. (61) list upper limit values for the emission, above 50 and 150 Mev, from a series of peculiar objects including Seyfert galaxies with an addition of galaxies identified through their X-ray emission, quasars arbitrarily selected with visual magnitude $m_v < 16.5$, BL Lac objects, assorted active galaxies including N galaxies and other emission line galaxies that are known as X-ray sources and some galaxies in the

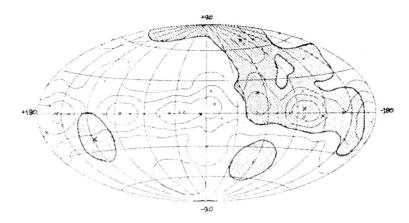

Fig. 15. Region of the sky used for the extragalactic search.

Local Group. The upper limits for the flux above 50 MeV range from 1 to 5 times 10^{-6} ph/cm^2 s confirming the conclusions obtained by SAS-2 (27) that the energy spectrum measured at X-ray energies does not extrapolate directly to gamma-rays but bends substantially much earlier.

REFERENCES

1. Hayakawa, S., 1952, Progr. Theor. Phys., 8, 571.
2. Hutchinson, G.W., 1952, Phil. Mag., 43, 847.
3. Morrison, P., 1958, Il Nuovo Cimento, 7, 858.
4. Frye Jr., G.M., et al., 1971, Nature, 231, 372.
5. Leray, J.P., et al., 1972, Astron. Astrophys., 216, 443.
6. Browning, R., Ramsden, D., Wright, P.J., 1972, Nature Phys. Science, 235, 128.
7. Parlier, B., et al., 1973, Nature Phys. Science, 242, 117.
8. Dahlbacka, G.H., Freier, P.S., Waddington, C.J., 1973, Ap. J., 180, 371.
9. Kraushaar, W.L., et al., 1965, Ap. J., 141, 841.
10. Kraushaar, W.L., et al., 1972, Ap. J., 177, 341.
11. Fichtel, C.E., et al., 1975, Ap. J., 198, 163.
12. Kniffen, D.A., Hartman, R.C., Thompson, D.J., Fichtel, C.E., 1973, Ap. J., (Letters), 186, L 105.
13. Fichtel, C.E., et al., 1978, NASA-GSFC Tech. Memorandum 79650.
14. Hartman, R.C., et al., 1979, Ap. J., 230, 597.
15. Fichtel, C.E., et al., 1977, Proc. 12th Eslab Symp.-ESA-SP-124, 191.
16. Fichtel, C.E., Simpson, G.A., Thompson, D.J., 1978, Ap. J., 222, 833.
17. Thompson, D.J., Fichtel, C.E., Hartman, R.C., Kniffen, D.A., Lamb, R.C., 1977, Ap. J., 213, 252.
18. Masnou, J.L., et al., 1977, Proc. 12th ESLAB Symp.-ESA-SP-124, 33.
19. Lamb, R.C., Fichtel, C.E., Hartman, R.C., Kniffen, D.A., Thompson, D.J., 1977, Ap. J., (Letters), 212, L 63.
20. Bennett, K., et al., 1977, Astron. Astrophys. 59, 273.
21. Ögelman, H.B., Fichtel, C.E., Kniffen, D.A., Thompson, D.J., 1976, Ap. J., 209, 584.
22. Manchester, R.N., 1980, private communication.
23. Kniffen, D.A., Fichtel, C.E., Thompson, D.J., 1977, Ap. J., 215, 765.
24. Fichtel, C.E., et al., 1977, Ap. J., (Letters) 217, L 9.
25. Fichtel, C.E., et al., 1977, Proc. 12th ESLAB Symp.-ESA-SP-124, 191.
26. Fichtel, C.E., Simpson, G.A., Thompson, D.J., 1978, Ap. J., 222, 833.
27. Bignami, G.F., Fichtel, C.E., Hartman, R.C., Thompson, D.J., 1979, Ap. J., 232, 649.
28. Scarsi, L., et al., 1977, Proc. 12th ESLAB Symp.-ESA-SP-124, 3.
29. Mayer-Hasselwander H.A., et al., 1979, Annals New York Ac. Sciences, 336, 211.

30. Strong, A.W., Worral, T.M., 1976, J. Phys. A. Math. Ge., $\underline{9}$, 823.
31. Black, J.H., Fazio, G.G., 1973, Ap. J., $\underline{185}$, L7.
32. Puget, J.L., Ryter, C., Serra, G., Bignami, G.F., 1976, Astron. Astrophys., $\underline{50}$, 247.
33. Lebrun, F., Paul, J.A., 1978, Astron. Astrophys., $\underline{65}$, 187.
34. Bignami, G.F., Morfill, G.E., 1980, Astron. Astrophys., $\underline{80}$, 1.
35. Lindblad, P.O., Grape, K., Sandqvist, A., Shober, J., 1973, Astron. Astrophys., $\underline{24}$, 309.
36. Rossano, G.S., 1978, Astron. J., $\underline{83}$, 241.
37. Caraveo, P., et al. (for the Caravane Collaboration), 1980, in preparation.
38. Hermsen, W., et al., 1977, Nature, $\underline{269}$, 494.
39. Swanenburg, B.N., et al., 1978, Nature, $\underline{275}$, 298.
40. Wills, R.D., et al., 1980, Advances in Space Expl. Vol. 7 (Pergamon Press).
41. Hermsen, W., 1980, Thesis University of Leiden.
42. Bennett, K., et al., 1977, Astron. Astrophys., $\underline{61}$, 279.
43. Kanbach, G., et al., 1980, Astron. Astrophys., in press.
44. Buccheri, R., et al., 1978, Astron. Astrophys. $\underline{69}$, 141.
45. Bignami, G.F., et al., 1980, Astron. Astrophys., in press.
46. Pollock, A.M.T., et al., 1980, Astron. Astrophys., to be published.
47. Sanduleak, N., 1978, IAU Circular n. 3170.
48. Hjellming, R., Hogg, D., Hvatum, H., Gregory, P., Taylor, R., 1978, IAU Circular N. 3180.
49. Perotti, F., et al., 1980, Ap. J., to be published.
50. Black, J.H., Fazio, G.G., 1973, Ap. J. Letters $\underline{185}$, L7.
51. Bignami, G.F., Morfill, G.E., 1980, Astron. Astrophys., $\underline{80}$, 1.
52. Wolfendale, A.W., 1980, Proc. of the IUPAP/IAU Symp. N. 94-Bologna, June 1980, in press.
53. Montmerle, T., Paul, J.A., M. Cassé, 1980, Proc. of the IUPAP/IAU Symp. N. 94-Bologna, June 1980,-in press.
54. Caraveo, P. and the Caravane Collaboration, 1980, in preparation.
55. Van den Bergh, S., 1979, Astron. J., $\underline{84}$, 71.
56. Montmerle, T., 1979, Ap. J., $\underline{231}$, 95.
57. Helmken, H.F., Weekes, T.C., 1979, Ap. J., $\underline{228}$, 531.
58. Bignami, G.F. and Caraveo, P.A., 1980, Ap. J., in press.
59. Bignami, G.F., Caraveo, P.A. and Maraschi, L., 1978, Astron. Astrophys., $\underline{67}$, 149.
60. Salvati, M., Massaro, E. and Panagia, N., 1980, Proc. of the IUPAP/IAU Symp. N. 94-Bologna, June 1980,-in press.
61. Pollock, A.M.T., et al., 1980, Astron. Astrophys., in press.

GAMMA RAYS FROM COSMIC RAYS

A.W. Wolfendale,
Physics Department, University of Durham, U.K.

It is shown that there is evidence favouring molecular clouds being sources of γ-rays, the fluxes being consistent with expectation for ambient cosmic rays interacting with the gas in the clouds for the clouds considered. An estimate is made of the fraction of the apparently diffuse γ-ray flux which comes from cosmic ray interactions in the I.S.M. as distinct from unresolved discrete sources. Finally, an examination is made of the possibility of gradients of cosmic ray intensity in the Galaxy.

1. INTRODUCTION

The nature and location of the sources of cosmic ray particles is one of the most important questions in contemporary cosmic ray physics. The reason for uncertainty is well known - the presence of the Galactic magnetic field causes particle trajectories to be torturous, and uncertain, for all but the most energetic particles. The first question to be asked is whether the bulk of the cosmic rays detected at the earth have come from sources within the Galaxy or outside it. For electrons, the answer is almost certainly the former because of the 'absorption' of most extragalactic electrons by the 2.7K radiation. For protons, however, there is still some uncertainty although it is often considered that our work on γ-rays from the Galactic anti-centre (Dodds et al., 1975) indicates that the bulk of the protons between 1 and 10 GeV are of Galactic origin and observations of anisotropies (see, for example, the summary by Kiraly et al., 1979) suggest a continuation of this type of origin to much higher energies.

The role of the γ-ray studies in this connection is that if those arising from particle interactions with the nuclei of the I.S.M. can be identified, and if the characteristics of the I.S.M. are known in detail then the intensity of particles can be determined at various locations in the Galaxy. Ideally one would hope to see the characteristic signature of γ's from π^o - mesons generated by protons near sources (S.N.R., pulsars ..., see, for example, Pinkau, 1975) but this condition has not yet been realised. Instead, attention has been directed towards searching for gradients of cosmic ray intensity in the Galaxy the detection of which would strongly suggest a Galactic origin for the particles.

Even this apparently modest goal is fraught with difficulties, however. These can be listed as follows: (i) with the poor angular resolution of contemporary detectors (a few degrees at the energies of interest here : 0.3 - 10 GeV), the contribution of unresolved discrete γ-ray sources is uncertain, (ii) the I.S.M. is not known in sufficient detail insofar as the column densities of the important H_2 component are not accurately known, (iii) most of the H_2 is in rather dense clouds and there is the possibility that some particles may not be able to penetrate them and (iv) although the electron to proton ratio at the earth is small, electrons are so efficient at producing γ-rays (and, furthermore the e/p ratio appears to be higher elsewhere than locally) that the important proton-contribution is hard to disentangle. Attention will be given to all these problems.

The form of the present paper is first to search for gas clouds which should give detectable γ-ray fluxes from ambient cosmic ray fluxes. The relevance of such detections as there are to the likely contribution of genuine discrete sources is then examined. Finally, a search is made for large scale cosmic ray gradients.

2. GAMMA RAYS FROM MOLECULAR CLOUDS

A test of several aspects of the diffuse γ-ray problem would be the observation of γ-rays from known molecular clouds, each with roughly the flux expected from the ambient cosmic ray flux acting on the known mass of the cloud. Preliminary estimates of the expected fluxes were made by Black and Fazio (1973); here, we make estimates using more recent I.S.M. data and compare with the observations from both the SAS II and COS B satellites.

Coverage of the sky in searches for 'large' molecular clouds is, of course, by no means complete and even for those clouds examined mass estimates are very imprecise. Most of our data have come from the survey by Blitz (1977) and Stark and Blitz (1978) but to this has been added data on ρ Oph (see our recent work, Issa et al., 1980a) on Cygnus X (Cong, 1978) and on the Galactic centre region (Scoville et al., 1974). Figure 1 gives a plot of the clouds referred to and also lines corresponding to particular γ-ray fluxes, integrated over the solid angle subtended by each cloud.

The value for the emissivity has been taken to be $\frac{q_\gamma}{4\pi}(E_\gamma > 100 \text{ MeV})$ = $2.2 \times 10^{-26} \text{s}^{-1} \text{H atom}^{-1}$ following our detailed analysis (Issa et al., 1980a). For those clouds within about 1 kpc of the sun, this q-value should be appropriate but further afield significant differences might well occur due to cosmic ray gradients.

The γ-ray data from the SAS II satellite are those tabulated by Fichtel et al. (1978a) and the COS B results come from Wills et al. (1980), this paper giving fluxes of detected 'sources' with $I_\gamma > 10^{-6}$ cm^{-2}s^{-1}, and from the flux contours of Mayer-Hasselwander et al. (1980).

The comparison of observed γ-ray intensities with prediction is rather difficult due to uncertainty in the contribution to the γ-flux from other regions not necessarily associated with the complex in question. This problem is particularly acute for the extended sources in Cygnus, Perseus and Orion, but we make the attempt nevertheless.

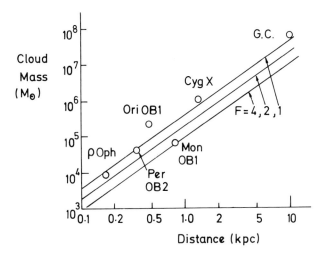

Figure 1. Estimated masses of molecular clouds versus approximate distance from the sun. The lines represent predicted γ-ray fluxes of 10^{-6} cm^{-2}s^{-1} ($E_\gamma > 100$ MeV) for the ambient cosmic ray particle fluxes of FX the local particle flux with F=1, 2 and 4. Only those clouds (from the references used) which lie above the line with F=4 are included. The references are given in the text.

ρ Oph. This cloud complex is particularly useful, being at a comparatively high latitude where confusion caused by background effects should be small. In addition the cloud is expected to be comparatively inert (i.e. it probably does not contain discrete γ-sources).

Issa et al. (1980a) have considered this cloud in detail and they conclude that there is reasonable consistency between observation and expectation for I_{CR} being close to the local value. Figure 2 shows the value of $F = I_{CR}(\rho\ Oph)/I_{CR}(local)$ needed (Note - the adopted value of I_γ is somewhat less than the COS B flux : 0.75 cf $1.1 \times 10^{-6} cm^{-2} s^{-1}$, because the SAS II upper limit was $\sim 0.7 \times 10^{-6} cm^{-2} s^{-1}$). As $\rho(Oph)$ is only $\simeq 160$ pc away, we would expect $F \simeq 1$.

<u>Per OB2.</u> This cloud is situated in the region of $\ell \simeq 160°$ and $b \sim -17°$ and has high gas column densities extending over about $5°$ in ℓ and $8°$ in b. Unfortunately this region is outside the published COS B coverage and it is necessary to fall back on the SAS II data. The cloud does not show up in the energy range 35-100 MeV, but for $E_\gamma > 100$ MeV a finite excess flux appears (the data were binned in $3.4°$ bins of latitude for the longitude range $155 - 165°$). The excess corresponds to a flux of $(1.4 \pm 0.7) 10^{-6} cm^{-2} s^{-1}$, in reasonable agreement with what we expect for $F = 1$ (Figure 2).

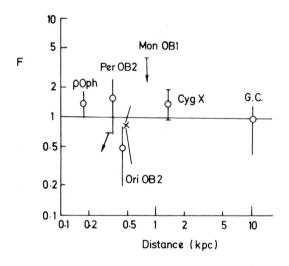

Figure 2. Enhancement factor for cosmic ray flux, F, versus distance of cloud from the sun for the clouds given in Figure 1. Vertical error bars correspond to $E_\gamma > 100$ MeV and oblique error bars to γ's in the range $35 - 100$ MeV.

<u>Ori OB 1.</u> This complex should show up in γ-rays in the region $\ell : 202° - 217°$, b : $-14°$ to $-22°$. Again, it is outside the COS B range and we have searched for it in the SAS II records. There is evidence for an excess in both energy ranges: the corresponding fluxes are $(1.7 \pm 0.8) 10^{-6} cm^{-2} s^{-1}$ for $35 - 100$ MeV and $(1.0 \pm 0.5) 10^{-6} cm^{-2} s^{-1}$ for $E_\gamma > 100$ MeV. These fluxes are a little less than we expect for $F = 1$ but not significantly so (see Figure 2).

<u>Mon OB 1.</u> This cloud is a 'long shot' in that it would require $F \simeq 4$ for detection. There is no COS B 'source' in this position and the SAS II data show no significant excess; an upper limit of $F \sim 4$ is therefore indicated. The nearest COS B 'source' is at $\ell = 195.1°$,

$b = 4.5°$ and is bright ($I_\gamma = 4.8 \times 10^{-6} \text{cm}^{-2}\text{s}^{-1}$); it is too far away to be identified with OB1 but it is interesting to note that there is a deficit of galaxy counts close by, at $\ell \simeq 196°$, $b \simeq +6°$, which may indicate the presence of a dense rather concentrated gas cloud which might account for this strong source.

Cygnus-X. The cloud complex marked 'Cyg-X' in Figure 1 comprises many clouds in the general direction of $\ell \sim 80°$, $b \sim 0°$. In an earlier paper (Protheroe et al., 1979a) we examined this region in some detail using quite extensive COS B data ($E_\gamma > 100$ MeV) and a brief resumé will be given here. The complex has been investigated in CO by Cong (1978) and this author estimates that the total mass of some 78 clouds in the range, $\ell : 75°$ to $85°$, $|b|<4°$ is $\sim 10^6 M_0$ and the distance range is 1 - 2 kpc. Protheroe et al. have derived column densities of H_2 from the CO results and combined these with the column densities of H from Weaver and Williams (1973) to predict the γ-ray flux. They give the longitude distribution of γ-ray flux for $|b|<6°$ and show that, with $q/4\pi = 1.8 \times 10^{-26} \text{s}^{-1}$ there is rough agreement. In fact, as the authors point out, a higher cosmic ray intensity would give a better fit. Inspection of the longitude plot indicates $F \simeq 1.7 \pm 0.4$ and if $q/4\pi$ is increased to the presently adopted value, $2.2 \times 10^{-26} \text{s}^{-1}$, F is reduced to 1.4 ± 0.5.

It is likely that the 'sources' quoted by Wills et al. at $\ell = 75.0°$, $b + -0.5°$ ($1.3 \times 10^{-6} \text{cm}^{-2}\text{s}^{-1}$) and $\ell = 77.8°$, $b = 1.5°$ ($2.5 \times 10^{-6} \text{cm}^{-2}\text{s}^{-1}$) are due to clouds within the complex. With a mass of $10^6 M_0$ and an effective distance of 1.3 kpc the expected net flux is $1.6 \times 10^{-6} \text{cm}^{-2}\text{s}^{-1}$, to be compared with the measured $3.8 \times 10^{-6} \text{cm}^{-2}\text{s}^{-1}$, i.e. we require $F \simeq 2.4$. The values, 1.5 and 2.4, are, understandably, not very different.

Galactic Center. The situation with respect to the flux of γ-rays from the G.C. region is confused. SAS II saw quite a respectable peak and Wolfendale and Worrall (1977) used the data to show that the flux above 100 MeV from within a few degrees of the G.C. was $\simeq 6.7 \times 10^{-6} \text{cm}^{-2}\text{s}^{-1}$. The COS B results appear to indicate a smaller source ($1.8 \times 10^{-6} \text{cm}^{-2}\text{s}^{-1}$) at $\ell = 359.5°$, $b = -0.5°$. However there is another source within $5°$ ($2.6 \times 10^{-6} \text{cm}^{-2}\text{s}^{-1}$) at $\ell = 356.5°$, $b = +3°$ and the summed intensity of $4.4 \times 10^{-6} \text{cm}^{-2}\text{s}^{-1}$ is not too far from that of the SAS II peak.

We are impressed by the evidence for the ring of molecular clouds round the G.C. subtending an angle of radius $\simeq 1.5°$ at sun (Kaifu et al., 1972 and many other references). The mass is very uncertain but here we adopt the range $(4-10)10^7 M_0$ quoted by Scoville et al. (1974) and treat this as a conventional cloud complex penetrated by the ambient cosmic ray flux (although we are mindful of the many problems in this region - the likelihood of genuine discrete sources, excess radiation density causing enhanced inverse Compton emission, etc.).

Adoption of the one source (359.5°, -0.5°) as being due to the ring gives F = 0.9 ± 0.5 (Figure 2).

Discussion of results on molecular clouds. The preceeding results give support to the idea that some, at least, of the so-called γ-ray sources are molecular clouds irradiated by the ambient cosmic ray flux and this is a feature which supports our contention that it is possible to derive information about the distribution of cosmic rays from an analysis of the diffuse flux in general.

The majority of the clouds in Figures 1 and 2 are near enough to the sun for $F \simeq 1$ to be expected (or at least for us to expect $F \underset{\sim}{<} 2$) and this is observed. The exception is the molecular ring at the G.C., where F might have been expected to be very high. Indeed, it is possible that the cloud mass is grossly over-estimated, in which case F can be large, but it is also possible that the magnetic field configuration is such that the cosmic rays generated there cannot escape (Wolfendale and Worrall, 1977). If this is the case then the injection rate will be much higher than locally to give the measured γ-ray flux. Our earlier estimate was an injection rate higher by a factor of 100; use of the new COS B data and the greater mass gives an enhancement factor of ~ 20.

3. GAMMA RAYS FROM DISCRETE SOURCES

Many attempts have been made to determine the fraction of the flux (f) from both resolved and unresolved discrete sources and estimates have varied from about 10% to near 100%. Such a spread is inevitable in view of the lack of identification of most of the apparent sources.

In an earlier work (Protheroe et al., 1979b) we examined the catalogue of 13 sources then available and used various arguments to determine values of f (for $E_\gamma > 100$ MeV) for various assumptions about the distribution of γ-ray sources in the Galaxy. The values derived were as follows: uniform slab model, $f \simeq 0.33$; distribution similar to that of X-ray sources, $f \simeq 0.19$; distribution similar to that of pulsars, $f \simeq 0.18$. The new catalogue of Wills et al.(1980) contains 29 sources of which 18 have $I_\gamma > 1.3 \times 10^{-6} cm^{-2} s^{-1}$ and represent a complete sample over a fraction of the Galactic plane (90° < ℓ < 300°). Riley et al. (1980) have analysed the new data using the uniform slab model and conclude that $f \simeq 0.25$; a similar result appears for the inner Galaxy (300° < ℓ < 90°) although here the method is not very accurate.

It seems likely that the value of f for the outer Galaxy is no higher than 0.25 and, if the distribution of source emissivity follows that of cosmic ray-induced emissivity throughout the Galaxy (a reasonable assumption), this value will pertain to the Galaxy as a whole.

The value of f will be somewhat of an overestimate because of the fact that some of the observed discrete sources are irradiated molecular clouds. Examination of the sources of Wills et al. which have $|b|>1.0°$ and $I_\gamma>1.3 \times 10^{-6} cm^{-2} s^{-1}$ shows 6 not associated with known clouds, 3(+2?) with clouds and 2 with SNR; in this region the visibility of clouds is probably reasonable so this means that probably $\sim$ 30% of the 'local' sources are irradiated clouds (of course, the uncertainty is considerable).

Riley et al. have used the data on the latitude distribution of sources to give a local source density of $\simeq 1.5$ kpc^{-2} for 'equivalent' sources with emission $5 \times 10^{38} \gamma$'s s^{-1} above 100 MeV (corresponding to a flux of $1.0 \times 10^{-6} cm^{-2} s^{-1}$ at 2 kpc). If 30% are clouds and if the distribution of genuine discrete sources follows roughly that of γ-emission in general the total Galactic flux will be $\sim 2 \times 10^{41} \gamma$'s s^{-1} above 100 MeV. This can be compared with the total emission of $\sim 1.3 \times 10^{42} \gamma$'s s^{-1} (Strong and Worrall, 1976) i.e. $f \simeq 17\%$.

It is useful at this stage to make a stock-taking of the various emission components. Prerequisites are a knowledge of the radial distribution of the densities of H and H_2 and that of the cosmic ray intensity. There are many permutations of these parameters but one which we prefer is as follows. For $I_{CR}(R)$ we take a distribution a little less rapid than that of SNR, viz $I_{CR}(R)/I_{CR}(local) \simeq 1.5$ at R = 3 kpc, 2.0 at R = 5 kpc, 1.6 at R = 8 kpc and 0.6 at R = 12 kpc. The H distribution is that given by Gordon and Burton (1976), the total mass of H in the Galaxy being $2 \times 10^9 M_\odot$. There is a problem with the Gordon and Burton H_2 distribution in that the densities are rather high. There are several reasons why densities just one half of those quoted are preferred: the formaldehyde analysis by Few (1979) gives such values, the local H_2 density measurements of Savage et al. (1977) lead to surface densities about 1/3 those of G and B and the ratio of CO to H_2 densities in molecular clouds may well be about twice the G and B value. Adopting the above parameters we find the following emission components: $H_2 : 6 \times 10^{41}$; H : 5×10^{41} and discrete sources, 2×10^{41} (all in γ's s^{-1} above 100 MeV).

If a division of this order is accepted for the time being, we can proceed to analyse the large scale gradients by neglecting the source contribution to the measured fluxes (with the exception of those from the CRAB and VELA, which are subtracted).

4. LARGE SCALE COSMIC RAY GRADIENTS

As was remarked in the Introduction, a demonstration of cosmic ray gradients in the Galaxy can be regarded as a first step along the road to identifying the sources of cosmic ray particles. Although there is no direct proof that protons, as distinct from electrons, are contributing to the γ-ray flux by way of π°-production there is circumstantial evidence from the spectral shape (e.g. Stecker,

1971) and insofar as the exponent of the γ-ray spectrum does not vary much with longitude (Hartman et al., 1979) it does seem that a gradient of 'cosmic ray intensity' would indicate a gradient of proton intensity (particularly at 'high' γray energies; $E_\gamma > 100$ MeV).

Our method of studying gradients is by way of an examination of the γ-ray emissivity, q, as a function of position in the Galaxy. Many studies have been made of the relationship between γ-intensity and column density of gas (usually, the C.D. of atomic hydrogen N_H). The q value determined from the usual relation $I_\gamma = (q|4\pi)N_H + I_b$ is clearly the average along the line of sight and is roughly representative of the q-value at the median value of N_H. More recent analyses (e.g. Protheroe et al., 1979a; Lebrun and Paul, 1979) have included the effect of molecular hydrogen and these analyses are continuing.

A problem occurs in that N_{H2} is only known in very restricted regions and in what follows q-values related to N_H will be of main concern. The SAS II results will be used because of their current availability.

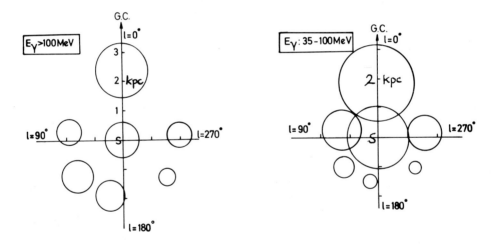

Figure 3. q-values at the median positions in the Galaxy. The radius of each circle is proportional to q. S = Sun's position.

Fichtel et al. (1978b) have made a comprehensive study of I_γ vs N_H using the N_H values of Daltabuit and Meyer (1972) and Heiles (1975) and have derived overall q-values. We have used their plots for individual ℓ- and b- ranges to derive appropriate q-values. The latitude distribution for each ℓ-range chosen has been used, together with the z-dependence of the gas, to determine the median linear distance appropriate to the q-values. An independent analysis has also been made of the q-values by taking the tabulated intensities of Fichtel et al. (1978a) together with the column densities of

Weaver and Williams (1973) and making a maximum likelihood fit to the I_γ, N_H expression. Finally, the q-values for the two methods have been averaged and the results are given in Figure 3. It should be remarked that the 'local' values are for 'high' latitudes ($|b|$: $12.8^\circ - 30^\circ$) and thus correspond to a radius of $\simeq 500$ pc round the earth.

The q-values given in Figure 3 can be used to make estimates of the likely cosmic ray gradients in the region of a few kpc from the sun although it must be remarked that there are some systematic errors present. One reason is that the q-values are over estimates because of the effect of molecular hydrogen and this overestimate is a function of both ℓ and b. The point is that there is some correlation between the column densities of H and H_2 (if only for geometrical reasons). Issa et al. (1980b and later work) find that the factor of overestimation, f, is $\simeq 1.22$ for $E_\gamma > 100$ MeV, $\ell : 10^\circ - 240^\circ$ and $|b| : 10^\circ - 51^\circ$. The overestimate will be larger at small latitudes but here the necessary H_2 data are sparse. An analysis of COS B and SAS II results for the Cygnus region, where H_2 data are available, indicates that, for $|b|<15^\circ$, $f_o \simeq 1.5$. Corrections have not been applied here for this effect for two

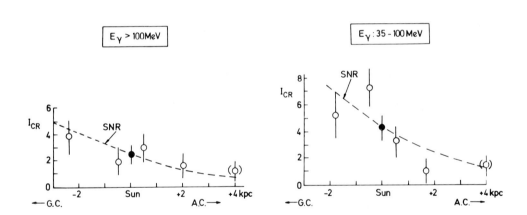

Figure 4. Very approximate values for the inferred 'cosmic ray intensity', I_{CR} (some combination of electrons in the range 100 MeV-1 GeV and protons etc. in the range 1-10 GeV) as a function of position in the Galaxy along the line $\ell=0$, 180°. Considerable averaging has been carried out. The points in parentheses are very approximate, having been derived from a subset of the γ-ray data (with $|b|<5.6^\circ$). The intensity units are arbitrary. The line marked SNR is the surface density of supernova remnants from the work of Kodaira (1974). The graph gives evidence for a gradient of the cosmic ray intensity (somewhat similar to that of SNR).

reasons: (a) the values of f_0 are not accurately known, (b) there is some measure of compensation as regards C.R. gradient because, due to the narrower thickness of the H_2 layer, the distance from the sun of the main producing layers is smaller than adopted. Later work will need to examine this problem, however.

Figure 4 shows the q-values, now designated as cosmic ray intensities, plotted as a function of radial distance from the sun, using average values from Figure 3 together with the dependence of q on ℓ at high latitudes ($|b|>12.8°$). The results are necessarily imprecise at this stage but there does seem to be evidence for a gradient of cosmic ray intensity in the local region of the Galaxy both for electrons (E_γ: 35-100 MeV) and what is probably a roughly equal mixture of electrons and protons (E_γ >100 MeV).

4. CONCLUSIONS

The foregoing can perhaps be regarded as an optimistic assessment of the role of γ-ray Astronomy in giving information about the distribution of cosmic rays in the Galaxy. As has been mentioned, a number of serious problems cause worry - notably uncertainty concerning the contribution to the γ-ray flux from discrete sources and uncertainties regarding the properties of the I.S.M. Doubtless a situation can be envisaged in which the bulk of the γ-rays are generated by electrons and the γ-rays then give no information at all about the proton component. However, it would be necessary to increase the e/p ratio considerably away from the sun and to keep protons (and presumably electrons) out of contact with the I.S.M. Such shielding might be possible in the inner Galaxy, where so much of the gas seems to be in dense clouds, but in the outer Galaxy shielding is unlikely. It is in the outer Galaxy too that an underestimate of the discrete source flux would increase rather than reduce the gradient.

In conclusion, then, the γ-ray evidence (particularly that away from the Galactic center) still seems to point to a Galactic origin for low energy cosmic rays. A reasonable case can be made for particle production in SNR (Figure 4) but any other sources having a similar distribution in the Galaxy would also be acceptable. A necessary consequence of SNR production would be that diffusive motion would be small by Galactic standards. If other evidence favoured much greater diffusive displacements then bigger source distribution gradients would be indicated. Considerable production at the Galactic Centre might not be ruled out.

ACKNOWLEDGEMENTS

The work described was carried out by the author's research group, comprising Dr. A.W. Strong, Dr. M.R. Issa and Dr. P.A. Riley and himself; he is very grateful to his colleagues for their

REFERENCES

Black, J.H., and Fazio, G.G.1 1973, Ap. J. Lett. 185, L7.
Blitz, L., 1977, Proc. Gregynog Workshop on Giant Molecular Clouds, Ed. P. Solomon.
Cong, H., 1978, Ph.D. dissertation, Columbia University (NASA Technical Memorandum 79590).
Daltabuit, E., and Meyer, S., 1972, Astr. Astrophys., 20, 415.
Dodds, D., Strong, A.W., and Wolfendale, A.W., 1975, Mon. Not. Roy. Astr. Soc. 171, 569.
Few, R.W., 1979, Mon. Not. Roy. Astr. Soc. 187, 161.
Fichtel, C.E., Hartman, R.C., Kniffen, D.A., Thompson, D.J., Ogelman, H.B., Turner, T., and Ozel, M.E., 1978a, NASA Tech. Mem. 79650.
Fichtel, C.E., Simpson, G.A., and Thompson, D.J., 1978b, Ap. J., 222, 833.
Gordon, M.A., and Burton, W.B., 1976, Ap. J., 208, 346.
Hartman, R.C., Kniffen, D.A., Thompson, D.J., Fichtel, C.E., Ogelman, H.B., Turner, T., and Ozel, M.E., 1979, Ap. J. 230, 597.
Heiles, C., 1975, Astron. Astrophys. Suppl. 20, 37.
Issa, M.R., Strong, A.W., and Wolfendale, A.W., 1980a (in the press).
Issa, M.R., Strong, A.W., and Wolfendale, A.W., 1980b (in the press).
Kiraly, et al., 1979, Rivista del Nuovo Cimento, 2, 1.
Kodaira, K., 1974, Publ. Astr. Soc. Japan, 26, 255.
Mayer-Hasselwander, H.A. et al., 1980, Proc. 9th Texas Symp. "Annals of the New York Academy of Sciences" (in the press).
Lebrun, F., and Paul, J.A., 1979, Proc. 16th Int. Cosmic Ray Conf., Kyoto, 12, 13.
Pinkau, K., 1975 "Origin of Cosmic Rays", Ed. J.L. Osborne and A.W. Wolfendale (D. Reidel Publ. Co.) 335.
Protheroe, R.J., Strong, A.W., and Wolfendale, A.W., 1979a, Mon. Not. Roy. Astr. Soc., 188, 863.
Protheroe, R.J., Strong, A.W., Wolfendale, A.W., and Kiraly, P., 1979b, Nature, 277, 542.
Riley, P.A., and Wolfendale, A.W., 1980 (in the press).
Savage, B.D., Bohlin, R.C., Drake, J.F., and Budich, W., 1977, Ap, J. 216, 291.
Scoville, N.Z., Solomon, P.M., and Jefferts, K.B., 1974, Ap. J., 187, L63.
Stark, A.A., and Blitz, L., 1978, Ap. J., 225, L15.
Stecker, F.W., 1971, Cosmic Gamma Rays, Mono. Book Co., Baltimore.
Strong, A.W., and Worrall, D.M., 1976, J. Phys. A., 9, 823.
Weaver, H., and Williams, D.R.W., 1973, Astron. & Astrophys. Suppl. 8, 1.
Wills, R.D. et al., 1980, Non Solar Gamma Rays (COSPAR) Ed. R. Cowsik and R.D. Wills, Pergamon Press, 43.
Wolfendale, A.W., and Worrall, D.M., 1977, Astron. Astrophys. 60, 165.

COSMIC RAYS FROM REGIONS OF STAR FORMATION
I. The Carina Complex

T. Montmerle, J.A. Paul and M. Cassé
Section d'Astrophysique
Centre d'Etudes Nucléaires de Saclay, France

Within the error circle of the COS-B gamma-ray source at $\ell = 288°$, $b = 0°$ (Wills et al., 1980) lies the Carina Nebula, one of the most active regions of star formation known, housing several OB associations and Wolf-Rayet stars (WRS), and perhaps also a supernova remnant (SNR). As a region containing intense mass-losing stars it belongs to the same species as the Rho Oph cloud (but much more active), suggested to be associated with the gamma-ray source at $\ell = 353°$, $b = +16°$ (Paul et al., this conference). As a group of OB association linked with a SNR, it belongs to the same species as "SNOB's" (Montmerle, 1979), possibly identified with about half of the COS-B sources. We suggest that the source at $\ell = 288°$, $b = 0°$ should be identified with the Carina complex. In this case the source would be at $\simeq 2.7$ kpc and its gamma-ray luminosity would be $\simeq 2.10^{35}$ erg s^{-1}. It is suggestive that the nearest aggregate of stars, gas and dust (Rho Oph) and the richest one (Carina) are both in the direction of a gamma-ray source. The Carina complex is noted in particular for the compact star clusters Tr 14, Tr 16 and Cr 228 (Humphreys, 1978), altogether comprising 6 of the 7 O3 stars observed in the Galaxy. It is also remarkable that 3 WRS are associated with the complex. All 3 WRS are of the WN7 type, having the highest mass-loss rate of WRS ($\simeq 10^{-4}$ $M_\odot$ yr^{-1}). Even more remarkable is the presence of the strange η Carina object which sheds mass at the extraordinary rate of 10^{-3} to 0.075 $M_\odot$ yr^{-1}, with a velocity of $\simeq 600$ km s^{-1}. Moreover, according to radio (Jones, 1973) and optical (Elliot, 1979) data, there seems to be a SNR buried in the Nebula. However, it is not seen at X-ray wavelengths by the Einstein observatory (Seward et al., 1979). This could be explained if the SNR has a luminosity $L_x < 10^{34}$ erg s^{-1}, since it could then be conceivably hidden by the unstructured, diffuse X-background. The molecular cloud associated with the Carina Nebula has been observed in the lines of H_2CO and OH (Dickel, 1974). The cloud has a derived mass of $\lesssim 10^5$ $M_\odot$, typical of other molecular clouds.

The inverse-Compton gamma rays emitted by the relativistic electrons in the SNR, inpinging on the UV and far-IR photons is, in the extreme $\simeq 30\%$ of the gamma-ray flux; the same electrons interact with interstellar matter to produce gamma rays by bremsstrahlung. In certain conditions (matter density averaged over the emitting volume $\simeq 600$ cm^{-3} and magnetic field $\simeq 10^{-5}$ G.) the totality of the flux would be produced by this process (Montmerle and Cesarsky, 1980). However, in view of the lack of spectral informations above

100 MeV, π^0 decay following proton collisions cannot be ruled out. Now if the SNR does not exist we may consider supersonic stellar winds (SSSW) as CR suppliers (Cassé and Paul, 1980).

The total mechanical power P_{car} injected by SSSW is as follows : i) From early type stars $P_{OB} \simeq 10^{38}$ erg. s^{-1}, ii) from WR stars, $P_{WR} \simeq 5.10^{38}$ erg. s^{-1} iii) from η Car, depending on the mass loss adopted, $P_\eta = 2.10^{38}$ to 2.10^{40} erg. s^{-1}. Therefore, at least $P_{car} = 8.10^{38}$ erg. s^{-1} i.e. at least 10^3 to 4.10^3 times the gamma-ray luminosity. If π^0 decay is the dominant gamma-ray production mechanism, the observed flux requires a CR proton density at least one order of magnitude higher than in the solar neighborhood. This implies that CR injected in the cloud by OB and WR stars, still very close to their birthplace, must be efficiently trapped there. This trapping in turn, allows CR to produce gamma rays very efficiently, since their lifetime against p-p collision is short in dense media (10^4 -10^5 yrs). Then altogether, and including the fact that low energy ($\lesssim$ 1 GeV) protons do not produce gamma rays by π^0 decay, the required acceleration efficiency is found to be $\lesssim 5.10^{-3}$.

REFERENCES

Cassé, M. and Paul, J.A. 1980, Ap.J., 237, 236.
Dickel, H.R. 1974, Astr.Ap., 31, 11.
Elliot, K.H. 1979, M.N.R.A.S., 186, 9p.
Humphreys, R.M. 1978, Ap.J.Suppl., 38, 309.
Jones, B.B. 1973, Austr.J.Phys., 26, 545.
Montmerle, T. 1979, Ap.J., 231, 95.
Montmerle, T. and Cesarsky, C.J. 1980, in "Non Solar Gamma Rays", eds.
    R. Cowsik and R.D. Wills (Oxford : Pergamon Press), 7.
Seward, F.D. et al. 1979, Ap.J. (Letters), 234, L55.
Wills, R.D. et al. 1980, in "Non Solar Gamma Rays", eds. R. Cowsik and
    R.D. Wills (Oxford : Pergamon Press), 43.

COSMIC RAYS FROM REGIONS OF STAR FORMATION
II. The OB associations

M. Cassé, T. Montmerle, J.A. Paul
Section d'Astrophysique
Centre d'Etudes Nucléaires de Saclay, France

Supersonic stellar winds have recently been proposed as active agents of cosmic-ray acceleration (Cassé and Paul, 1980). We try to insert this potential acceleration mechanism in its general astrophysical context. Among galactic objects, OB associations have the narrowest latitude distribution, resembling to that of gamma-ray sources. Here we focus on the bulk rate of kinetic energy deposition in molecular clouds through stellar winds of individual stars pertaining to OB associations. Assuming that a minute fraction of this mechanical energy can be transferred to suprathermal particles (Montmerle et al., 1980), we examine whether OB associations are detectable high-energy gamma-ray sources, owing to the interaction of accelerated particles with the dense molecular cloud still present close to young and massive stars. The 3 factors that govern the gamma-ray "visibility" of a given OB association are i) the rate of kinetic energy deposition $\dot{E} = \Sigma\, 1/2\, \dot{M}_i\, V_i^2$, where the summation is done on all mass-losing stars; $\dot{M}_i$ is the mass-loss rate for the star i and V_i is the terminal velocity of its stellar wind;(ii) the distance of the OB association;(iii) the angular extent of the association.

We have examined the relative detectability of the 72 OB associations listed by Humphreys (1978), supplemented by newly discovered associations. Wolf-Rayet stars in OB associations (Van der Hucht et al., 1980) have been included. Since the wind parameters are only measured in a limited number of cases, we used empirical laws derived from recently published bolometric corrections, mass-loss rates and terminal velocities (Montmerle et al., this conference). Table 1 presents the expected gamma-ray surface brightness for the 10 brightest OB associations, normalized to that of the Carina complex (excluding ηCar and ηCar-like objects, and the contribution of SNRs when existing). It can be seen that Wolf-Rayet stars bring a dominant contribution to the mechanical energy injected.

The expected gamma-ray flux from only a few associations (e.g. Cyg OB2, Sco OB1) is comparable to that of the Carina complex possibly associated with the COS-B source at $\ell = 288.5°$, b = -0.5° (Montmerle et al., 1980). Most of the remaining OB associations must be undetectable as individual gamma-ray sources by the COS-B satellite, but they must contribute significantly to the diffuse galactic background. The large scale correlation between galactic gamma radiation and Lyman continuum photons (Pinkau, 1979) is naturally explained here, as well as the dominance of the spiral structure on the galactic gamma-ray emission (Caraveo and Paul, 1979).

Table 1

Rank	Association	number of WN7, WN8 stars (a)	number of other WR stars (b)	number of stars	gamma-ray surface brightness
1	Cyg OB2	0/1	0/2	3	0.45/1.71
2	Cr 121	0	0/1	0	0/1.34 (c)
3	Sco OB1	1	1	3	1.13
4	Carina	3	0	1	1.00
5	H.M. (d)	2	0	2	0.76
6	Cyg OB1	0	4/5	3	< 0.74
7	Sct OB2	0	1	2	< 0.39
8	Sct OB3	0	0	0	< 0.14
9	Cyg OB9	0	0	0	< 0.14
10	Cyg OB3	0	1	1	< 0.13

(a) $\dot{M} = 10^{-4} M_\odot \, yr^{-1}$
(b) $\dot{M} = 3 \, 10^{-5} M_\odot \, yr^{-1}$
(c) WR star probably foreground (Moffat and Seggewiss, 1979)
(d) Havlen and Moffat (1977)

REFERENCES

Caraveo, P. and Paul, J.A. 1979, Astr.Ap., 75, 340.
Cassé, M. and Paul, J.A. 1980, Ap.J., 237, 236.
Havlen, R.J. and Moffat, A.F.J. 1977, Astr.Ap., 58, 351.
Humphreys, R.M. 1978, Ap.J. Suppl., 38, 309.
Moffat, A.F.J. and Seggewiss, W. 1979, Astr.Ap., 77, 128.
Montmerle, T., Paul, J.A. and Cassé, M. 1980, this conference.
Pinkau, K. 1979, Nature, 227, 17.
Van der Hucht, K.A., Conti, P.S., Leep, E.M. and Wary, J.D. 1980, submitted
    to Space Sci. Rev.

COSMIC RAYS FROM REGIONS OF STAR FORMATION
III. The role of T-Tauri stars in the Rho Oph cloud

J.A. Paul, M. Cassé, T. Montmerle
Section d'Astrophysique
Centre d'Etudes Nucléaires de Saclay, France

1. The Rho Oph cloud as a cosmic-ray source

It has long been expected that gamma-ray astronomy will visualize the mysterious cosmic-ray (CR) sources. Indeed, on the basis of COS-B observations (Wills, 1980), it has been proposed that CR acceleration within the Rho Oph cloud complex (ROCC) is required to account for its gamma-ray luminosity (Cassé and Paul, 1980, herein after CP, Bignami and Morfill, 1980). However, Issa et al. (1980) have suggested that the size of the ROCC has been underestimated by a factor $\simeq 2$, and that the cloud mass is a factor $\simeq 4$ higher than given by Myers et al. (1978), making unnecessary CR acceleration and trapping in the cloud interior.

Counter arguments are presented in the following. The radial density distribution from the center of the cloud to a distance $r = 2.1$ pc can be parametrized as $n(r) \sim r^{-\alpha}$ (1) (Myers et al., 1978). Let us define the cloud boundary radius, R, as the radius at which relation (1) ceases to be valid. Assuming perfect spherical symmetry and taking $\alpha = 1$ for simplicity (still compatible with star counts, Myers et al., 1978), the column density depends on the radial distance r through $N_H = 1.7 \, 10^{22} \, \text{Ln} \, (R/r + ((R/r)^2 -1)^{\frac{1}{2}})$ H-atom cm^{-2} (2). The constant corresponds to an integrated mass of 1800 $M_\odot$ within a radius of 2.1 pc (Myers et al., 1978). R can be determined from the column density on the line of sight towards stars observed in the direction of the ROCC. Take, for instance, HD 148605 located at a distance of $\simeq 217$ pc from the Sun, i.e. $\simeq 57$ pc behind the ROCC. Its line of sight intercepts the ROCC region at 2.2 pc from the cloud center. The total column density toward this star is estimated to be $9.1 \, 10^{20}$ H-atom cm^{-2} (Bohlin et al., 1978). This value is much lower than expected from (2) with R > 2.1 pc. Other examples lead to the same conclusion. Consequently, it does not seem that the radius of the dense part of the ROCC has been underestimated, and we confirm the picture depicted by Myers et al. (1978) and hence the necessity of CR acceleration in the cloud.

2. The role of T-Tauri stars

The gamma-ray emission of the ROCC likely results from CR interactions with the cloud material. CR would be accelerated at the boundary between supersonic stellar winds and the circumstellar medium, as suggested by CP. In CP it was proposed that the wind from the B1 III star HD 148165 was the main accelerating agent, but Copernicus measurements (Snow, private communication) tend to show that the mass-loss rate of this star is actually much less than expected on the basis of its spectral type. However, the contribution of T-Tauri stars (TTS) may in fact account for a substantial release of mechanical power. In the survey of Cohen and Kuhi (1979), 13 TTS are apparently associated to the ROCC. For these 13 TTS the visual absorption A_V never exceeds 5 mag (mean value ~ 1.8). Since A_V up to 15 mag are noticed in the densest part of the ROCC, only the brightest and/or less obscured stars are visible (Cohen and Kuhi, 1979). Assuming a cloud of uniform density, ~ 40 TTS would be obscured by the cloud material. This crude estimate is sufficient since the wind parameters from TTS are still debated. Typical values as $\dot{M} = 10^{-7} M_\odot \text{ yr}^{-1}$ and wind velocity $\sim 200 \text{ km s}^{-1}$ have been proposed, leading to a mechanical power injected in the cloud $\sim 10^{33} \text{ erg s}^{-1}$ per TTS. Including hidden TTS, the whole T-association would release $\sim 10^{34} \text{ erg s}^{-1}$, to be compared with the gamma-ray luminosity ($\sim 8 \cdot 10^{32} \text{ erg s}^{-1}$) of the ROCC. This power has to be considered as a lower limit since it is expected that the mass-loss rate is higher for younger TTS, supposed to be in the densest part of the cloud (Cohen and Kuhi, 1979). It is worth noting that most of the mechanical power released by TTS is not used to accelerate CR, but remain available, for instance, to excite the nebula (Schwartz and Dopita, 1980).

REFERENCES

Bignami, G.F. and Morfill, G.E. 1980, Astr.Ap. (in press).
Bohlin, R.C., Savage, B.D. and Drake, J.F. 1978, Ap.J., 224, 132.
Cassé, M. and Paul, J.A. 1980, Ap.J., 237, 236.
Cohen, M. and Kuhi, L.V. 1979, Ap.J.Suppl., 41, 743.
Issa, M.R., Strong, A.W. and Wolfendale, A.W. 1980, preprint.
Myers, P.C. et al. 1978, Ap.J., 220, 864.
Schwartz, R.D. and Dopita, M.A. 1980, Ap.J., 236, 543.
Wills, R.D. et al. 1980, in "Non Solar Gamma Rays" (COSPAR), eds. R. Cowsik and R.D. Wills (Oxford : Pergamon Press), 43.

HIGH ENERGY γ-RAYS FROM THE DIRECTION OF THE CRAB PULSAR

T. Dzikowski, B. Grochalska, J. Gawin, J. Wdowczyk,
Institute of Nuclear Research and University of Lodz,
Lodz, ul. Uniwersytecka 5.

A search has been made for very high energy photons from the direction of the Crab Pulsar using the Lodz extensive air shower array. This device is particularly suitable for such a study because it consists of a large muon detector which can be used to search for the characteristic muon poor showers.

The method involved selecting those showers falling within $15°$ of the Crab direction, the observation time being chosen in the interval when the Crab was not lower than $40°$ from the zenith. For the Lodz geographical latitude ($51.6°N$) this corresponds to 8 hours observation per day. The Crab has been observed for 5600 hours from 1975 to 1979. The sample from the general direction of the Crab has been compared with three background samples taken from the points on the sky located at the Crab declination but with R.A. displaced by $90°$, $180°$ and $270°$. The background samples were taken in this way to make sure that these showers were observed from the same direction with respect to the apparatus as the showers coming from the direction of the Crab.

The results of observation are summarised in the Table.

	Crab	RA+90°	RA+180°	RA+270°	Excess
$N_e > 4 \times 10^5$	313	292	290	310	52.3 ± 20.0
$N_e > 10^6$	156	108	115	120	41.7 ± 13.9
$N_e > 2 \times 10^6$	52	28	36	36	18.7 ± 7.9

It is seen that a clear excess of showers from the direction of the Crab is observed and that the fractional excess is increasing with shower size.

The result is confirmed by preliminary analysis of an earlier set of data taken in the period 1968 - 72. For that period we have compared only the excess in the intensity at very high densities corresponding to large showers, $N_e > 10^6$ particles. Here, there exists an excess amounting to 20.0 ± 7.4 showers. The joint probability of obtaining the two results by chance is 2.8 x 10^{-4} (3.6 σ for a Gaussian distribution).

The excess showers seem to be deficient in muons. The average muon content in the Crab showers amounts to (0.60 ± 0.12) of that in the normal showers. Normal showers contain ∼ 7 muons per 55 m^2 detector, whereas the excess showers contain 4 ± 1 per detector (we note that photon-induced showers should contain ∼ 2 muons per detector: Wdowczyk, 1965, not too far from observation).

The intensity of the excess showers observed in the present experiment can be evaluated taking into account the fact that the collection area of our device is approximately 1000 m^2 and is virtually energy independent; however it is dependent on angle and this gives a significant uncertainty in the absolute magnitude of the intensity. Including all uncertainties the overall flux from the Crab direction (in excess of our background level) is (3 ± 2) x $10^{-13} cm^{-2} sec^{-1}$ above an energy of ∼ 10^{16} eV.

In view of the large angular region round the Crab direction considered the evidence that the excess events are due to the Crab itself is circumstantial. A number of possibilities need to be considered including: a general enhancement over several degrees in the direction of the Crab, a discrete source within a few degrees of the Crab etc. (we are grateful to Dr. K.J. Orford for discussions on this point). Further, more detailed work is necessary to distinguish the possibilities.

If the excess flux is indeed to be identified with photons from the Crab, a possible model is one where protons accelerated by the pulsar interact with visible light. The results would not conflict with the measurements summarized by Weekes, 1979, if the spectrum between 10^{12} eV and 10^{16} eV was rather flat (or indeed if there were a negligible flux of photons below 10^{15} eV). A flat spectrum would be expected from the model referred to.

The authors are grateful to Dr. J.L. Osborne and Prof. A.W. Wolfendale for helpful discussions.

REFERENCES

Wdowczyk, J., 1965, Proc. IXth Int. Cosmic Ray Conf. London II, 691.
Weekes, T.C., 1979, Preprint No. 1260 - Center for Astrophysics,
  Cambridge, Massachusetts 02138.

A NEW KIND OF GAMMA RAY BURST?

G. Pizzichini
Istituto TESRE/CNR
Via De' Castagnoli, 1
40126 Bologna, Italy

One of the most recently detected (Cline et al., 1980) Gamma Ray Bursts (GRB) appears to have very unusual properties.

We recall here briefly the main features of the time history and of the spectral data, as given by Cline (1979). The time history has a very fast initial rise, less than 200 μsec., a smooth, large but very short initial peak, with a maximum intensity of several x 10^{-3} ergs cm^{-2} sec^{-1} and a 150 msec duration, followed by an oscillating decay phase, with at least 22 compound 8 second pulses. The spectrum of the initial phase of the event corresponds to a steep power law with possibly a line at 420 keV. The total spectrum of the decay phase is even steeper and shows no lines (Mazets and Golenetskii,1979).The location of the event (Evans et al., 1980) corresponds to N49, a supernova remnant in the Large Magellanic Cloud, which gives a total (isotropic) emission of $\approx 10^{45}$ erg, one half of it in the initial spike, the rest in the decay phase. Three later events, apparently with no special properties, are attributed to the same source (Mazets and Golenetskii, 1979), with increasing delays (0.6, 29 and 50 days) and decreasing peak intensities (3%, 1% and 0.5% of the first event), because their locations are all consistent with the much smaller March 5 error box.

We wish to discuss here whether this event should be assigned to a new class of GRB as suggested by Cline (1979), on the base of its observed properties. Let us compare these properties with those of other bursts:

a) short rise time and duration of the initial peak: for at least 2 more events it cannot be excluded that they had the same properties. They are events 69-3 and 72-5 in Strong et al. (1974);

b) structureless initial spike: the event of March 22, 76 (Cline et al., 1979) had only one simple shape peak, albeith with much larger duration and slow rise;

c) slow decay phase, with a $\approx$ 50 sec time constant: the July 8, 77 event (Cline et al., 1979) probably had an even larger time constant;

d) pulsations in the decay phase: they would not have been detected by the early GRB detectors, but event 69-3 had a "weak continuing flux" (Strong et al., 1974);

e) steep power law spectra: also observed in other events, even if more

typically in their final stages (i.e.: Mazets and Golenetskii, 1979);
f) evidence of a line at ≈ 420 keV: also present for example in the November 19, 1978 event (Teegarden and Cline, 1980; Mazets and Golenetskii, 1979);
g) repeated bursts and association with a known object: except possibly for Cyg X-1 (Strong, 1975), this is the only instance of burst repetition and identification of a GRB source with a known celestial object, and it is based on a small (1'x2') error box;
h) peak flux at the earth: this is indeed at least one order of magnitude larger than for any other event detected.

For the energy output at the source, a somewhat longer comment is necessary. It has been suggested, both on the base of the log N-log S curve (Fishman, 1979) and of limitations to the energy density required at the source (Schmidt, 1978; Cavallo and Rees, 1978), that GRB are of Galactic origin. If, at the same time, we accept N49 as the source of the March 5 event, then both its peak intensity and its total energy output are several orders of magnitude (10^5-10^6) larger than those of the other bursts. This fact alone puts it in class by itself. On the other hand, the identification of the source with N49, means that at least one burst mechanism does not obey those restrictions to the energy density and weakens the case for the Galactic origin of the other events. Such a mechanism has been suggested by Ramaty et al. (1980a and 1980b).

We conclude that, while many of the properties of the March 5 event have been seen in at least one more event, they certainly have not been observed together in any other GRB. Although we must be cautious in considering this event as unique, because of instrumental limitations in early detections, we should at least separate GRB into "slow" and "fast" (March 5 type) events, assigning tentatively to the latter class also the two events 69-3 and 72-5. The suggestion is not entirely new: in fact it was noticed very early that some bursts were unusually short (Strong et al., 1974). Future GRB monitors should be flexible enough to detect both kinds.

References

Cavallo, G. and Rees, M.J.: 1978, MNRAS 183, 359.
Cline, T.L.: 1979, NASA Tech. Mem. 80630.
Cline, T.L. et al.: 1979, Ap. J. L 232, L1.
Cline, T.L. et al.: 1980, Ap. J. L 237, L1.
Evans, W.D. et al.: 1980, Ap. J. L 237, L7.
Fishman, G.J.: 1979, Ap. J. 233, 851.
Mazets, E.P. and Golenetskii, S.V.: 1979, preprint Akad.Sci. USSR N.632.
Ramaty, R. et al.: 1980a, Nature 287, 122.
Ramaty, R., Lingenfelter, R.E. and Bussard, R.W.:1980b,NASA TM 80674.
Schmidt, W.K.H.: 1978, Nature 271, 525.
Strong, I.B., Klebesadel, R.W. and Olson, R.A.: 1974, Ap. J. L 188, L1.
Strong, I.B.: 1975, Proc. XIV ICRC, München 1, 237.
Teegarden, B.J. and Cline, T.L.: 1980, Ap. J. L. 236, L67.

THE COMPONENTS OF THE GALACTIC γ-RAY EMISSION

M. Salvati,
Istituto di Astrofisica Spaziale CNR, Frascati, Italy
E. Massaro,
University of Rome and Istituto di Astrofisica Spaziale CNR, Frascati, Italy, and
N. Panagia,
Istituto di Radioastronomia CNR, Bologna Italy

The galactic luminosity for $0.1 < E_\gamma < 2$ GeV can be evaluated directly from observational data and the commonly accepted value is $5 \; 10^{38}$ erg s^{-1}. Various contributions to the γ-ray emission can be recognized, namely i) strong sources, ii) pulsars, and iii) diffuse processes. Their relative importance is discussed in the following.

A recent list of γ-ray sources (Caravane Collaboration, 1979) includes 29 entries, 3 of which have $|b|$ much larger than average and are disregarded here. The main features of the distribution are a strong concentration toward the galactic plane and a marked asymmetry between the regions inside and outside the solar circle. This property can be measured by the ratio $r = (N_i - N_o)/(N_i + N_o)$, with $N_{i,o}$ = detected sources having $-90° < \ell < +90°$ and $+90° < \ell < +270°$, respectively. To avoid biases deriving from different sensitivity at different longitudes, we retain only sources with flux $\geqslant \Phi_m = 1.1 \; 10^{-6}$ ph cm^{-2} s^{-1} = $5.2 \; 10^{-10}$ erg cm^{-2} s^{-1}. Thus, we obtain $N_i + N_o = 20$ and $r = .65$. Such high ratio, and the evolutionary constraints discussed by Panagia and Zamorani (1979) suggest a Pop I distribution $\exp(-\alpha R)$, with $\alpha = (2.2 \text{ kpc})^{-1}$ and an inner cutoff at 5 kpc. Under the assumption that all the sources have the same luminosity L_γ, a second order expansion around the solar position gives

$$(N_i + N_o) = (\sigma_\odot/4 \Phi_m) \{ 1+(1-1/\alpha R_\odot)(\alpha^2 d^2/8)\}$$

$$r = (4/3\pi) \alpha d \{1+(1-1/\alpha R_\odot)(\alpha^2 d^2/8)\}^{-1}$$

here $\sigma_\odot$ = luminosity surface density at Sun due to the strong sources and $d^2 \doteq L_\gamma/(4\pi\Phi_m)$. By integrating σ over the galactic plane with the same Pop I distribution, we find a source contribution of L(strong sources) = $(2.58 \pm 0.85) \; 10^{38}$ erg s^{-1}.

Another established class of γ-ray sources are the radio pulsars, including both very young objects and more typical ones with ages $\geqslant 10^5$ yrs. In order not to count twice strong sources, such as those associated with the Crab and the Vela pulsars, in the following we consider only the latter subclass. Thus, we can apply the findings of Buccheri et al. (1978) that the conversion efficiency from rotational energy loss $\dot{E}$ to γ-ray luminosity L_γ increases with age and is about unity for most objects. We use the radio data from Gullahorn and Rankin (1978) and Taylor and Manchester (1975); rather than $\dot{E}$ we calibrate $\dot{P}$ vs L_r, and convert it to $\dot{E} \equiv L_\gamma$ by means of the average P^{-3} and the standard moment of inertia 10^{45} g cm^2. The scatter diagram in the log L_r - log $\dot{P}$ plane appears as an approximately uniform distribution within a strip, whose upper and lower envelopes define $L_{\gamma max}$ and $L_{\gamma min}$ for any L_r. This result combined with the pulsars' radio luminosity function (Taylor and Manchester, 1977), uniquely determines their bivariate distribution as a function of L_γ and L_r. Integrating this distribution over the galactic plane with allowance for the limited observable bandwidth and the radio beaming factor (the γ beaming factor is irrelevant), the final result is L(pulsars) = (1.25 ± 0.50) 10^{38} erg/s.

The remaining fraction of the γ-ray luminosity may be then ascribed to cosmic-ray interactions with the interstellar gas. Current estimates indicate that π^0-decay is the only relevant mechanism for E_γ > 0.1 GeV, and its contribution to the galactic luminosity is

$$L(\pi^0) = L(total) - L(strong\ sources) - L(pulsars) =$$
$$= (M_g/m_p)\ s_\odot\ <f>$$

where $M_g \approx 10^{43}$ g is the total mass of the interstellar gas (Mezger, 1978), m_p the proton mass, $s_\odot$ the energy yield in the solar neighbourhood ($\approx$ 5.5 10^{-29} erg/s), and $<f> = \int fn_g\ dV / \int n_g dV$ is the ratio of the cosmic-ray density to the local value weighted with the gas distribution. So we are able to evaluate $<f>$, and with $L(\pi^0)$ = (1.2 ± 1.0) 10^{38} erg s^{-1} we obtain $<f>$ = 0.1 - 0.6 . $<f>$ can also be computed "a priori" by assuming a gas distribution and a relation between f and n_g. For example, using the data of Gordon and Burton (1976) and f = $n_g/n_\odot$ as suggested by Fichtel et al. (1976) gives $<f>$ = 2.6, much higher than the above value. Therefore, such a positive correlation of f with n_g can be excluded. Indeed, this provides evidence that π^0-decay emissivity is reduced in the high density regions.

References
Buccheri, R. et al. : 1978, Nature 274, 572.
Caravane Collaboration: 1979, Proc. 16th ICRC, Kyoto, 12, 36.
Fichtel, C.E. et al.: 1976, Astrophys. J. 208, 211.
Gordon, M.A. and Burton, W.B.: 1976, Astrophys. J. 208, 346.
Gullahorn, G.E. and Rankin, J.M.: 1978, Astron. J. 83, 1219.
Mezger, P.G.: 1978, Astron. Astrophys. 70, 565.
Panagia, N. and Zamorani, G.: 1979, Astron. Astrophys. 75, 303.
Taylor, J.H. and Manchester, R.N.: 1975, Astron. J. 80, 794.
Taylor, J.H. and Manchester, R.N.: 1977, Astrophys. J. 215, 885.

GAMMA RAYS FROM GALAXY CLUSTERS

M. Giler,
University of Lodz, Poland.
J. Wdowczyk,
Institute of Nuclear Research, Lodz, Poland.
A.W. Wolfendale,
University of Durham, U.K.

In our recent work (Giler et al., 1980a) we have shown that a good model for the highest energy cosmic rays is one where they are generated in clusters of galaxies with a relatively flat production spectrum and then propagate with an energy dependent diffusion coefficient. There is evidence for a similar flat production spectrum ($\gamma \simeq 2.1$) for both protons and electrons in our own Galaxy (Giler et al., 1980b). Here we assume that the spectrum for clusters (specifically the VIRGO cluster) is independent of energy and an evaluation is made of the likely flux of γ-rays.

Figure 1 shows the derived γ-ray flux from the VIRGO cluster assuming that the mean gas density in the cluster is $5 \times 10^{-4} cm^{-3}$ and that p|e at production is 10. Also shown are upper limits from the SAS II experiment of Fichtel et al. (1975) and various Cerenkov EAS experiments. Insofar as the angular diameter of the whole VIRGO

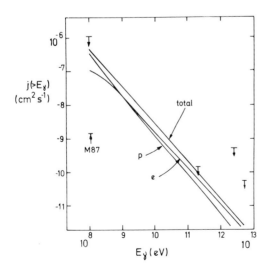

Figure 1. Gamma ray flux from the VIRGO cluster.

source is larger than the resolution of the Cerenkov detectors the
upper limits should be raised (by ~ 5). The expected flux from M87
alone has also been calculated using its radio emissivity. Inspection
of Figure 1 indicates that the VIRGO cluster should be detectable in
γ-rays with the next generation of γ-ray detectors.

If all clusters of galaxies are similar sources of cosmic rays to
the VIRGO cluster, a significant contibution to the intensity of
diffuse γ-rays will result. Our estimate is given in Figure 2.

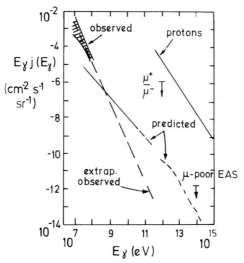

Figure 2. Diffuse isotropic background of γ-rays. The
shaded area denotes experimental measurements (Fichtel et
al., 1977).

The γ-rays above 10^{12} eV come mainly from the e-γ cascade resulting
from high energy cosmic ray interactions with the 2.7K radiation (for
details see Wdowczyk et al., 1972). It is interesting to note that
the prediction crosses an extrapolation of low energy measurements
by 10^9 eV and should thus be amenable to experimental check quite soon.

REFERENCES

Fichtel, C.E., et al., 1977, Ap. J., 217, L9.
Giler, M., Wdowczyk, J., and Wolfendale, A.W., 1980a (these
  Procedings); 1980b, Astron. Astrophys. 84, 44.
Wdowczyk, J., Tkaczyk, W., and Wolfendale, A.W., 1972, J. Phys. A.,
  5, 1419.

HIGH ENERGY GAMMA RAYS FROM ACCRETION DISC

F. Giovannelli
Istituto di Astrofisica Spaziale/CNR, Frascati, Italy
S. Karakula, W. Tkaczyk
Dept. of Physics, University of Lodz, Poland

In the case of spherical symmetric accretion into a black hole, the matter may be heated up to the temperature $KT = 0.1\ m_p c^2$ (Kolykhalov and Sunyaev, 1979). In such a hot plasma inelastic collisions of protons may produce Π° which is the gamma quantum source (Dahlbacka et al., 1974; Kolykhalov and Sunyaev, 1979).

In this work we determined γ-rays production spectrum in the comoving plasma reference frame, expected γ-rays spectrum for the case of spherical symmetric accretion of matter into a black hole and the upper limit to the number of black holes in Galaxy is evaluated.

In the calculations we made the following assumptions: 1) the plasma is fully ionized; 2) the proton momentum distribution is described by the relativistic Maxwell distribution; 3) the characteristics of interactions $p + p \to \Pi^\circ +$ anything were derived from an approximation of the experimental data (Barashenkov et al., 1972).

Fig. 1 shows the γ-rays production energy spectrum in the comoving plasma reference frame.

To evaluate the temperature, concentration and velocity of the plasma near a black hole, the system of equations describing the plasma motion should be solved (Michel, 1972)

$$\left(\frac{x}{x-1}\Theta + 1\right)^2\left(1 - \frac{1}{r} + u^2\right) = \text{const},\ nur^2 = \text{const},\ \frac{\Theta}{n^{x-1}} = \text{const},$$

where: $\Theta = KT/m_p c^2$, n – temperature and concentration of plasma in its own reference system, $r = R/R_g$, R – distance from black hole, R_g – its gravitation radius, u – R component of four velocity, $x = 5/3$.

To determine the temperature as a function of the distance from black hole, different values of u_0^2 were taken for given r_0. Calculations were done for $r_0 = 10^4$, $u_0^2 = 2.6\ 10^{-5}$ and $6\ 10^{-6}$ corresponding to Mach numbers 1.0266 and 2.1213, respectively.

Fig. 2 shows expected energy spectrum $Q(E)$ from accretion disk multiplied by $R_g/\dot{M}^2$ ($\dot{M}$ – accretion rate) which is determined in Schwarzschild metric with regard to relativistic effects (curves a, b – for $u_0^2 = 2.6\ 10^{-5}$ and $6\ 10^{-5}$, respectively).

Assuming the black hole mass equal $10\ M_\odot$ and accretion rate $\dot{M} = 10^{-8} M_\odot/$year, we found its luminosity $L(E > 100\ \text{MeV})$ and emissivity $N(E > 100\ \text{MeV})$:

L = 6.3 10^{33} erg/sec and 2.8 10^{32} erg/sec, N = 2.1 10^{37} phot/sec and 7.3 10^{35} phot/sec for u_o^2 = 2.6 10^{-5} and 6 10^{-5} respectively.

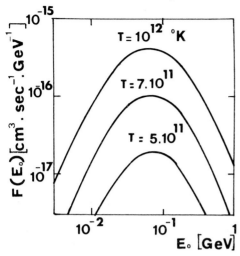

Fig. 1. The γ-rays production spectrum in the comoving plasma system.

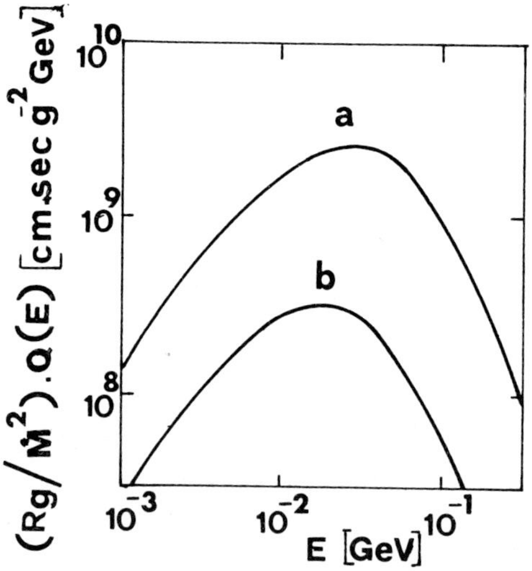

Fig. 2. The expected energy spectrum Q(E) from accretion disc multiplied by $R_g/\dot{M}^2$.

From the experimental measurements SAS-II, Galaxy emissivity of photons of energies E > 100 MeV is 1.3 10^{42} sec^{-1} (Strong et al., 1976). Assuming that all the Galaxy emissivity arises from considered objects, their number should be about 10^5 that gives 10^{-5} of the total star population. Taking into account another accretion parameter for instance u_o^2 = 6 10^{-5} we derive the contribution of these objects of about 10^{-4}.
We conclude that such objects may give a significant contribution to the total emissivity of our Galaxy.

References

Kolykhalov, P.I., Sunyaev, R.A.: 1979, Soviet Astron. J. 56, 338
Dahlbacka, G.H. et al.: 1974, Nature 250, 36
Barashenkov, W.S. et al.: 1972, Vzajmodestvija vysokoenergeticeskih castic i atomnyh jader s jadrami, Moscow
Michel, F.C.: 1972, Astrophys. and Space Sci. 15, 153
Strong, A.W. et al.: 1976, Mon. Not. R. Astr. Soc. 175, 23P

SEARCH FOR X-RAY SOURCES IN THE COS-B GAMMA-RAY ERROR BOXES

Giovanni F. Bignami
Laboratorio di Fisica Cosmica e Tecnologie Relative del CNR
via Bassini 15/A, 20133 Milano, ITALY

The publication of the first (Hermsen et al, 1977) and preliminary second catalogue of the COS-B sources (Wills et al, 1980), has brought to the attention of the astronomical community the new reality of the high-energy Galactic Gamma-Ray Sources (GGRS). It is worth recalling here the definition of such objects, in the COS-B sense: "A GRS is a significant excess of photon counts, compatible with the instrument's angular resolution, or, more clearly, with the instrument's Point Spread Function". This definition of an unresolved (as it is the case for the vast majority of the GGRS) GRS is, to some extent, dependent on the shape of the source spectrum; however a general shape of the PSF for the case of the source associated with 3C273 can be found in Bignami et al, 1980. Naturally, the absolute flux (i.e. the total number of photons) from a source is also important in determining the positional error, especially when usage is made of the cross-correlation method (Hermsen, 1980). Typical photon numbers (100 Mev) for GGRS range from 50 to few hundreds. The COS-B catalogue error boxes are defined taking into account both the PSF and the photon statistics; it is then apparent that the search for candidate counterparts of the GGRS should be carried out inside such boxes, or in their immediate vicinity, and that, for examples, such loose positional coincidences as obtainable by increasing a quoted error radius by 50%, let alone doubling it, are totally unacceptable.
Systematic searches for counterparts at other wavelengths have not yet been carried out, except for a few cases of X-ray and radio observations, as for example in the case of CG 135+1 (the only COS-B GRS erbox compatible with a 4U erbox) which led to the discovery of a new X-ray QSO by Apparao et al, 1978.
A truly systematic approach for a search in the X-ray band has been rendered possible by the advent of the Einstein Observatory, with its Guest Observer Program. The size of the IPC instrument field of view is comparable with the COS-B error radius, so that with few (from three to five) mosaic-arranged pointings it is normally possible to cover a GRS erbox. An IPC coverage of the COS-B erboxes has been foreseen, and so far about 11 GGRS regions have been, or are being, mapped, as a result of a collaborative effort involving R.C. Lamb, (ISU), R. Hartman and D.Thompson (GSFC), T. Markert(MIT), P. Caraveo (LFCTR) and, for some cases, the

whole COS-B Caravane Collaboration.

From a pre-Einstein X-ray astronomy point of view, these IPC pointings (a total of about 50) represent a small galactic survey, totally unbiased and quite random. It has, however, been limited to regions as far away from the center as possible: the sensitivity of exposure is that achievable with $1.5 - 2.0 \times 10^3$ secs of useful IPC time, yielding sources down to 5×10^{-13} erg/cm^{2}sec, or 3×10^{-2} UEFU (1 UEFU $\equiv$ 1 UFU in the Einstein 1 - 3 keV range).

The data are slowly coming in, showing preliminarly that the number of sources found is compatible to that obtained by extrapolating the Uhuru-Einstein LogN - LogS of Giacconi et al (1979), i.e. (7 ± 2.6) $S^{-1.5}$ sources/steradian, down to 3×10^{-2} UEFU, or ≤ 1 source/square degree ($\equiv 1$ IPC field). Thus the survey will yield a number of new IPC sources which will be totally manageable from the point of view of individual follow-up observations, already under way with HRI and optical data. Last, but not least, the problem of what to look for in the new X-ray sources: obviously time variability and/or common properties between sources in different GRS erboxes could be powerful tools to single out potential candidate counterparts of the GGRS ; in general, it is clearly useful to explore the boxes down to an energy flux limit which is 10^{-2} of that associated to the gamma-ray emission, since even a negative result could be very important. In any case, new interesting results of galactic X-ray astronomy per se are being obtained, one for all the "case of the twin pulsars" (see Lamb et al, 1980).

References

Apparao, K. et al, 1978 Nature, 273 450.
Bignami, G.F., et al, 1980 Astron. and Astrophys., in press.
Giacconi, R. et al, 1979 Ap. J. 234, L1.
Hermsen, W., 1980, Thesis, Univ. of Leiden.
Hermsen, W. et al, 1977 Nature, 269, 494.
Lamb, R.C. et al, 1980 Ap. J., in press.
Wills, R.D. et al, 1980 Proc. XXII COSPAR Pergamon Press, pag. 43.

THE ACCELERATION OF GALACTIC COSMIC RAYS

W.I. Axford
Max-Planck-Institut für Aeronomie, D-3411 Katlenburg-Lindau 3,
Germany

1. INTRODUCTION

Although it is clear that cosmic rays are of considerable astrophysical significance we know surprisingly little about their origin(s), their mode(s) of propagation and escape from the galaxy, and even of their general properties. Practically nothing is known of the composition and energy spectra of different species at energies below about 200 MeV/nucleon as a consequence of solar modulation, which has a significant effect on the observed spectra at energies up to about 10 GeV/nucleon. The energy spectra of ions (especially protons) appear to be well approximated by a power law (differential spectral index $\gamma \sim 2.6$) in the range $10^{10}-10^{15}$ eV/nucleon. At higher energies (up to $\sim 10^{20}$ eV/nucleon) both the spectrum and the composition of the ion component are less well known, however it is significant that the anisotropy, as evidenced from the sidereal diurnal variation, increases substantially, suggesting that perhaps an extragalactic origin must be considered (Wolfendale, 1977).

As far as the composition is concerned, it appears that there is a noticeable enhancement of the heavier primary ions relative to alpha-particles and (to a greater extent) protons, although the latter are still the dominant species at, say, a given energy/nucleon. One should note that relative abundances, instead of referring to the whole spectrum, are usually expressed on the basis of equal energy/nucleon, total energy, energy/charge or rigidity and the choice among these is largely a matter of taste. As far as the heavier ions are concerned, it seems to be the case that the composition is more or less consistent with what might be expected for the interstellar medium at the present epoch with due allowance for propagation of the primary beam through ~ 3 g/cm² after injection. Ignoring the effects of modulation, the total energy density of cosmic radiation in the solar neighbourhood is of the order of 10^{-12} ergs/cm³ which is comparable to the energy density of the interstellar magnetic field and the total (thermal plus kinetic) energy density of the interstellar gas. This fact is presumably a consequence of the nature of the mechanisms of containment and escape of cosmic rays from the galaxy rather than to their sources and modes of acceleration. However, with this result and guessing the total volume occupied by galactic cosmic rays to be $2\times10^{67} cm^3$ we can use the "age" of 2×10^7 years deduced from the Be^{10} flux observed at (modulated) energies of ~ 100 MeV/nucleon (Garcia-Munoz et al., 1977) to obtain a total source power of $3\times10^{40} ergs/sec$.

Measurements of cosmic ray induced radioactivity in meteorites indicate that the average intensity of the cosmic radiation in the vicinity of the sun has remained constant to within a factor of 2 or 3 over the last 10^8-10^9 years (Forman and Schaeffer, 1979; Honda, 1979). This result does not exclude the possibility of short-lived variations, however, recent measurements of the amounts of Be^{10} in deep sea sediments suggest that there have been no significant enhancements during the last 2×10^6 years (Somayajulu, 1977). This may be an especially important result in view of the fact that there is some evidence that the sun is situated close to, or even inside, a very large supernova remnant which is of the order of 10^6 years old (Frisch, 1979, and private communication).

The situation with respect to electrons is rather less clear since their flux at a given energy is considerably less (~ 100 times) than that of protons and electrons are experimentally more difficult to measure. The spectral index in the range 1-10 GeV is similar to that of the protons and the bending that should result from synchrotron and inverse Compton losses gives an age which is roughly consistent with that obtained from Be^{10} in the cosmic radiation (Giler and Wolfendale, 1979). The positrons, which are clearly secondaries, provide in principle a better determination of both path length and age but again due to experimental difficulties the results are still somewhat uncertain.

Synchrotron emission associated with the interstellar magnetic field and inverse Compton scattering of starlight and microwave background radiation photons provide means of investigating the intensities of electrons throughout large regions of interstellar space, especially, in the case of synchrotron radiation, at energies below the range in which solar modulation effects become severe (e.g. Webber et al., 1979). As the observations refer to line-of-sight averages and involve several unknown factors such as the strength of the interstellar magnetic field and its variations, it is difficult to reach any very definitive results. However, it is clear from the distribution of non-thermal radio emissions in the galaxy that the cosmic ray electron flux is far from uniform. There are large variations, especially associated with supernova remnants and hence sources of energetic electrons may exist in which the present intensity is relatively high. Similar conclusions can be drawn from observations of the gamma ray emissions induced by protons (Kanbach, 1979).

2. ACCELERATION MECHANISMS

An excellent summary of the current view of various possible cosmic ray acceleration mechanisms has been given by Lingenfelter (1979). It seems likely, on the basis of the rather scanty evidence outlined above, that the most important energizing processes take place in the interstellar medium rather than in unusual objects such as supernovae, pulsar magnetospheres, flare stars, and the like. The mechanisms which are available in interstellar space are all dependent on mass motions of the medium, notably small-scale motions or waves (Fermi acceleration and magnetic pumping), and large scale motions involving shocks which are associated with supernovae, novae, expanding HII regions, stellar winds, and so on. In fact, these may all play a role but on energetic grounds at least the mass motions due to supernovae are likely to be the most important source. Shock waves provide the most direct and effective way of accelerating high energy particles and most of this review is concerned with this particular topic. However, the other processes are not negligible and may play a role at least in providing "seed" particles.

Of the possibilities which will not be discussed here in detail the supernova mechanism of Colgate and Johnson (1960, see also Colgate, 1979) is perhaps the most interesting. According to this idea the supernova shock wave, which tends to speed up on reaching the outer envelope of the star, eventually becomes relativistic. If this material were able to escape freely and diffuse into the interstellar medium without losing any significant amount of energy an energy spectrum could be achieved which is a power law not very different from that observed. There appears to be an essential difficulty, however: the particles must lose energy by doing work on the gas surrounding the star (either the pre-existing stellar wind or the interstellar medium) and there is no reason to believe that they retain their initial spectrum or even remain relativistic. Rayleigh-Taylor type instabilities might help overcome this difficulty to some extent but do not obviously save it.

A second process involving supernova remnants, namely acceleration in pulsar magnetospheres, has been considered in some detail during the last ten years (see Arons, these Proceedings). Current ideas suggest however, that the particles accelerated in such regions are essentially all electrons and positrons and so it is clear that, as presently envisaged, the mechanism has nothing to do with cosmic rays. Of course the power associated with slowing down of the pulsar rotation should eventually appear as kinetic energy of the associated supernova remnant and may in turn be used to accelerate cosmic rays but the mechanisms involved would then be those discussed later in this paper.

A third mechanism which is sometimes invoked involves magnetic field line reconnection, i.e. stellar and galactic "flares". It is not easy to evaluate the effectiveness of this mechanism in view of our relative lack of knowledge of flare processes other than those occurring on the sun and in the earth's magnetosphere. One can say, however, that the maximum energy obtainable is determined by the voltage drop occurring along a magnetic "neutral" line which should exist in the flare site. This voltage drop is of the order of $V_A BL$, where V_A and B are the characteristic Alfvén speed and magnetic field strength, respectively, and L is the length of the neutral line. With typical values of these quantities appropriate to the earth's magnetosphere, the sun and the galaxy as a whole, one finds maximum energies of the order of 10^6eV, 10^{10}eV, and 10^{13}eV per charge, respectively. It seems doubtful that the total energy available in stellar flares is adequate to account for the cosmic radiation and even though large solar flares are quite efficient in producing high energy particles, one must remember that if the particles do not escape freely they loose energy by doing work on the surrounding medium. Galactic flares may not suffer from these disadvantages but on the other hand their existence is questionable, since there is no obvious reason why magnetic field line reconnection on a galactic scale should take place abruptly in a flare-like manner.

It is instructive in considering possible cosmic ray acceleration mechanisms to take account of the "ground truth" provided by in situ observations of particle acceleration made in the terrestrial and Jovian magnetospheres and in the interplanetary medium. Shock acceleration is commonly observed both in magnetospheric bow shocks and in interplanetary propagating shocks; acceleration associated directly with magnetic field line reconnection is observed in the tail of the earth's magnetosphere (Sarris and Axford, 1979, and references therein); adiabatic acceleration and deceleration occurs as a result of magnetospheric convection (Axford, 1969); semi-adiabatic acceleration is associated with the break-down of one but not all adiabatic invariants of particle motion in the presence of large-scale, time-varying electric fields (Schulz and Lanzerotti, 1974); stochastic acceleration due to interactions with waves appears to occur throughout the magnetosphere and possibly the magnetosheath; acceleration due to the presence of electric fields parallel to the magnetic field occurs in the auroral zone magnetosphere where electric currents become too large to be carried otherwise by the ambient plasma (Swift, 1979; Johnson, 1979); and finally, in the case of Jupiter, magnetospheric rotation and the effects of neutral particle pick-up are not insignificant acceleration mechanisms at least at low energies. All of these processes can be considered as candidates for cosmic ray acceleration and, except perhaps for the semi-adiabatic mechanisms mentioned, all may have a role to play. Probably, however, the processes which are most important are shock acceleration together with acceleration/deceleration effects due to compressions/expansions of the medium as a whole and also stochastic acceleration due to the presence of turbulence. In any case, it cannot be too strongly emphasized that the processes available to us should not be considered as being in any sense speculative or without observational foundation – indeed, any theoretical schemes we may construct should be tested against in situ observations available in interplanetary space, if they are to be taken at all seriously.

3. STOCHASTIC ACCELERATION

There are many possibilities for particle acceleration involving plasma/magnetohydrodynamic wave turbulence. For moderately energetic particles (i.e. those with speeds large compared with the characteristic Alfvén and magnetoacoustic speeds in the medium) the most interesting are those involving Alfvén wave turbulence (second order Fermi acceleration, Fermi 1949, 1954; Hasselmann and Wibberenz, 1968) and magnetoacoustic turbulence (magnetic "pumping", Thompson, 1955; Fisk, 1976). With suitable assumptions the source strength for these two mechanisms can be written

$$Q = \frac{\partial^2}{\partial T^2} (D_1 U) - \frac{\partial}{\partial T} (D_2 U/T) ,\qquad(1)$$

where $U(T)$ is the cosmic ray number density in the kinetic energy range $(T, T+dT)$. $D_1(T)$ and $D_2(T)$ are the sum of two components $D_A \propto \langle V_A^2 \rangle / \kappa$ and $D_S \propto \langle V_S^2 \rangle / \kappa$, where $\langle V_A^2 \rangle$ and $\langle V_S^2 \rangle$ are the mean square speeds due to Alfvén and acoustic mode turbulence, respectively, and κ is the spatial diffusion coefficient which is associated largely with pitch angle scattering. Note that both processes are stochastic in nature, involving a second derivative with respect to energy and therefore diffusion in energy space. One may regard the Fermi acceleration mechanism as one in which light test particles (i.e. cosmic rays) try to come into thermal equilibrium with a field of randomly moving heavy particles (i.e. Alfvén waves). The magnetic pumping mechanism may be regarded as a process in which acoustic waves are damped by the equivalent of heat conduction by cosmic rays.

It is difficult to evaluate the effectiveness of these mechanisms as far as accelerating cosmic rays is concerned, as we have little information concerning the intensities of these modes of turbulence in the interstellar medium (see Jokipii, 1977; Cesarsky, 1980). Obviously, most of the energy available in the interstellar medium is involved in large scale mass motions which even if they are superficially chaotic cannot be regarded as representing turbulence in the sense implied above. Of course, such large scale motions may eventually degenerate into turbulence, so that on energetic grounds these stochastic processes should not be ruled out. However, if acceleration in connection with the large scale motions is very efficient as we will argue, one would expect that stochastic processes play only a secondary role. Nevertheless, since shock waves must inevitably be associated with large scale mass motions and hydromagnetic shocks can generate quite intense wave turbulence (as observed in interplanetary space) it seems quite likely that stochastic acceleration is important in producing "seed" particles from the high energy tail of the background plasma velocity distribution which could be ultimately accelerated to much higher energies by the shock acceleration mechanism.

As randomly moving shock waves interact with each other and the inhomogenieties that inevitably occur in the interstellar medium one would expect the strong shocks to degenerate into a random assembly of weak shock waves with associated expansions and compressions of the medium. The effect of such a situation on energetic particles has been considered by Bykov and Toptygin (1979) who show that it leads to acceleration in a manner similar to that expressed in equation (1) above. This is perhaps not surprising since weak shock waves are in every other sense essentially equivalent to sound waves.

4. GENERAL ASPECTS OF SHOCK ACCELERATION

Shock acceleration with regard to cosmic rays was first discussed theoretically by Parker (1958), Hoyle (1960), and Schatzmann (1963). Direct evidence for the occurrence of shock acceleration in interplanetary space was first found in the case of moderately energetic (1-100 GeV) cosmic rays by Dorman and his colleagues (see Dorman, 1963) in the form of a small ($\sim 1\%$) increase in intensity of cosmic rays observed by ground stations immediately before the occurrence of a geomagnetic sudden commencement. The latter is of course the signature of an interplanetary shock wave interacting with the earth's magnetosphere. Dorman and Freidman (1959) formulated a single reflection theory to account for these observations, which was the first of a series of papers treating this topic (see section 5).

Much clearer evidence for shock acceleration in interplanetary space was obtained by Axford and Reid (1962, 1963) from observations of polar cap absorption events using riometer networks which respond to the influence of solar energetic particles with energies of the order of 10 MeV. They found a number of cases in which large quasi-exponential increases of the solar energetic particle flux occurred prior to geomagnetic storm sudden commencements. Since a single reflection would be inadequate to explain the effect, these authors invoked multiple reflection on looped magnetic field lines which intersect a shock in two places or between two approaching shocks to obtain a many-fold increase in energy.

These ground-based observations were confirmed by in situ spacecraft observations, notably by Explorer 12 (Bryant et al., 1962) and later by numerous other US, Soviet, and German spacecraft. The observations are by now quite detailed in terms of mass, energy, directional and time resolution and show that in addition to the relatively slow, quasi-exponential increases mentioned above there are also short-lived, highly anisotropic bursts ("shock spikes") which occur near the time of but not necessarily coincident with the shock passage (Sarris and Van Allen, 1974). Perhaps the most impressive examples of shock acceleration in interplanetary space are those obtained by Helios (Richter and Keppler, 1977), and from Pioneers 10 and 11 (Barnes and Simpson, 1976). The latter show distinct double-peaked structures associated with forward-reverse shock pairs which occur as part of corotating interaction regions in the solar wind. The peaks are evidently the result of shock acceleration but the characteristic depression in the middle is probably the result of adiabatic deceleration in the expanding region between the two shocks (Skadron and Axford, 1976; Axford, 1977).

Isolated bursts and a generally structured distribution of energetic electrons was observed by several of the first spacecraft to pass out of the earth's magnetosphere on the sunwards side (Van Allen, 1959). These electron fluxes were considered to be part of the evidence for the existence of a bow shock in front of the magnetosphere and to be the result of acceleration in shock-associated turbulence (Axford, 1962). Jokipii and Davies (1964) interpreted the electrons found upstream of the shock as being the result of Fermi acceleration between the shock wave and scattering fields carried in the approaching solar wind (for a detailed discussion, see Jokipii, 1966). This suggestion, which is in principle similar to that of Schatzmann (1963) could not be fully worked out in the absence of suitable transport equations for the energetic electrons. However, it is essentially the concept involved in present-day scattering theories of shock acceleration as outlined in section 6 of this paper (see also Van Allen and Ness, 1967). In fact, it is possible that a third quite different mechanism may be responsible for many of the electron bursts seen upstream of the earth's bow shock, namely an electrostatic field in regions of the bow shock which are approximately tangential to the external interplanetary magnetic field and which increases the electron energy by a factor of the order of 10-40 times the solar wind proton energy of about 1 keV (e.g. Sonnerup, 1969). It would require only modest scattering in pitch angle for electrons which have transiently achieved such energies to escape upstream across the curved shock surface.

5. ACCELERATION BY LAMINAR SHOCKS

In the absence of pitch angle scattering energetic particles with large gyro-radii which traverse abrupt discontinuities in magnetic field strength and/or direction are found to conserve their magnetic moment, at least approximately (Parker, 1958; Alexeyev and Kropotkin, 1970). This has the immediate consequence for shocks propagating perpendicular to the upstream magnetic field that the increase in perpendicular kinetic energy (non-relativistic) is proportional to the increase in magnetic field strength and the parallel kinetic energy remains unchanged. The energy gain can be viewed as being the result of drift in the inhomogeneous magnetic field near the shock wave in the direction parallel or anti-parallel to the ambient electric field and in the case of strong shocks, amounts to a factor 4 for non-relativistic particles. This is sufficient to enhance the synchrotron emissivity per unit volume of the medium by a factor $\sim$ 500-1000 so it is clear that shocks are likely to produce significant inhomogeneities in the galactic non-thermal radio emission (cf. Hoyle, 1960; Chevalier, 1977).

In considering the reflection of particles from non-perpendicular shocks it should be noted that a particle must have a certain minimum kinetic energy T_m if it is required to move upstream:

$$T_m = \frac{1}{2} mV_1^2 \cot^2\theta_1 , \tag{2}$$

where the magnetic field line makes an angle θ_1 with respect to the shock surface and V_1 is the flow speed normal to the shock. For example, in the case of the shock wave terminating the supersonic solar wind near the ecliptic plane at a distance of 100 AU, say, $\theta_1 \sim 0.5°$ and $T_m \sim 10$ MeV/nucleon: galactic particles with lower energies are unable to penetrate into the supersonic region.

Assuming that particles interact with an oblique shock in such way that their magnetic moments are conserved one finds that for reflection to occur the incident particles must have a certain minimum energy T_1 and the reflected particles have a minimum energy T_2 in the shock frame:

$$T_1 = \frac{B_1}{(B_2 - B_1)} T_m , \qquad T_2 = (4 + \frac{B_1}{(B_2 - B_1)}) T_m , \tag{3}$$

where B_1 and B_2 are the magnetic field strengths ahead and behind the shock respectively. The transmitted particles tend to have a flat pitch angle distribution and there is no restriction on their energies, whereas reflected particles tend to be beamed along the magnetic field and must have energies greater than T_2. As in the case of perpendicular shocks, the acceleration can be regarded as being the result of drift in the direction of the electric field which occurs when particles interact with the inhomogeneous magnetic field at the shock front.

Numerous analytic and numerical studies have been made of energetic particle interactions with laminar shock waves (Dorman and Freidman, 1959; Parker, 1958, 1963; Shabansky, 1962; Wentzel, 1963, 1964; Hudson, 1965; Sonnerup, 1969; Alexeyev and, Kropotkin, 1970; Sarris and Van Allen, 1974; Chen and Armstrong, 1975; Pesses, 1979; Terasawa, 1979a). The results obtained must be correct if the particle gyroradius R_g is large compared with the shock thickness d, provided only that the scattering mean free path λ is very much larger than R_g. These are conditions which should certainly prevail in shock waves occurring in the interstellar medium and in the solar wind, so that this type of acceleration must be taken into account. If $R_g \sim d$ similar acceleration effects should occur but in this case the particles are likely to be affected by turbulent fields in the shock itself, so that the analysis would not be strictly valid.

One should note that a model based on purely laminar reflection and transmission of particles at shock waves cannot explain features such as the quasi-exponential increases which occur ahead of interplanetary shock waves. Such effects require that scattering be present, which in turn produces a qualitative change in the acceleration mechanism and can lead to acceleration to much higher energies. Attempts have been made to include scattering in numerical simulations of the laminar reflection process but in some cases the authors appear to have done this incorrectly (see, however, Terasawa, 1979b). It seems to be the case, nevertheless, that interplanetary shock spike events can be explained entirely on the basis of direct acceleration at the shock front itself without recourse to pitch angle scattering (Sarris et al., 1976).

6. ACCELERATION BY SHOCKS IN SCATTERING MEDIA

In order to understand the effects of scattering on the acceleration of energetic particles by shock waves it is necessary to have suitable transport equations describing the effects of convection and diffusion of particles in the scattering medium as well as energy changes due to various causes. The appropriate Fokker-Planck equation

$$\frac{\partial U}{\partial t} + \frac{\partial}{\partial x}(VU) = \frac{\partial}{\partial x}\left(\kappa \frac{\partial U}{\partial x}\right) + \frac{1}{3}\frac{\partial V}{\partial x}\frac{\partial}{\partial T}(\alpha T U) + Q, \qquad (4)$$

was given independently by Parker (1965) and Dolginov and Toptygin (1966). Here V is the speed of the scattering medium, Q represents energy gains and losses other than those due to compression and expansion, and $\alpha = (T+2T_0)/(T+T_0)$ where T_0 is the particle rest energy. The equation is written in its one-dimensional form appropriate to the treatment of plain shocks and accordingly κ is to be regarded as the coefficient for diffusion normal to the shock front. In order to make progress with the problem of shock acceleration it is necessary to use a pair of first order equations which are equivalent to (4)

$$\frac{\partial U}{\partial t} + \frac{\partial S}{\partial x} = -\frac{1}{3} V \frac{\partial^2}{\partial x \partial T}(\alpha T U) + Q, \qquad (5)$$

$$S = V\left(U - \frac{1}{3}\frac{\partial}{\partial T}(\alpha T U)\right) - \kappa \frac{\partial U}{\partial x}, \qquad (6)$$

obtained originally by Gleeson and Axford (1967). Here S is the particle current in (T, T+dT) and $C = (1 - (\partial(\alpha TU)/\partial T)/3U)$ is the "Compton-Getting" coefficient. In deriving these equations it is assumed that (1) the distribution function is nearly isotropic ($S/U \leq 0.1$ v, where v = particle speed); (2) the Compton-Getting transformation is linear (V << v); and (3) inertia effects are negligible ($\partial S/\partial t << v^2 S/3\kappa$). It should be noted that these assumptions can easily be violated in shock acceleration problems and conclusions should accordingly be drawn only with care. The equations are readily generalized to take into account anisotropic diffusion, three-dimensional flow, etc. Furthermore, it should be noted that the "adiabatic" acceleration/deceleration terms appearing in (4) and (5) are first order Fermi effects being in principle reversible and non-statistical in nature. However, if the velocity of the scattering medium has a random component with wave-lengths large compared with the scattering mean free path, it is possible, with suitable averaging, to derive the acceleration rate due to magnetic pumping given previously (Bykov and Toptygin, 1979).

Anticipating the use of the above equations in shock acceleration problems Gleeson and Axford (1967) gave the following "jump" conditions for the energetic particle density and current for a shock wave described as a rapid or discontinuous change of V with respect to x:

$$U_1 = U_2, \quad S_1 = S_2, \tag{7}$$

where subscripts 1 and 2 refer to conditions ahead of and behind the discontinuity (shock), respectively. These conditions are obtained by integrating equations (5) and (6) across the transition, assuming uniform conditions on either side, that V, U, S, C, Q and $\partial U/\partial t$ are finite and that $\kappa > 0$. Note that although (7) refers to simple normal shock waves, a generalization to the case of oblique shocks with anisotropic diffusion coefficients is trivial. The procedure outlined is in fact invalid for the general case of oblique shock waves where the magnetic field lines are sharply kinked since we expect that $\lambda \gg d$ and the above equations may be invalid if used on such a scale. Furthermore, the possibility of reflection by the shock front itself is completely neglected; indeed, since particle inertia is neglected in the transport equations the requirement for a minimum energy for motion upstream and all the other considerations discussed in section 5 are also absent. Consequently, results obtained on the basis of the above equations and jump conditions should be considered as lacking an important physical element in all but the cases of perpendicular, parallel, and very weak oblique shocks where the magnetic field lines remain straight.

Consider a situation in which a shock wave is situated at $x = 0$ facing in the negative x- direction so that $V = V_1$ in $x < 0$ and $V = V_2$ in $x > 0$, with $V_1 \geq V_2$. We assume that $U(T) \to U_1(T)$ as $x \to -\infty$ and U remains finite ($U \to U_2(T)$) as $x \to +\infty$. The appropriate solutions of equations (5) and (6) are

$$U(x,T) = U_1(T) + \left[U_2(T) - U_1(T)\right] \exp \int_0^x (V/\kappa)dx, \quad x < 0,$$
$$U = U_2(T), \quad x > 0, \tag{8}$$

where $U_2(T)$ is to be determined by requiring continuity of $S(T)$ at $x = 0$:

$$\frac{1}{3}(V_1 - V_2) \frac{\partial}{\partial T}(\alpha T U_2) + V_2 U_2 = V_1 U_1. \tag{9}$$

Hence

$$U_2(T) = \frac{3V_1}{\alpha(V_1 - V_2)} T^{-(\lambda+1)} \int U_1(T) T^\lambda dT, \tag{10}$$

where $\lambda = 3V_2/\alpha(V_1-V_2)$.

The following are examples of the types of accelerated spectra that can be obtained:

(a) $U_1(T) = U_1 \delta(T-T_1)$,

$U_2(T) = (V_1 \lambda/V_2)(U_1/T_1) H(T-T_1)(T/T_1)^{-(\lambda+1)}$; (11)

(b) $U_1(T) = U_1(T/T_1)^{-\mu} H(T-T_1)$,

$U_2(T) = \beta U_1 \left[(T/T_1)^{-\mu} - (T/T_1)^{-(\lambda+1)}\right] H(T-T_1)$; (12)

(c) $U_1(T) = U_1(T/T_1)^{-(\lambda+1)} (\log_e T/T_1)^n H(T-T_1)$,

$U_2(T) = (V_1 \lambda/V_2) U_1 (T/T_1)^{-(\lambda+1)} \dfrac{(\log_e T/T_1)^{n+1}}{(n+1)} H(T-T_1)$; (13)

where $H(x)$ is the Heaviside step function and $\beta = 1/(1-C_1(V_1-V_2)/V_1)$. The basic result (11) was obtained independently by Axford et al. (1977), Krimsky (1977), Blandford and Ostriker (1978), and Bell (1978). The most notable features of these solutions are that the pre-shock increase is exponential (if κ_1 is independent of x) with a scale length κ_1/V_1 and that a special power law appears with index $(\lambda+1)$, which is dependent only on the shock strength and not on the form of the diffusion coefficient.

A number of generalizations of this result are possible:
(a) The diffusion coefficient may be taken to be anisotropic and the shock oblique. One finds that in the case where the magnetic field makes a small angle with the shock surface the anisotropy can become very large and the transport equations may then cease to be valid.

(b) The effects of sources and sinks can be taken into account in certain special cases. For example, if $Q = -U/\tau(T)$ in $x > 0$ to simulate ionization and nuclear interaction loss for example, one finds that the effectiveness of the acceleration mechanism is drastically reduced if $\tau \leq \kappa_2/V_2^2$ (Völk et al., 1979; Bulanov and Dogiel, 1979). The effects of losses due to synchroton and inverse Compton emission have been considered by Krimsky et al. (1979b) and the effects of acceleration by turbulence have been considered by Tverskoi (1978) and, in a Monte-Carlo treatment, by Scholer and Morfill (1975).

(c) Time variations, namely the inclusion of the term $\partial U/\partial t$ in (8), have been treated by Fisk (1971), Toptygin (1978) and Forman and Morfill (1979). It is found that the equilibrium spectrum is achieved only after a time of order $\kappa_1 \ln(T/T_1)/V_1^2$, which, for the more energetic particles, may be excessively long in comparison with other characteristic time scales so that the equilibrium spectrum is never achieved at high energies.

(d) Various solutions have been given for the case of flow with spherical symmetry. These include: the familiar solar modulation problem with a shock (Fisk, 1969; Axford, 1972; see also Jokipii, 1968); a Forbush increase/decrease model with a shock wave moving outwards with constant speed (Fisk, 1969); a driven double shock/contact surface combination representing a corotating interaction region (Skadron and Axford, 1976); a Sedov blast wave solution (Krimsky et al., 1979a), and an accretion shock model (Cowsik and Lee, 1980). Some of the above results are in need of revision as part of the solution has been overlooked. A quite realistic model of a corotating interaction region has been treated by Fisk and Lee (1980) who have shown that the effects of adiabatic deceleration on both sides of an expanding shock wave in the solar wind are significant in shaping the spectrum, as is also the form of the diffusion coefficient, and that it is possible to account for the observed spectral form of an exponential in rigidity (see also Forman, 1980).

The essential conclusion of this type of analysis is that, provided scattering occurs, energetic particles can be accelerated very efficiently by shock waves without recourse to the laminar reflection mechanism discussed in section 5. Power law spectra can be achieved fairly easily and in the case of strong shocks ($V_1 \to 4V_2$) we obtain $(\lambda+1) \to 2$ for relativistic particles ($\alpha \to 1$) which is rather close to the value 2.6 found for protons at energies of the order of 10 GeV or more (Krimsky, 1977; Blandford and Ostriker, 1978). In fact, not too much should be made of this result since more general analyses show that the spectrum obtained can be affected by such factors as adiabatic acceleration/deceleration, energy sources and sinks, time-dependence and non-linear effects (see section 7), all of which permit the diffusion coefficient (and especially its energy dependence) to have some influence on the spectrum of the accelerated particles. Perhaps the most important conclusion is that the acceleration is non-adiabatic (i.e. irreversible) in the sense that much more energy is given to energetic particles than would be achieved by adiabatic compression with the same compression ratio. This possibility, which was first noted by Hoyle (1960), contrasts with the common assumption that the energetic particles are simply compressed adiabatically (e.g. Newman and Axford, 1968; Chevalier, 1977).

7. THE SELF-CONSISTENT PROBLEM

If one calculates the pressure p_{c2} of cosmic rays behind a strong shock wave according to equation (11), it is found that it can become very large:

$$P_{c2} = \int_0^\infty (\tfrac{1}{3} \alpha U_2 T) dT \propto \left[\log_e T/T_1\right]_{T_1}^\infty$$

In reality such a divergence would be easily supressed by effects such as time-dependence and energy losses, however it suggests that we should seriously take into account the effects of cosmic ray pressure on the background plasma flow. This problem has been considered by Axford et al. (1977) (see also Leer et al., 1976) for the case of a steady state one-dimensional flow, thus:

$$\rho V = \rho_1 V_1 = A_1 , \tag{15}$$

$$p + p_c + \rho V^2 = p_1 + p_{c1} + \rho_1 V_1^2 = A_2 , \tag{16}$$

$$\rho V \frac{d}{dx}\left[\frac{1}{2} V^2 + \frac{\gamma}{\gamma-1} \frac{p}{\rho}\right] = - V \frac{dp_c}{dx} , \tag{17}$$

where ρ is the plasma mass density, p and p_c the plasma and cosmic ray pressures, respectively, and V the speed of the plasma in the x direction. In addition we make use of the cosmic ray transport equation (4), integrated over energy and assuming $\kappa = \kappa(x)$ for convenience:

$$V p_c - \kappa \frac{dp_c}{dx} + \frac{\alpha}{3} \int_{-\infty}^{x} \frac{dV}{dx} p_c \, dx = V_1 p_{c1} . \tag{18}$$

It is easily shown, using (15)-(17) that the background plasma behaves isentropically if there are no shock waves in the flow (i.e. p/ρ^γ = constant), and also that the total energy flux is constant:

$$\rho V(\tfrac{1}{2} \rho V^2) + \rho V(\tfrac{\gamma}{\gamma-1} p/\rho) + \int_0^\infty S(T) T dT = F_K + F_T + F_c = A_3 . \tag{19}$$

Numerous solutions of the above set of equations for the case $\alpha = 2$ have been given by Leer et al. (1976) and Axford et al. (1977) with various initial conditions. To understand the nature of the solutions it is sufficient to consider the very simple case in which the plasma pressure is neglected everywhere ($p = 0$), so that ρ and p_c can be eliminated to yield

$$\kappa \frac{dV}{dx} = (1 + \alpha/6)(V_1 - V)(V_2 - V) , \tag{20}$$

where $V_2 = [(1+\alpha/3)A_2/(1+\alpha/6)A_1] - V_1$. Note that for $p_{c1} = 0$, $V_2 \to V_1/4$ if $\alpha = 2$, $V_2 \to V_1/7$ if $\alpha = 1$. We see immediately that the only solutions which are finite over the whole range of x and which begin with $V \to V_1$ as $x \to -\infty$ are such that V decreases smoothly and asymptotically to $V = V_2$ as $x \to \infty$, with a characteristic length scale of the order $L = \kappa/(V_1-V_2)$, provided $V_1^2 \geq (1+\alpha/3)p_{c1}/\rho_1$ (i.e. the incident flow is super supersonic, since $1+\alpha/3$ is the specific heat ratio for the cosmic ray gas).

This result shows quite clearly that shock-like transitions are possible in which the pressure is provided entirely by the cosmic ray component and cosmic ray diffusion plays a role similar to that of heat conduction in ordinary shock waves with vanishing Prandtl number (see Illingworth, 1953, section 5). In this case, we find that the change in cosmic ray energy flux as a fraction of the initial kinetic energy flux is

$$(F_{c2} - F_{c1})/F_{K1} = 1 - (\alpha/(6+\alpha))^2 = 0.92 - 0.98 . \qquad (21)$$

That is, the plasma kinetic energy can be converted to cosmic ray energy with 92-98% efficiency.

It is found that for highly supersonic shocks the transition is smooth as in the above example. In general, however, a plasma shock must be included in the transition using the usual Rankine-Hugoniot conditions and also requiring that p_c and F_c are continuous in accordance with equation (17) (Drury and Völk, these Proceedings).

To obtain the cosmic ray spectrum and current and thereby close the loop self-consistently we must next solve the transport equations (4) and (5) using the form $V(x)$ obtained in the above manner. The result is presumably not very sensitive to the precise form of $V(x)$ but only to the length scale of the transition (L). Furthermore we may permit κ to be energy dependent in this second step since the κ which determines L must be defined by the particles which provide most of the pressure. Analytic solutions of the equations are unfortunately difficult to obtain. However, the results to be expected from a smooth transition, for example, are obvious: (1) low energy particles ($\kappa \ll (V_1-V_2)L$) do not diffuse readily and are simply adiabatically compressed; (2) very high energy particles ($\kappa \gg (V_1-V_2)L$) should see the transition as a discontinuity in flow speed and should accordingly be accelerated more or less as in the non-self-consistent case treated in section 6; (3) a 'special' spectrum may develop and, since there is a length scale in the problem, its form will depend on κ as well as V_1 and V_2. It is unfortunately not easy to solve the one-dimensional transport equation analytically with V a function of x and κ a function of T (and x), however attempts are in progress. Blandford (1980) has carried out a perturbation analysis for the case $p_c \ll \rho_1 V_1^2$ (presuming a shock wave to be present) and finds that the spectrum tends to flatten at high energies, which is not inconsistent with the above remarks. Eichler (1979) has argued that the incident kinetic energy flux is divided equally between relativistic ($\sim$ 1 GeV) particles and plasma thermal energy, however, this is in contradiction to the existence of smooth transitions as shown above.

8. CONCLUSIONS

The essential conclusion of this review is that provided they are scattered effectively and provided energy losses are not too severe, cosmic rays can be very efficiently accelerated by shock waves in the interstellar medium. Indeed, in favourable circumstances, most of the kinetic energy associated with the shock wave can be converted to cosmic ray energy. The provisos are significant since the Alfvén waves which are necessary to scatter the particles can only be produced effectively by the same shock waves and by strong anisotropies in the cosmic rays themselves. In the cooler regions of the interstellar medium (which in any case are regions which do not contain strong shocks) since Alfvén waves tend to be strongly damped by ion-neutral collisions and cosmic ray energy losses due to ionizing collisions and nuclear interactions are also relatively important, one would expect little acceleration to occur. We are therefore directed towards the hot, low density, fully-ionized regions (i.e. large HII regions and supernova remnants) which occupy most of the interstellar medium by volume. In such regions, ionization losses are very low and Alfvén waves are effectively damped only by viscosity, Landau damping, non-linear decay, expansion of the medium, and as a result of second order Fermi acceleration of cosmic rays. In particular, the very low density, high temperature (10^5-10^6K) interstellar medium is especially important since shock waves should tend to speed up and strengthen in such regions and, depending on the geometry of the situation, may propagate over large distances without too much attenuation. [See Cesarsky (1980) for a review of the problem of Alfvén wave generation and damping in the interstellar medium.]

In considering the behaviour of cosmic rays in the galaxy, one should note that during the time $\tau_c \sim 2 \times 10^7$ years, which is of interest, some 5×10^5 supernova occur, each of which expands to a radius of the order of 100 parsecs in a period of 10^6 years or so. This means that every point in the interstellar medium is passed by a rather strong shock wave (~ 100 km/sec) about ten times during the period in question, excluding reflections. From the point of view of the cosmic rays therefore, the interstellar medium is a violent place in which they are continually accelerated or decelerated or accelerated again, with of course a strong net acceleration because the effect is statistically irreversible (Ostriker, 1979). One should also note that the cosmic rays, at least in the low density medium, must continually be transported in one direction or another by mass motions and indeed produce these mass motions in part as a result of their own contribution to the pressure of the medium.

If one adopts the above picture of the interstellar medium, noting of course that it is complicated by the existence of stellar winds and expanding HII regions, there appears to be no strong reason to doubt that galactic cosmic rays are accelerated by shock waves with some assistance from second order Fermi effects and magnetic pumping. Furthermore, the cosmic rays play a major role in the energy balance and dynamics of the medium, and to some extent control their own destiny. That is, cosmic rays are in no sense to be regarded as passive riders on a more or less static interstellar magnetic field which scatters them occasionally so that they eventually by chance find their way out of the galaxy. Instead, the cosmic rays leave the galaxy because they are forced out and because they force their own way out. As a result of this process, a galactic halo is formed and eventually a galactic wind in which the cosmic rays are energetically dominant (Johnson and Axford, 1971). Shock waves which pass into the halo should tend to strengthen as they move into regions of rapidly decreasing density and in doing so further heat the halo gas and accelerate the cosmic rays in a manner analogous to the process suggested by Biermann (1948) for heating the solar corona.

It has been argued occasionally that the diffusion coefficient in the interstellar medium, at least in the direction normal to the plane of the galaxy, should lie in the range 10^{27}-10^{28}cm²/sec since the cosmic ray scale height is of the order of 10^2-10^3 parsecs and the escape time 2×10^7 years. Such large diffusion coefficients, if typical, would tend to make all but nearly perpendicular shocks ineffective in accelerating cosmic rays since the time scale for acceleration by shock waves with speeds of the order of 100 km/sec would be at least 10^{13}-10^{14}sec, which is too long. In fact, this diffusion coefficient has little physical significance beyond the two parameters used in deriving it as it contains the implicit assumption that the interstellar medium is more or less static, non-convective, and lacking the highly dynamic mass motions which are implied in the description given above. If it means anything at all, this diffusion coefficient is related to the supersonic turbulent diffusion of the hot interstellar medium as a whole rather than to the diffusion of the cosmic ray gas it carries along with it. Except in the outer regions of the halo, where the cosmic rays tend to separate out, the diffusion coefficient for the cosmic ray gas relative to the plasma could be very much smaller.

The question of seed particles is an important but difficult question in view of our lack of knowledge of the microscopic properties of the interstellar medium. A number of suggestions for favouring the medium and heavy nuclei relative to hydrogen and helium have been made but it is difficult to find critical tests for any of them except perhaps by observing what happens in the interplanetary medium. One suggestion is that the ionization potential is the significant parameter since this would clearly discriminate against hydrogen and helium (Cassé et al., 1975). If this were the case, however, one would have to rely on the acceleration taking place in relatively cool and low density HI regions which, on the basis of the discussion given here, is most unfavourable since Alfvén waves damp rapidly, ionization losses are relatively important and if the shock waves are strong they would fully ionize the medium in any case. A second suggestion uses the condensation of solids as the filter, namely by invoking sputtering from interstellar grains following passage by a shock wave (Meyer et al., 1979; Cesarsky et al., these Proceedings). The sputtered atoms would have speeds of the order of the shock speed and hence, on becoming ionized, would have some chance of being selectively accelerated. An obvious difficulty with this idea is that helium and neon would be essentially absent.

Since we have advocated the hot intercloud regions of the interstellar medium as being the most likely sites for acceleration it is perhaps more reasonable to seek a seeding mechanism which is consistent with the properties of such a hot plasma. Thus we suggest that the parameters mass (A) and mass per charge (A/Z) could lead to the favoured injection of heavier species into the cosmic radiation. Since it is likely that in a collision-free shock ions tend initially to have the same energy per nucleon rather than the same total energy, the heavy ions are immediately favoured. Ions with large A/Z ratios receive further favourable treatment since their gyro-periods are larger and they therefore tend to resonantly interact with lower frequency components of the shock-induced turbulence where more wave power is available. This hypothesis, or variants of it, could in principle be put to test in the interplanetary medium (e.g. Hamilton et al., 1979). The existence of the anomalous component is perhaps some evidence in its favour since the particles possibly have large A/Z (Fisk et al., 1974; Klecker, 1977) and may result from a combination of stochastic (Fisk, 1976) and shock acceleration (Axford et al., 1977). In addition of course it must be recognized that the shock acceleration mechanism is also selective to the extent that the diffusion coefficient plays a role as it must in the non-linear interaction described in section 7. At a given energy per nucleon heavy ions tend to have higher rigidities and larger diffusion coefficients than protons; consequently, the heavy particles tend to "see" the transition differently and may be preferentially accelerated (cf. Eichler, 1979). It should of course be remembered that seed particles produced by stochastic acceleration in a shock transition must have at least the minimum energy T_m (equation (2)) if they are to be subsequently accelerated in the upstream region.

As far as electrons are concerned, once they are sufficiently relativistic they should behave more or less like protons and be almost as easily accelerated. However, the seeding processes for electrons are quite another matter since they respond to a quite different frequency domain in shock-induced electromagnetic turbulence and also to electrostatic fields in the shock, so that a separate discussion is required. However, it is clear from observations of energetic electrons associated with the interplanetary shocks that there is no principle difficulty in accelerating electrons (Armstrong and Krimigis, 1976).

We have emphasized expanding supernova remnants as the most likely means of accelerating cosmic rays, because more than enough energy is available ($\sim 10^{42}$ ergs/sec) and because they can interact with the hot intercloud medium on a very large scale. However, there are other sources of mass motion in the interstellar medium which are not insignificant from the point of view of energetics, namely novae, expanding HII regions and stellar winds. Novae produce about 10^{39} ergs/sec. In the case of HII regions, the energy available is $\sim 6 \times 10^{41} \eta$ ergs/sec, where η is the efficiency (about 1%) with which the energy of ionising photons is converted to kinetic energy (Kahn and Dyson, 1965). There are of course shock waves around HII regions and cosmic rays must be affected by them (Newman and Axford, 1968); however the shock speeds are relatively low (10-50 km/sec) and they tend to occur in relatively dense HI regions where the conditions for efficient cosmic ray acceleration are unfavourable. Stellar winds would be negligible contributors of kinetic energy ($\sim 3\times10^{37}$ergs/sec) if the solar wind were a typical example; however, it appears that the distribution of stellar wind energy fluxes is very non-uniform and is dominated ($\sim 10^{41}$ergs/sec?) by the same O and B stars which are responsible for the large HII regions (Cassé and Paul, 1979). On energetic grounds alone one could therefore not rule out stellar winds as a significant cause of cosmic ray acceleration (Cassé and Paul, 1979). Dorman (1979) has made this point in a somewhat different way by noting that solar modulation is in fact an acceleration mechanism since the solar wind must do work against the cosmic ray pressure gradient; one can readily show that if the typical stellar wind region were only about ten times larger than the 100 AU expected in the case of the sun the necessary power required to maintain the galactic cosmic radiation would be available. Dorman's calculation is in fact conservative since he neglects the enhancement due to acceleration at the shock wave terminating the supersonic wind.

Finally, it should be noted that intergalactic acceleration of cosmic rays by shocks is a perfectly feasible proposition and may account for the highest energy cosmic rays we observe. Inverse Compton and adiabatic expansion losses must of course be overcome but there is no shortage of seed particles and intergalactic shocks, at least those associated with radio galaxies, certainly exist. On this basis, one is entitled to ask whether galactic winds modulate the intergalactic cosmic radiation and also contribute to their acceleration.

REFERENCES

Alexeyev, I.I. and A.P. Kropotkin: 1970, *Geomag.Aeron.* 10, 755.
Armstrong, T.P. and S.M. Krimigis: 1976, *J. Geophys. Res.* 81, 677.
Axford, W.I.: 1962, *J.Geophys.Res.* 67, 3791.
Axford, W.I. and G.C. Reid: 1962, *J. Geophys. Res.* 67, 1692.
Axford, W.I. and G.C. Reid: 1963, *J. Geophys. Res.* 68, 1793.
Axford, W.I: 1969, *Rev. Geophys.* 7, 421.
Axford, W.I.: 1972, *"Solar Wind"*, NASA SP-308, 609.
Axford, W.I.: 1977, *'Study of Interplanetary Travelling Phenomena'*, (ed. M.A. Shea et al.), Reidel, 145.
Barnes, G.W. and J.A. Simpson: 1976, *Astrophys. J.* 210, L91.
Bell, A.R.: 1978, *M.N.R.A.S.* 182, 147 and 443.
Biermann, L.: 1948, *Z. Astrophys.* 25, 161.
Blandford, R.D. and F.R. Ostriker: 1978, *Astrophys. J.* 221, 229.
Blandford, R.D.: 1980, in press.
Bryant, D.A., T.L. Cline, U.D. Desai, and F.B. McDonald: 1962, *J. Geophys. Res.* 67, 4983.
Bulanov, S.V. and V.A. Dogiel: 1979, *Proc. 16th Internat. Cosmic Ray Conf. (Kyoto)* 2, 70.
Bykov, A.M. and I.N. Toptygin: 1979, *Proc. 16th Internat. Cosmic Ray Conf. (Kyoto)* 3, 66.
Cassé, M., P. Goret, and C.J. Cesarsky: 1975, *Proc. 14th Internat.*

Cosmic Ray Conf. (Munich) 2, 646.
Cassé, M. and J.A. Paul: 1979, *Proc. 16th Internat. Cosmic Ray Conf. (Kyoto)* 2, 103.
Cesarsky, C.J.: 1980, *Ann. Rev. Astron. Astrophys.*, in press.
Chen, G. and T.P. Armstrong: 1975, *Proc. 14th Internat. Cosmic Ray Conf. (Munich)* 5, 1814.
Chevalier, R.A.: 1977, *Astrophys. J.* 213, 52.
Colgate, S.A. and M.H. Johnson: 1960, *Phys. Rev. Lett.* 5, 235.
Colgate, S.A.: 1979, *Proc. Symp. on 'Very Hot Plasmas in the Universe'*, IAU (Montreal), in press.
Cowsik, R. and M.A. Lee: 1980, in press.
Dolginov, A.Z. and I.N. Toptygin: 1966, *JETP* 51, 1771.
Dorman, L.I.: 1963, *Prog. Elem. Part. and C.R. Phys.* 7.
Dorman, L.I. and G.I. Freidman: 1959, *"Problems of MHD and Plasma Physics"* (Riga), 77.
Dorman, L.I.: 1979, *Proc. 16th Internat. Cosmic Ray Conf. (Kyoto)* 2, 49.
Eichler, D.: 1979, *Astrophys. J.* 229, 419.
Fermi, E.: 1949, *Phys. Rev.* 75, 1169.
Fermi, E.: 1954, *Astrophys. J.* 119, 1.
Fisk, L.A.: 1969, *Ph.D. Thesis*, U.C.S.D.
Fisk, L.A.: 1971, *J. Geophys. Res.* 76, 1662.
Fisk, L.A., B. Kozlovsky, and R. Ramaty: 1974, *Astrophys. J.* 190, 235.
Fisk, L.A.: 1976, *J. Geophys. Res.* 75, 1169.
Fisk, L.A. and M.A. Lee: 1980, *Astrophys. J.*, in press.
Forman, M.A. and G. Morfill: 1979, *Proc. 16th Internat. Cosmic Ray Conf. (Kyoto)* 5, 328.
Forman, M.A. and O.A. Schaeffer: 1979, *Rev. Geophys. Space Phys.* 17, 552.
Forman, M.A.: 1980, *Proc. COSPAR Symp. on the Heliosphere*, in press.
Frisch, P.C.: 1979, *Astrophys. J.* 227, 474.
Garcia-Munoz, M., G.M. Mason, and J.A. Simpson: 1977, *Astrophys. J.* 217, 859.
Giler, M., T. Wdowczyk, and A.W. Wolfendale: 1979, *Proc. 16th Internat. Cosmic Ray Conf. (Kyoto)* 1, 507.
Gleeson, L.J. and W.I. Axford: 1967, *Astrophys. J.* 149, L115.
Hamilton, D.C., G. Glockler, T.P. Armstrong, W.I. Axford, C.D. Bostrom, C.Y. Fan, S.M. Krimigis, and L.J. Lanzerotti: 1979, *Proc. 16th Internat. Cosmic Ray Conf. (Kyoto)* 5, 363.
Hasselmann, K. and G. Wibberenz: 1968, *Z. Geophys.* 34, 353.
Honda, M.: 1979, *Proc. 16th Internat. Cosmic Ray Conf. (Kyoto)* 14, 159.
Hoyle, F.: 1960, *M.N.R.A.S.* 120, 338.
Hudson, P.D.: 1965, *M.N.R.A.S.* 131, 23.
Illingworth, C.R.: 1953, *Chapter 4 of "Modern Developments in Fluid Dynamics"* (ed. L. Howarth), Oxford.
Johnson, H.E. and W.I. Axford: 1971, *Astrophys. J.* 165, 381.
Johnson, R.G.: 1979, *Rev. Geophys. Space Phys.* 17, 696.
Jokipii, J.R. and L. Davis, Jr : 1964, *Phys. Rev. Lett.* 13, 739.
Jokipii, J.R.: 1966, *Astrophys. J.* 143, 961.
Jokipii, J.R.: 1968, *Astrophys. J.* 152, 799.
Jokipii, J.R.: 1977, *Proc. 15th Internat. Cosmic Ray Conf. (Plovdiv)* 1, 429.
Kahn, F.D. and J.E. Dyson: 1965, *Ann. Rev. Astron. Astrophys.* 3, 47.

Kahn, F.D. and J.E. Dyson: 1965, Ann. Rev. Astron. Astrophys. 3, 47.
Kanbach, G.: 1979, Proc. 16th Internat. Cosmic Ray Conf. (Kyoto) 14, 105.
Klecker, B.: 1977, J. Geophys. Res. 82, 5287.
Krimsky, G.F.: 1977, Dok. Akad. Nauk, SSR, 234, 1306.
Krimsky, G.F., A.I. Kuzmin, and S.I. Petukhov: 1979a, Proc. 16th Internat. Cosmic Ray Conf. (Kyoto) 2, 44.
Krimsky, G.F., A.I. Kuzmin, and S.I. Petukhov: 1979b, Proc. 16th Internat. Cosmic Ray Conf. (Kyoto) 2, 75.
Leer, E., G. Skadron, and W.I. Axford: 1976, EOS 57, 780.
Lingenfelter, R.E.: 1979, Proc. 16th Internat. Cosmic Ray Conf. (Kyoto) 14, 135.
Meyer, P., R. Ramaty, and W.R. Webber: 1974, Phys. Today 27, 23.
Meyer, J.P., M. Cassé, and H. Reeves: 1979, Proc. 16th Internat. Cosmic Ray Conf. (Kyoto) 12, 108.
Newman, R.C. and W.I. Axford: 1968, Astrophys. J. 153, 595.
Ostriker, J.P.: 1979, Proc. 16th Internat. Cosmic Ray Conf. (Kyoto) 2, 124.
Parker, E.N.: 1958, Phys. Rev. 109, 1328.
Parker, E.N.: 1963, Interplanetary Dynamical Processes, (New York: Interscience Publishers).
Parker, E.N.: 1965, Planet. Space Sci. 13, 9.
Pesses, M.E.: 1979, Proc. 16th Internat. Cosmic Ray Conf. (Kyoto) 2, 18.
Richter, A.K. and E. Keppler: 1977, J. Geophys. 42, 645.
Sarris, E.T. and J.A. Van Allen: 1974, J. Geophys. Res. 79, 4157.
Sarris, E.T., S.M. Krimigis, and T.P. Armstrong: 1976, Geophys. Res. Lett. 2, 133.
Sarris, E.T. and W.I. Axford: 1979, Nature 277, 460.
Schatzmann, E.: 1963, Ann. d'Astrophysique 137, 135.
Scholer, M. and G. Morfill: 1975, Solar Phys. 45, 227.
Schulz, M. and L.J. Lanzerotti: 1974, 'Particle Diffusion in the Radiation Belts', Springer, New York.
Skadron, G. and W.I. Axford: 1976, EOS 57, 980.
Sonnerup, B.U.O.: 1969, J. Geophys. Res. 74, 1301.
Swift, D.W.: 1979, Rev. Geophys. Space Phys. 17, 681.
Somayajulu, B.L.K.: 1977, Geochim. et Cosmochim. Acta 41, 909.
Terasawa, T.: 1979a, Planet. Space Sci. 27, 193.
Terasawa, T.: 1979b, Planet. Space Sci. 27, 365.
Thompson, W.B.: 1955, Proc. Roy. Soc. London A223, 402.
Toptygin, I.N.: 1979, Isv. Akad. Nauk SSR 43, 755.
Tveskoi, B.A.: 1978, Proc. 10th Leningrad Symposium, 137.
Van Allen, J.A.: 1959, J. Geophys. Res. 64, 1683.
Van Allen, J.A. and N.F. Ness: 1967, J. Geophys. Res. 72, 935.
Volk, H.J., G. Morfill, and M.A. Forman: 1979, Proc. 16th Int. Cosmic Ray Conf. (Kyoto) 2, 38a.
Webber, W.R., R.A. Goeman, and S.M. Yushak: 1979, Proc. 16th Internat. Cosmic Ray Conf. (Kyoto), 1, 495.
Wentzel, D.G.: 1962, Astrophys. J. 137, 135.
Wentzel, D.G.: 1964, Astrophys. J. 140, 1013.
Wolfendale, A.W.: 1977, Proc.15th Int. Cos. Ray Con. (Plovdiv) 10, 235.

COSMIC RAY ACCELERATION IN THE PRESENCE OF LOSSES

H.J. Völk, G.E. Morfill and M. Forman
Max-Planck-Institut für Kernphysik
Postfach 10 39 80
6900 Heidelberg - 1, W. Germany

Diffusive Fermi acceleration on hydromagnetic shock fronts (Axford et al., 1977; Bell, 1978a,b; Blandford and Ostriker, 1978) is a fairly slow process: most scatterings are energetically neutral, only those across the velocity jump yield a first order acceleration. Thus energy losses of cosmic rays (CR) due to e.g. adiabatic cooling, ionising, Coulomb-, or nuclear collisions, bremsstrahlung, synchrotron radiation, should play a role in limiting the acceleration, since most shocks are coupled with a loss region (HII regions, SNR's, galactic density wave, stellar wind shocks). In addition, for this process to be a local one in the galaxy, the magnetic irregularities must either be excited by the accelerated CR's or produced by a downstream source. This implies a finite wave build-up time during the shock life time. Nevertheless, in the loss free case, the time-asymptotic amplification is independent of the mean free path λ (or the diffusion coefficient κ) which only appears in the spatial scale for the CR intensity. To investigate the effects of losses, the CR diffusion equation is amended by a simple loss term f/τ, with an (energy dependent) loss time τ, f being the CR momentum distribution. For τ spatially homogeneous, a distributed source is required. Then acceleration is only effective if $X \equiv 4\kappa/V_s^2 \cdot \tau \lesssim 1$, i.e. if the acceleration time $t_{acc} = 4\kappa/V_s^2$ is smaller than τ, on either side of the shock. As the Figure shows, also the spatial intensity profile is modified. Wave excitation in dense clouds is prohibitive (Cesarsky and Völk, 1978). Even in a "warm" (T $\sim 10^4$K) intercloud medium shock speeds $V_s \gtrsim 3 \times 10^7$ cm/sec are required to accelerate mildly relativistic particles. Waves from an upstream source (star) inside clouds should frictionally dissipate after distances $L \lesssim 5 \times 10^{14}$ cm $<< \lambda$ (10 MeV) if a solar wind scaling is adopted. Thus, presumably in such a case there is no acceleration of stellar or shock-injected particles through a stellar wind shock by scattering within the cloud, but possibly by reflection from beyond the cloud, the condition being $1/\tau \cdot V_s \lesssim 1$, where l is the linear cloud size. Diffusive approach from outside to a standing shock (Jokipii, 1968) appears to be very energy-selective, even in a loss free medium.

Extended SN shocks in a hot medium appear very efficient accelerators. However the waves are subject to severe nonlinear Landau damping (Lee and Völk, 1973). Neglecting trapping of thermal ions (Kulsrud, 1978), this leads to a wave cutoff resulting in strongly suppressed acceleration below that particle momentum where the CR intensity turns down to a spectrum less steep than p^{-2} thus possibly requiring injection from behind. Also t_{acc} exceeds the typical SN life time of 8×10^5 yr. for energies beyond about 10 GeV in this limiting case.

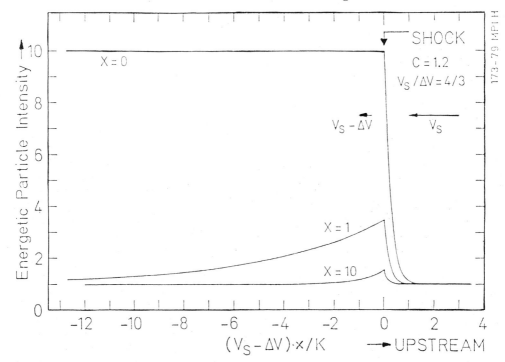

Figure: Relative intensity vs distance in units of $\kappa/V_s - \Delta V)$ near a strong shock. A power law source $\sim p^{-3C}$ is assumed. $X = 4\kappa/V_s^2 \tau$, for various uniform loss rates $1/\tau$.

References

Axford, W.I., et al.: 1977, Proceedings 15th International Cosmic Ray Conference, Plovdiv, 11, 132.
Bell, A.R.: 1978a, MNRAS, 182, 147; 1978b, MNRAS, 182, 443.
Blandford, R.D., and Ostriker, J.P.: 1978, Ap.J.Lett., 221, L29.
Cesarsky, C.J., and Völk, H.J.: 1978, Astron.Astrophys. 70, 367.
Jokipii, J.R., 1968, Ap.J., 152, 799.
Kulsrud, R.M., 1978, Copenhagen Univ. Obs. pp. 428, 317.
Lee, M.A. and Völk, H.J.: 1973, Astrophys.Space Sci., 24, 31.

COSMIC-RAY INJECTION INTO SHOCK-WAVES

Catherine J. Cesarsky
Section d'Astrophysique
Centre d'Etudes Nucléaires de Saclay, France

Jean-Pierre Bibring
Laboratoire René Bernas, Orsay, France

When corrected for the effects of propagation in the interstellar medium (i.s.m.), the observed composition of galactic cosmic rays can give us some clues as to the origin of these particles. It is noteworthy that the main pecularities of the cosmic ray source composition (CRS), as compared to normal i.s.m. abundances (Meyer 1979), bear some resemblance to that of i.s. grains, as inferred from i.s. absorption line measurements (e.g. York 1976) : (1) the refractory elements Al, Si, Mg, Ni, Fe and Ca, which in i.s. clouds are almost completely locked into grains, are present with normal abundance ratios in the CRS. (2) normalized to Si, the volatile and reactive elements C, N, O, S and Zn are underabundant in CRS by factors of 2.5 to 6 ; these elements are only partially depleted in the i.s.m. (3) at a given rigidity the ratios H/Si and He/Si are lower than in the i.s.m. by a factor of ~ 25 ; while H and He atoms are virtually absent in i.s. grains. (1) implies that cosmic rays originate in astrophysical sites where the grains have either not condensated as yet, or where they have been (at least partially) destroyed. Then , to account for (2) and (3), one might consider that an unspecified mechanism selects the particles to be accelerated, possibly according to their first ionization potential (Cassé 1979 and references there-in).

We present here an alternative point of view : we propose that cosmic rays are accelerated out of grain material freshly released in the i.s.m., and of ambient particles which have interacted with the grains. A more specific model, which can account for (1), (2) and (3), is based on the dynamical and physical behaviour of i.s. grains embedded in clouds swept by fast ($v_s \gtrsim 100$ Km s^{-1}) i.s. shock waves.

a) Gas-grain dynamics

The interaction of a shock wave of velocity v_c with a gas cloud is radiative if the gas behind the shock cools down in a time shorter or comparable to that of passage of the shock [i.e. if the gas column N > v_s (n t_c), where n is the density and t_c the cooling time ; e.g. if N $\gtrsim 10^{18}$ cm^{-2} for $v_s = 100$ Km s^{-1}] . Behind the shock, the gas cools rapidly and gets compressed ; the grains, whose interaction length with the gas is much longer than the mean free path of thermal particles, do not "see" the shock, so that they have a velocity of order v_s with respect to shocked gas. If the magnetic field has a component parallel to the shock front, the grains can be accelerated by a betatron mechanism, which offsets the braking effect of ambient particles. Consequently, gas-grain interactions lead to chemical and physical implantation effects, including sputtering ; this is probably why refractory elements are less depleted in fast interstellar clouds (Spitzer 1976).

b) Sputter-induced suprathermal composition

Particles with energy < 600 ev/n, corresponding to relative velocities <350 Km/s, are implanted into the grains at depths < 300 $\overset{o}{A}$. If the grain diameter exceeds ~ 500 A°, no transmission sputtering occurs ; then, for the dominant sputtering species, He, the net sputtering yield at relevant energies is $\sim 1-2 \times 10^{-1}$ at/at for most refractory materials. If the grain temperature is high enough, most of the implanted H and He atoms diffuse out. Thus, sputtering in the cooling layers of shocked gas creates a population of particles with grain-like velocities ($v_g \simeq v_s >>$ thermal velocity), which is a mixture of ambient particles (mostly H and He) and of heavier atoms detached from the grains. The number ratio of H to (Z > 2) is consistent with the CRS value of 57 (CRS from Meyer 1980, private communication).

c) The Ne problem

In this framework, Ne atoms would be implanted together with H and He, and subsequently thermally released. One would then get a ratio Ne/H $\simeq$ cosmic value (10^{-4}), which is 8 times smaller than the CRS value. On the other hand, it is possible that i.s. grains, prior to the passage of the shock, have suffered adsorption and/or irradiation sequences that have loaded them with gaseous species, which are locked in the lattice or adsorbed on active sites at the surface. The very top atomic monolayers would then be saturated with C, N, O, Ne and Ar atoms mainly (this picture is consistent with the depletion of Ar derived from Copernicus results, York 1971). Thus, in the most vulnerable part of ordinary silicate grains, the Ne/Mg ratio can be , like in CRS, of order 1.

d) Shock induced injection mechanism

Shock waves have been invoked by many authors as a possible site of cosmic ray acceleration ; this hypothesis is based on theoretical and on energetical arguments, and is supported by interplanetary observations. However, as in most theories of statistical acceleration, the injection problem remains open (e.g. Blandford 1979). We have shown that a suprathermal population of a composition akin to that of CRS is generated behind radiative shock waves. The particles released by the grains will be somewhat boosted by betatron acceleration in the layers which are undergoing compression ; however, in most cases, this mechanism cannot compensate for the ionization losses. Nevertheless, if we postulate the existence of magnetic turbulence in the troubled, recently shocked gas, the suprathermal particles preserve or increase their velocity, so that some of them can overtake the shock front and be injected in a shock acceleration mechanism.

REFERENCES

Blandford, R.D., 1979, Workshop on Particle Acceleration Mechanisms in Astrophysics, AIP Conf. Proc. n° 56, p. 333.
Cassé, M., 1979, ibidem, p. 211.
Meyer, J.P., 1979, 16th Int. C.R. Conf., Kyoto, 2, 115
Spitzer, L., 1976, Comments Ap., 6, 157.
York, D.G., 1976, Mem. S.A. It., 493.

SHOCK STRUCTURE INCLUDING COSMIC RAY ACCELERATION

L. O'C. Drury and H.J. Völk,
Max-Planck-Institut für Kernphysik,
Postfach 10 39 80
6900 Heidelberg 1, W. Germany

The backreaction of shock accelerated cosmic rays (CR) on the hydrodynamic flow is studied in a simple macroscopic model introduced by Axford et al. (1977): the fluid is isentropic except at discontinuities and the energy of the scattering wave field is neglected. With gas density ρ, velocity u, pressure p_G and a constant specific heat ratio γ_G we have in the one dimensional steady state:

$$\rho u = A; \quad A u + p_G + p_C = B \qquad (1)$$

$$u \frac{\partial p_C}{\partial x} + \gamma_C \cdot p_C \cdot \frac{\partial u}{\partial x} = \frac{\partial}{\partial x} \int dp \cdot \kappa' \cdot p^3 \cdot v \cdot \frac{\partial f}{\partial x} \qquad (2)$$

where the CR pressure $p_C = 4\pi \int p^2 dp \cdot p \cdot v \cdot f(x,p)$ and γ_C equals 4/3 (5/3) in the (non-)relativistic case but is assumed constant; with an effective diffusion coefficient (assumed positive) determining the spatial structure $\kappa(x) = \int p^3 v dp \cdot \kappa' \cdot (\partial f/\partial x) / \int p^3 v dp \partial f/\partial x$, given in terms of the CR momentum distribution $f(x,p)$ and the diffusion coefficient κ', (2) integrates to

$$\frac{1}{2} A u^2 + \frac{\gamma_G}{\gamma_G - 1} u p_G + \frac{\gamma_C}{\gamma_C - 1} u p_C = C + \frac{\kappa}{\gamma_C - 1} \frac{\partial p_C}{\partial x} \qquad (3)$$

with A,B,C constant. The figure is a (u,p_C) diagram for studying (1) and (3); it shows the physically allowed triangular region, wherein the upstream and downstream states lie on the hyperbola $\partial p_C/\partial x = 0$. Those adiabats that do not cross the line $Au = \gamma_G p_G$ between these states correspond to smooth transitions. Adiabats having intersection points with the sheared reflection of the hyperbola in that line may continue from one of them as shock transitions along p_C = const., with $\gamma_C u p_C - \kappa (\partial p_C/\partial x)/(\gamma_C - 1)$ continuous. In general there exist 3 or 1 solutions, the extension of the test particle solution (Axford et al., 1977; Bell 1978a,b; Blandford and Ostriker, 1978) corresponds to that with smallest downstream p_C. The two other possible solutions have finite p_C downstream and finite or zero p_C upstream. In the latter case they should be interpreted as CR confinement and acceleration by a CR-free supersonic wind. They have no correspondent test particle solution of CR accelerated at a stellar wind terminal shock.

The relation between upstream and downstream pressures and the magnitude of discontinuities can be (at least time-asymptotically) determined within the model which also shows that a shock can put almost all its flow energy into cosmic rays.

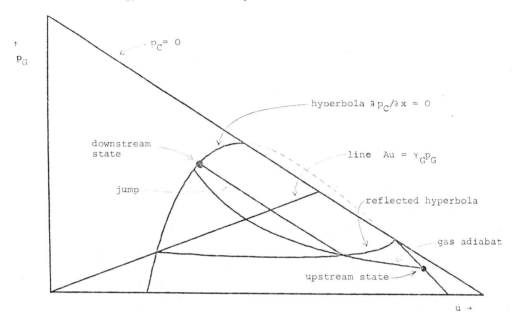

Fig   The shock construction for the case M=2.0, N=0.3, γ_G=5/3, γ_C=4/3.

References

Axford, W.I., et al.: 1977, Proceedings 15th International Cosmic Ray Conference, Plovdiv, 11, 132.
Bell, A.R.: 1978a, MNRAS, 182, 147
          1978b, MNRAS, 182, 443.
Blandford, R.D., and Ostriker, J.P.: 1978, Ap.J.Lett., 221, L29.

SELECTIVE EFFECTS IN COSMIC RAYS INDUCED BY COULOMBIAN INTERACTIONS WITH FINITE TEMPERATURE PLASMAS

J. Pérez-Peraza
Tata Institute of Fundamental Research, Bombay, India and
Instituto de Astronomia, U.N.A.M., Mexico, D.F., Mexico, and
S.S. Trivedi
Tata Institute of Fundamental Research, Bombay, India

The role of Coulombian energy losses in cosmic ray physics is generally over simplified by using the Bethe-Block formulation which does not depend explicitly on the temperature of the medium. The role of low energy particles is usually neglected, as a result of the over estimation of losses when the temperature of the medium is ignored. A deep analysis of Coulombian losses may raise the importance of these particles in the dynamics of the Galaxy. In fact, the deceleration of these particles is determined by charge interchange processes with the target ions and electrons, which energy dependence is roughly the inverse of ionisation losses. Even high energy particles may be subject to this kind of deceleration if the temperature is very high. The consideration of Coulombian losses through all energy ranges with explicit dependence on the temperature has been discussed by Perez and Lara (1979): a fully ionized medium of hydrogen has been assumed to prevail in most of cosmic ray sources. One kind of the implications is the determination of particle composition. It is claimed that a given kind of ion is preferentially accelerated or depleted depending on whether the acceleration is higher or lower than the deceleration rate at the beginning of the acceleration of thermal material. Species which undergo depletion are accelerated only if their energy is higher than that for which both rates are equated (E_c, E_c', E_c'') in such a way that only those of the hot tails of their thermal distributions are effectively accelerated. These will appear depleted relative to other species which are free accelerated because their deceleration rates at low energies are lower than the acceleration rate. It can be noted in the next figures, that if both rates would not intersect at the beginning of the acceleration, they would not join at higher energies because the acceleration rate grows faster with energy than the deceleration rate. Three arbitrary acceleration rates are used for illustration: Fermi-2nd order ($\alpha\beta W$), Betatron or adiabatic heating ($\alpha\beta^2 W$) and shock wave acceleration (αW), where α, β and W are the

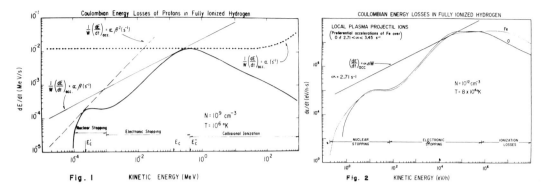

Fig. 1   KINETIC ENERGY (MeV)

Fig. 2   KINETIC ENERGY (eV/n)

efficiency, the particles velocity and the total energy per nucleon respectively. In Fig.1 it can be seen that this selective acceleration relative to Coulombian losses is defined at different energy levels depending on the kind of acceleration involved. Since the main effect of the temperature on the losses at the beginning of the acceleration is through the local charge states of the ions, the sequence of energy losses among different species is highly assorted. This is translated in a great amount of possibilities of particle enhancements and depletions according to the temperature of the source and the kind of acceleration operating therein. If particles under go acceleration in a fully stripped state, the sequence of losses at all energy levels is such that the heavy elements are depleted in relation with the lighter ones; same is the situation, what-ever the initial charge state, for high energy particles in the range of ionisation. It may be concluded, on basis to the observational enhancement of heavy cosmic rays, that hot regions are not likely sources, and that acceleration initiates from thermal energies. On Fig.2 it is illustrated the enhancement of Fe over O in solar flare conditions, on basis to the charge states as given by Jordan (1969). If $\alpha < 2.71$ s^{-1} both elements would be depleted, whereas if $\alpha > 3.45$ s^{-1} both would be preferentially accelerated.

The temperature effect must also be considered in the demodulation of particles fluxes to derive the cosmic ray source abundances within the Galoctogenic or Autogenic hypothesis. The ability of KeV particles to escape from their sources and propagate in the inhomogeneous interstellar medium must be quantified on the same grounds.

Acknowledgments:
J.P.P. wish to thank Prof. S. Biswas for his kind hospitality and to the TIFR for financial support.

References:
Jordan, C: 1969, Mon. Not. R. Ast. Soc., 142, 501
Pérez-Peraza, J. and Lara, R: 1979, Proc. of 16th Int. Conf. Rays Conf., 5, 153

HIGH ENERGY PHENOMENA IN THE SUN

Claudio Chiuderi
Istituto di Astronomia, Università di Firenze, Italy.

INTRODUCTION

High energy phenomena in the solar physics context, simply means solar flares. To be sure, the energies attained during flares are certainly not very impressive on a cosmic-ray scale. The most energetic particles belong the GeV range, the highest temperatures are of the order of 10^7 K, γ-ray emission is occasional and the total energy emitted remains below 10^{33} ergs for all the flares so far observed. Apart from an absolute energy scale, flares are also energetically irrelevant on a solar scale. In fact in a large flare a few units in 10^{32} ergs are emitted, with a total duration of about one hour and a total surface area involved of a few units in 10^{-4} of the solar surface. Recalling the values of the luminosity, $L_\odot \simeq 4 \times 10^{33}$ erg s^{-1} and the solar flux $F_\odot \simeq 6.3 \times 10^{10}$ erg cm^{-2}s^{-1}, we see that

$$L_{flare} \simeq 10^{-5} L_\odot \quad \text{and} \quad F_{flare} \simeq (10^{-1} - 10^{-2}) F_\odot \ .$$

In spite of their irrelevance to the Sun's global energetics, flares are an important and fascinating phenomenon, not only for solar physics but for the whole of astrophysics as well. The reason is that they most likely are a small-scale model of a class of phenomena widespread in the astrophysical Universe: the violent conversion of some form of energy into heat and kinetic energy of accelerated particles. As we shall see, it can be argued that in flares the primary energy supply comes from previously stored magnetic energy. Flares occur relatively nearby and can be observed to a reasonable degree of detail. We can hope, therefore, to be able to identify the key processes responsible for the transformation of magnetic energy in other forms of energy. Nobody can be sure that the information gained through the study of flares can be safely scaled to other astrophysical situations, but this appears to be a sensible first step toward a proper quantitative understanding of one of the basic astrophysical processes.

In the following I shall briefly outline the most important observational aspect of solar flares as we know them today, and concentrate then on the mechanisms of primary energy release.

FLARE OBSERVATIONS

In the following the word flare is meant to indicate the set of high-energy events that involve the presence of suprathermal particles, the emission of radiation whose brightness temperature exceeds the local kinetic temperature and the ejection of large masses of plasma from the solar atmosphere.

Flares are observed over the whole electromagnetic spectrum, which implies that a large vertical portion of the solar atmosphere is involved. However, they are seldom seen in white light, a direct result of the low flare luminosity as compared with the total solar output. Observations show that the continuum of the solar spectrum is hardly affected by flares, while the most prominent absorption lines show strong temporal variations, thus proving the chromospheric rather than photospheric nature of the optical flare. In the temporal development of a flare, three phases are usually recognized. A precursor phase, where a slowly rising intensity is observed mainly in EUV and soft X-rays and at microwave radio frequencies. Then an impulsive or flash phase follows, characterized by intense emission at practically all wavelengths. Finally we have the gradual or main phase, where the activity decays over varied timescales. Typically the impulsive phase may last 10^2 s and the gradual phase 10^3 s. Emission in the EUV, hard X-rays and microwave is mainly concentrated in the impulsive phase.

As a result of observations from space our global knowledge of the flare phenomenon has considerably improved. A major achievement of the Skylab Mission has been the recognition of the importance of the coronal aspect of the flares and their close association with the solar magnetic field structure. In the light of these new observations the chromospheric flare, the only one once known, must be considered rather a consequence of processes that are initiated in the corona. Coronal flares occur almost exclusively in arch-like structures (coronal loops). The loop is filled with coronal plasma ($T \simeq 1\text{-}2 \times 10^6$K) and links two low-temperature ($T \simeq 1\text{-}5 \times 10^4$ K) chromospheric "ribbons", belonging to oppositely polarized magnetic regions. During the flare's impulsive phase a hot plasma cloud ($T \simeq 2 \times 10^7$ K) is formed at the apex of the loop. The plasma then cools and becomes clearly visible in many intermediate temperature ($T \simeq 2 \times 10^6$ K) coronal lines (post-flare loops). The loops supposedly delineate the gross magnetic field structure. Sometimes they can be observed before the occurence of the flare. They appear to be relatively long-lived structures, with lifetimes of hours to more than one day. Apart from electromagnetic radiation, a major manifestation of flares is the emission of high-energy particles. Acceleration of electrons and ions to 100 KeV is a standard feature, but the so-called "proton flares" produce protons up to 100 MeV and the "cosmic-ray flares" may boost heavy particles in the GeV range.

By interacting with the ambient nuclei in the solar atmosphere, these high energy particles can initiate nuclear reactions. This may result in γ-ray emission via the radiative decay of excited nuclei, or positron annihilation, or deuterium formation. Observations have been successfully performed of the 2.23 MeV line from $p + n \rightarrow d + \gamma$, of the 0.51 MeV line from $e^+ + e^- \rightarrow \gamma + \gamma$ and of the 4.43 MeV and 6.14 MeV de-excitation lines of C^{12} and O^{16}. Another energetically important aspect of flares is the ejection of large masses of plasma ($M \simeq 10^{16}$ g). The kinetic energy associated with the ejecta represents about one half of the flare energy.

This sketchy description of the most relevant observable features of flares, is already sufficient to show the complexity of the phenomenon. The theoretical interpretations do not attempt a detailed description but rather try to identify the basic physical processes responsible for the great variety of the observed effects.

FLARE PHYSICS

The minimum set of questions that <u>any</u> flare theory should answer is the following:

i) Which are the energy build-up mechanisms?
ii) Which are the energy release mechanisms?

As already anticipated, there is a general consensus on the fact that energy is stored in magnetic form prior to the flare. Alternative sources, such as gravitational or thermal energy, fall short of orders of magnitude, especially in the case of large flares. On the contrary, there are in principle no difficulties in fulfilling the energy requirements with magnetic energy: a field of 160 G, occupying a volume of 10^{29} cm^3, has an energy of 10^{32} ergs. Of course the <u>type</u> of field must be specified. In fact potential (i.e. current-free) fields are of no interest here, since they constitute the minimum energy configuration. Thus, if the energy storage mode is magnetic, electrical currents must flow in the solar corona. The presence of currents is also to be expected on different grounds. In fact, the magnetic field lines threading the solar corona must be rooted in the much denser photospheric layers. Given the high electrical conductivity of the solar atmosphere, the field is effectively frozen-in. Thus any motion in the high-β photospheric plasma inevitably induces magnetic stresses, or equivalently currents, that propagate in the low-β coronal plasma. According to this picture, the ultimate energy source for flares resides in those (turbulent) photospheric or sub-photospheric motions that are responsible for the generation of currents. The subject of the dynamical evolution of magnetic fields, especially force-free fields, has recently received a great deal of attention, in spite of the considerable degree of mathematical difficulty involved. Solutions can be found that prove the existence of a critical stress: if the initial magnetic configuration is stressed beyond a certain level a catastrophic rearrangement of the field takes place. This is highly suggestive of the abrupt onset of a physical flare, but a convincing correlation between this rather abstract treatment and the observed properties remains to be made. The above mentioned studies have however shown one thing: the accumulation of magnetic energy appears to be a plausible assumption on general ground, but a direct verification of the validity of the proposed mechanisms requires a knowledge of the magnetic field structure <u>and</u> evolution well beyond present-day possibilities.

If the stored energy is magnetic, we must still explain how this energy is released, and how it is released on a rather fast timescale. The main problem here comes from the fact that the decrease of magnetic energy implies the destruction of part of the magnetic field and this is by no means easy in a highly conductive plasma, like the solar one. The basic equation that governs the changes of the magnetic field is:

$$\frac{\partial \tilde{B}}{\partial t} = \tilde{\nabla} \times (\tilde{v} \times \tilde{B}) + \eta \nabla^2 \tilde{B} \qquad (1)$$

where $\eta = c^2/(4\pi\sigma)$, σ being the electrical conductivity. The above equation displays two characteristic timescales, depending on the ratio of the two terms on the right hand side, i.e. on the magnetic Reynolds number S :

$$S = L\,v/\eta$$

At large S the first term dominates, the fields are frozen-in and the typical timescale is the Alfvén transit time

$$\tau_a = L/v = L/c_A$$

since in this situation the velocity scale is given by the Alfvén speed, c_A. At low S, the field changes are due to resistive diffusion, or in other words to the ohmic dissipation of the sustaining currents. The timescale is then

$$\tau_r = L^2/\eta \quad.$$

In solar situations $S > 10^9$, τ_a is of the order of seconds and τ_r of the order of 10^2 years. Neither of these two timescales is suitable for the description of a flare, whose characteristic time lies in between. It is clear however that the resistive dissipation of the magnetic field is the most direct, and possibly the only, mechanisms for magnetic energy conversion. We are thus faced with a two-fold problem: to locate in the high-S solar corona the places where the resistive term can play a role and to reduce τ_r by many orders of magnitude.

To answer the first question let us consider an incompressible case and rewrite the magnetic induction equation in the form

$$\frac{\partial \underset{\sim}{B}}{\partial t} = (\underset{\sim}{v}\cdot\nabla)\underset{\sim}{B} - (\underset{\sim}{B}\cdot\nabla)\underset{\sim}{v} + \eta \nabla^2 \underset{\sim}{B} \qquad (2)$$

In plane geometry, $\underset{\sim}{B} = B_y(x)\underset{\sim}{e}_y + B_z(x)\underset{\sim}{e}_z$ and the time variation of the initially vanishing component B_x (the "reconnecting" field) is given by

$$\omega B_x = -i(\underset{\sim}{k}\cdot\underset{\sim}{B})v_x + \eta(\frac{\partial^2}{\partial x^2} - k^2) B_x \quad . \qquad (3)$$

Here we assumed $B_x \sim \exp(i \underset{\sim}{k}\cdot\underset{\sim}{x} + \omega t)$ and $\underset{\sim}{k} = k_y \underset{\sim}{e}_y + k_z \underset{\sim}{e}_z$. From Eq. (3) we see that the resistive term will dominate wherever $\underset{\sim}{k}\cdot\underset{\sim}{B} \simeq 0$. Thus the location of the "resistive" layer depends on $\underset{\sim}{k}$, which in turn is controlled by the boundary conditions. The width of the resistive layer is determined by the rate of change of $\underset{\sim}{k}\cdot\underset{\sim}{B}$, or in other words by the degree of <u>shear</u> of the field. Growing perturbations ($\omega > 0$) can be found when the last term of Eq. (3) is positive. For a highly sheared field, the resistive sheet width, ℓ, will be much smaller than the typical linear dimension of the system, L. Thus the resistive time $\tau_r \simeq \ell^2/\eta \ll L^2/\eta$. Outside the resistive layer the field is effectively frozen-in and is convected towards the dissipative region by the fluid motion, where the plasma and the magnetic field can decouple. The magnetic lines then tear and reconnect, forming elongated cells or "islands".

There are a number of features of this scheme of dynamical reconnection that makes it particularly attractive for application to solar flares, especially when compared with the more familiar process of stationary reconnection at neutral points. First of all the field need not vanish anywhere, but must simply possess a certain degree of shear. Second, the tearing-mode instability is a spontaneously growing perturbation: the flow pattern and the field distorsion result from the mutual plasma-field interaction without any need to impose artificial external conditions. Finally, the location of the reconnecting process is not fixed <u>a priori</u> but is chosen by the perturbation itself. From this considerations we see that it is rather difficult to make general statements about the efficiency or the speed of the tearing-mode instability, since they depend crucially on the detailed geometry of the magnetic field.

Again, the solution of the primary energy release in flares seems to be tied to the measurement of coronal magnetic fields. Such measurements have not been possible so far. The only chance in a near future seems to reside in polarimetry in the EUV range. First-generation instruments of this kind are part of the instrumentation of the Solar Maximum Mission, second-generation are being planned to fly on Spacelab. As a conclusion, it is fair to say that flare studies have not yet succeeded in pinning down the mechanism that governs this complex phenomenon. However, even if the final answers to the basic questions listed at the beginning of this Section have not been given, it has been possible to identify a few firm physical points that will certainly be the basis for the theory of solar flares.

References

The literature on solar flares is very vast. We list here a few recent review papers or books to which we refer for an extensive bibliography.

Kane, S.R.: 1975, Solar Gamma, X and EUV Radiation, Reidel Publ. Co. Dordrecht, Holland.
Priest, E.R. (Ed.): 1980, Solar Flare MHD, Gordon and Breach, New York, USA (to appear).
Rust, D.M.: 1980, Solar System Plasma Physics, (C.F. Kennel, L.J. Lanzerotti and E.N. Parker, Eds.),North-Holland, New York,USA.
Sakurai, K.: 1974, Astrophys. Space Sci. 28, 375.
Sturrock, P.A. (Ed.): 1980, Solar Flares, Proceedings of the Second Skylab Workshop, Univ. of Colorado Press, Boulder, USA.
Svestka, Z.: 1976, Solar Flares, Reidel Publ. Co. Dordrecht, Holland.
Tandberg-Hanssen, E.: 1967, Solar Activity, Blaisdell, Waltham, USA.

MAGNETOSPHERIC PROCESSES POSSIBLY RELATED TO THE ORIGIN OF COSMIC RAYS

Gerhard Haerendel
Max-Planck-Institut für Physik und Astrophysik
Institut für extraterrestrische Physik
8046 Garching , W-Germany

ABSTRACT

Two processes are discussed which violate the frozen-in condition in a highly conducting plasma, reconnection and the auroral acceleration process. The first applies to situations in which $\beta = \frac{8\pi p}{B^2} \approx 1$. It plays an important role in the interaction of the solar wind with the Earth's magnetic field and controls energy input into as well as energetic particle release from the magnetosphere. Detailed in situ studies of the process on the dayside magnetopause reveal its transient and small-scale nature. The auroral acceleration process occurs in the low magnetosphere ($\beta \ll 1$) and accompanies sudden releases of magnetic shear stresses which exist in large-scale magnetospheric-ionospheric current circuits. The process is interpreted as a kind of breaking. The movements of the magnetospheric plasma which lead to a relief of the magnetic tensions occur in thin sheets and are decoupled along the magnetic field lines by parallel electric potential drops. It is this voltage that accelerates the primary auroral particles. The visible arcs are then traces of the magnetic breaking process at several 1000 km altitude.

1. INTRODUCTION

There are three ways in which a cosmical plasma can interact with a self-gravitating system. They depend on the existence of an extended atmosphere and/or an appreciable intrinsic magnetic moment. If the system is lacking both of these qualities completely, the plasma will impinge directly on it, as, for instance, the solar wind does on the moon. In case of a zero or weak magnetic moment, but with the existence of an extended atmosphere, the plasma will interact with the latter via its ionized component. In the solar system, examples of this class are the comets and the planet Venus.

The existence of a strong magnetic field shifts the location of interaction outside the atmosphere. It is typically collision-free in nature, although the magnetic field can transport the forces towards the central body where collisions are important. A *magnetosphere* is being created. It is the region in which the magnetic field linked to the self-gravitating system dominates the forces acting on the plasma component. Examples in the solar system are provided by the Earth, Mercury, perhaps Mars, Jupiter, Saturn, possibly Neptune and Uranus.

What is the possible relevance of the study of planetary magnetospheres to the origin of cosmic rays? It can be threefold. (1) Planetary magnetospheres are emitters of energetic particles and thus contribute - although on a modest scale - to the low energy end of the cosmic ray spectrum. (2) Since we have direct access to the regions where the particles are accelerated, we can probe into the processes and the detailed plasma and field configurations in which they operate. (3) We can study the electromagnetic radiation emitted from acceleration regions and thus try to calibrate the emissions with respect to the properties of the exciting particle beams, with the aim to apply this knowledge to cosmical sources.

In this contribution, I want to deal with the second aspect, and among all possible acceleration processes only with two of them, reconnection and the auroral acceleration process. Both are of universal importance. Although the first one has been known conceptually since more than twenty years and has been studied extensively, we are only now beginning to collect conclusive direct observations of this process in the magnetosphere. There are several unexpected features emerging that should be taken into consideration when studying other reconnection situations.

The second process is much less known. It does not even have a name apart from its auroral context. But there are good reasons to believe that it is rather universal. In purely physical terms, it could be briefly described as the set-up of large voltages parallel to the magnetic field direction in the presence of strong field-aligned currents and as a consequence of current-driven instabilities. The parallel voltage is, however, only one aspect of the process. Large transverse interchange motions are accompanying it. Disruptions as known to occur in Tokomaks may be of similar nature. For the time being, we may just call it "the auroral acceleration process" keeping in mind that it does not underlie all kinds of aurora, but rather the structured arcs. In the light of a new theory which I will characterize later we will find another, more descriptive name for this process.

2. RECONNECTION

Reconnection or merging of two oppositely directed magnetic fields in a highly conducting fluid has been first studied in the astrophysical context by Sweet (1958) and Parker (1957). Dungey (1961) applied it to the magnetosphere and proposed the general scheme of magnetic flux transport between the dayside and nightside magnetosphere. Although it has undergone substantial modifications, this scheme is still believed to be valid. Depending on their orientation, interplanetary magnetic field lines carried past the magnetopause by the solar wind can connect temporarily with the Earth's field. Such field lines are then called "open". They give the solar wind an easy means to transfer flow momentum to the magnetospheric plasma via magnetic shear stresses and drag it into the antisolar direction. Thereby, the field is being stretched and a magnetic tail is formed. The central region of the tail which contains a layer of hot plasma, the so-called plasma-sheet, separates stretched fields of anti-parallel direction. They are subject to the merging or reconnection process under circumstances which are not yet fully understood. This occurs in events of typically 1 hour duration which are called "substorms". As a result, open field lines are transformed into closed ones (i.e. dipole-like field lines) plus field lines that are completely disconnected from the Earth and are carried away by the solar wind flow. Internal convection motions carry the closed magnetic flux-tubes back towards the dayside where the process can eventually start again.

The physics of reconnection events in the tail, the substorms, is subject to intense study by magnetospheric physicists and much could be said about it that might have a bearing on the origin of cosmic rays. A good overview of substorms is contained in a recent text-book by Akasofu (1977). However, we are still lacking fully convincing data sets on the very process, much of the inferences are indirect.

On the front-side of the magnetopause, we are on somewhat safer grounds. Data have become available which allow quantitative checks of the predictions of the macroscopic theory of reconnection. Temporal and spatial scales of the process can as well be inferred. Therefore, and because our laboratory has been strongly involved in this area of research, I will deal in detail with the reconnection at the front-side of the Earth's magnetopause.

Before we enter this discussion, we may pause for a moment and ask ourselves what role reconnection could play in the production of cosmic rays. It can contribute to it in three ways. (1) Some of the magnetic energy released in the merging process is being transferred directly to suprathermal particles which appear at the lower end of the energy spectrum of cosmic rays. (2) Reconnection can modulate the energy input into a magnetic configuration where part of it is stored as magnetic energy. This can be used to some extent for the production of high-energy particles by unspecified internal processes. (3) Reconnection may then lead to a modulation of the storage of such high-energy particles, i.e. it may control their eventual leakage from the "accelerator" into interstellar space. At the present state of knowledge we cannot say even for the Earth's magnetosphere what dominates, the direct production of low energy cosmic rays during the reconnection process or the acceleration in the strong field of closed configurations and subsequent release from them. But there is no doubt that reconnection is an important contributor to the origin of cosmic rays.

In view of the great interest in that process and of the many satellites launched with the aim of investigating the magnetosphere, it is rather surprising that it took so long to identify it unambiguously by in situ measurements. All what is needed is the combination of a magnetometer and a plasma detector capable to establish the flow of the dominant plasma component in three dimensions. As it turns out that the process is rather short-lived, both measurements must be made sufficiently fast and afford a high telemetry rate. The International Sun-Earth Explorers, ISEE 1 and 2, were the first satellites with the right orbits and instruments to meet these conditions. Paschmann et al. (1979) report an event of a few minutes duration at the dayside magnetopause near noon in which the observed changes of flow momentum at the magnetopause are consistent with the magnetic stresses that would exist if the internal and external fields were connected through a rotational discontinuity.

Whatever the detailed structure of a reconnection region may be (Petschek, 1964; Vasyliunas, 1975), it should contain a rotational discontinuity in which the magnetic field changes direction by a large angle and has a finite component normal to the plane of the discontinuity. This is sketched in Figure 1 for the simple case of almost anti-parallel fields. The plasma that transits through the discontinuity should undergo little change of its thermodynamic properties, but be accelerated by an amount that is related to the balance of magnetic and mechanical stresses which (in an isotropic plasma) reads:

$$[\rho \underline{v}_t v_n] = [\frac{\underline{B}_t B_n}{4\pi}] \quad . \tag{1}$$

Index "n" and "t" refer to the normal and tangential components, respectively. Square brackets indicate the jump across the discontinuity. Since B_n = const. and $\rho \simeq$ const., we have the simple relations:

$$[\underline{v}_t] = \frac{[\underline{B}_t]}{\sqrt{4\pi\rho}} \tag{2}$$

and

$$v_n = \frac{B_n}{\sqrt{4\pi\rho}} \quad , \tag{3}$$

which the plasma should obey. The jump in tangential flow velocity (Equation 2) is easily observable, in contrast to the quantities of Equation 3. The reason is that on the one hand most theories predict small normal components, and on the other hand one or even two satellites are not sufficient to establish with sufficient confidence the normal vector of an observed discontinuity.

In earlier studies of the plasma flow at the magnetopause (Heikkila, 1975; Paschmann et al., 1976; Haerendel et al., 1978), it was quite

Figure 1. Reconnection situation at the dayside magnetopause with southward pointing interplanetary field. The shaded area shows a layer of accelerated plasma flow after transition of the solar wind through the rotational discontinuity (Paschmann et al., 1979).

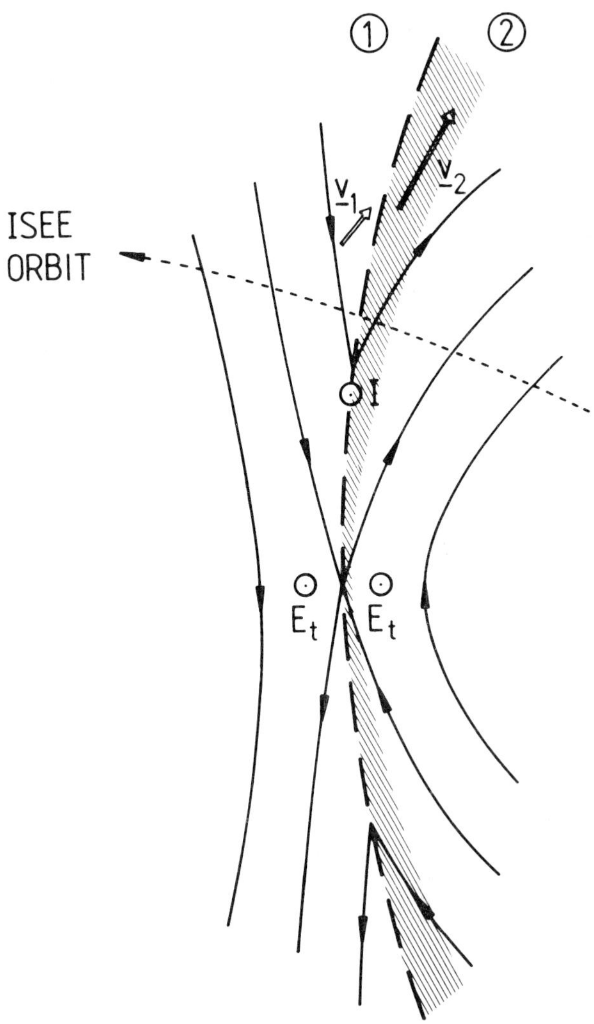

disturbing that the predictions of Equation 2 were not encountered even when the magnetic fields inside and outside the magnetopause were almost oppositely directed. It became evident that if reconnection occurred at all in the explored regions it should be transient and small-scale in nature and thus escape detection. The recent measurements of Paschmann et al. (1979) were, however, sufficiently fast to cope with this difficulty. Figure 2 shows a set of data on three subsequent transitions through the magnetopause along a pass as sketched by the dashed line in Figure 1. The displayed data are total plasma density, N_p, magnitude of the flow velocity, v_p, the component of B perpendicular to the ecliptic, B_z, the plasma pressure P and magnetic pressure, B, and finally, the sum of gas and magnetic pressures, P_T. The units are, respectively: cm^{-3}, km/sec, nT, and 10^{-7} N m^{-2}. The magnetopause (MP) undergoes usually radial oscillations; hence there are three transitions as revealed most clearly by the jumps of B_z from positive to negative values. The most important feature is the large increase of plasma flow velocity by several 100 km/sec just inside the magnetopause. Via the total density measurements which the same instru-

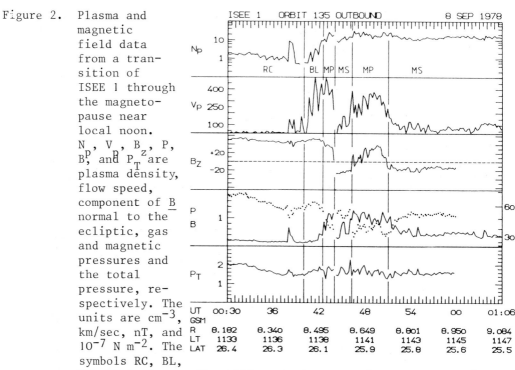

Figure 2. Plasma and magnetic field data from a transition of ISEE 1 through the magnetopause near local noon. N_p, V_p, B_z, P_B, and P_T are plasma density, flow speed, component of $\overline{B}$ normal to the ecliptic, gas and magnetic pressures and the total pressure, respectively. The units are cm^{-3}, km/sec, nT, and 10^{-7} N m^{-2}. The symbols RC, BL, MP, MS designate the different plasma regimes encountered, namely ring current, boundary layer, magnetopause layer, magnetosheath (i.e. shocked solar wind) (Paschmann *et al.*, 1979).

ment yields and the magnetometer data, Equation 2 can be checked. Agreement is found within 10%. This must be considered as rather good in view of several sources of experimental error.

A special technique developed by Sonnerup (1971) and co-workers, which is called "minimum variance technique", allows the derivation of B_n if the orientation of the discontinuity is sufficiently stable during the transit time of the satellite. For the event contained in Figure 2 an inward pointing normal component of 5.4 nT was found. An inward flow component of 28 km/sec would go along with this value (Equation 3). The existence of two closely spaced spacecraft (ISEE 1 and 2) allows the determination of the speed of normal motion of the magnetopause. This is needed to correct the value found for v_n by a similar minimum variance technique. Though affected by large error bars both values are in good agreement.

I have chosen to discuss this particular measurement in detail in order to give the reader some feeling for the difficulties involved in establishing with great confidence the existence of a fundamental plasma process in space, even under rather favorable circumstances. Meanwhile about 10 events of this kind have been identified, approximately 30% of the total number of cases in which the orientation of the interplanetary field was favorable for reconnection. Strong anti-parallel field components are apparently not sufficient for the process to occur. What a sufficient condition could be, is not known at this moment.

Earlier studies of the plasma near the magnetopause (Hones et al., 1972; Akasofu et al., 1973; Rosenbauer et al., 1975; Paschmann et al., 1976; Haerendel et al., 1978) had revealed an important feature of the magnetopause. It is covered on its inside by a boundary layer which covers it down to the distant tail. It is particularly thick and dense in the region of the polar cusps, which is shown in Figure 3. Here it

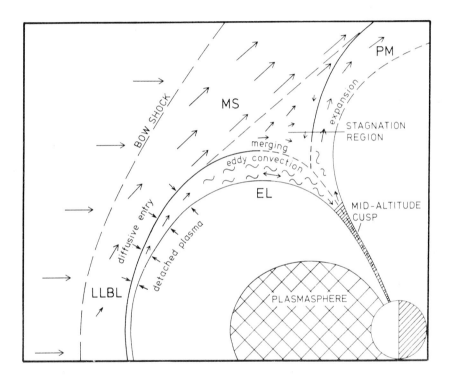

Figure 3. Meridian cut through the frontside boundary layers with indication of the dominant processes (MS = magnetosheath, LLBL = low latitude boundary layer, EL = entry layer, PM = plasma mantle) (Haerendel et al., 1978).

has been given the name "entry layer" (EL), since it is believed that this is the region of dominant plasma entry into the magnetosphere. Only a small fraction of the plasma in this part of the boundary layer penetrates along the field lines into the polar region. Most of it flows along the tail boundary layer into the distant tail. This part has been called "plasma mantle" (PM). The boundary layer on the low-latitude dayside (LLBL) is rather thin, of lower density then in the adjacent solar wind (magnetosheath (MS)) and exhibits strong temporal modulations.

If the cusp regions play a dominant role in plasma entry, they should also be the location of frequent reconnection events (Haerendel, 1978). From the observation of very irregular flows in the entry layer it was concluded that these events should be transient and small-scale in nature. Scales of only 1000 km and 20 sec have been deduced from measurements with insufficient temporal resolution on the ESA satellite HEOS 2. This feature was probably also responsible for our inability to identify, in the same manner as discussed before, the signature of reconnection. However, the irregularity of the flow gives ground to the hypothesis that the mass transport inside the boundary layer is a kind of eddy convection process. An order of magnitude estimate of its efficiency is consistent with the implications of the drainage of this region by the mantle flow (Haerendel, 1978).

A more recent study of the low latitude part of the dayside boundary layer by Sckopke et al. (1980) has revealed its transient nature rather clearly. Here the time-scales are, however, much longer. They are of the order of a few minutes and the spatial scales of several Earth's radii. Of the three possible interpretations of the observations which are shown in <u>Figure 4</u>, case (c) appears as the most likely. This means

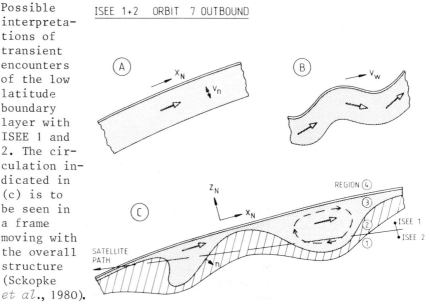

Figure 4. Possible interpretations of transient encounters of the low latitude boundary layer with ISEE 1 and 2. The circulation indicated in (c) is to be seen in a frame moving with the overall structure (Sckopke et al., 1980).

that plasma is carried in form of separate "blobs" along the inside of the magnetopause in the downstream direction. The expected rotational motion of the plasma inside and outside the boundary layer has been confirmed observationally. In the event that has been extensively studied by Sckopke et al. (1980), the bending of the magnetic field inside the boundary layer is in a sense as if the plasma were pulled from above, i.e. the cusp region. This is again consistent with dominant reconnection in the cusps. The periodicity of the events on the low latitude side may be related to the separation of "vortices" from the stagnation region (shown in Figure 3) outside the cusp magnetosphere, as it would happen in an ordinary fluid streaming around a corner (Haerendel, 1978). It could also be the consequence of a Kelvin-Helmholtz instability of the boundary layer flow (Sckopke et al., 1980).

The consequences of reconnection in the cusps on the gross topology of the field is shown in Figure 5 taken from Haerendel et al. (1978).

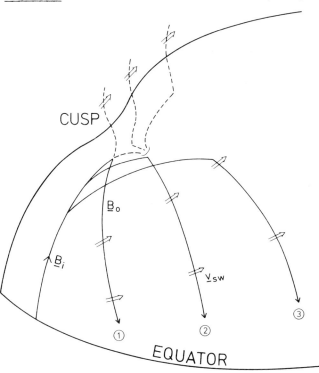

Figure 5. Sketch of the erosion of magnetic flux from the frontside of the magnetosphere initiated by reconnection in the cusp regions (Haerendel et al., 1978).

Magnetic flux is being eroded from the frontside of the magnetosphere. On the low latitude side, the magnetic field becomes stretched, i.e. its magnitude increases, as it is often observed.
This means that reconnection does not necessarily imply a release of magnetic energy everywhere. Part of the space involved may experience a growth of magnetic energy at the expense of kinetic energy of flow. The erosion of magnetic flux occurs in short-lived events of tens of seconds to a few minutes. Sometimes such flux-tubes can be identified outside the magnetopause in the solar wind plasma (Russell and Elphic, 1979). In addition to the magnetic signature (increase of $|\underline{B}|$) one finds

also hot electrons streaming away from the magnetopause. They may have been accelerated at the magnetopause or released from the interior. Whatever the origin of these energetic electrons is, we seem to observe an important step in the production of energetic particles by a magnetosphere.

A more important source of energetic particles is actually the geomagnetic tail (Anderson, 1965; Baker and Stone, 1976; Hones *et al.*, 1976; Sarris *et al.*, 1976). Again it seems that part of the energization is due directly to the tail reconnection process and part to a leakage of the energetic trapped particles from the outer radiation belt when the tail recovers after a reconnection event ("recovery phase" of a substorm) (Belian *et al.*, 1980).

In summary, we find the following properties of the reconnection process in the Earth's magnetosphere. It is transient and small-scale, i.e. the spatial scale is much smaller than the size of the overall magnetic configuration. These scales may be connected to the hydrodynamic properties of the plasma flow around the object (turbulence). The cusp regions seem to be the primary site of the reconnection process. Boundary layers inside the magnetopause are set up as a consequence. The short duration of the reconnection events leads to the erosion of magnetic flux in form of rather discrete flux-tubes, which provide paths for the escape of energetic particles from the interior. At the same time, direct acceleration of energetic particles is observed.

When applying theoretical models of reconnection to an astrophysical system, we should be warned that it may be dangerous to use a stationary picture and scales of the size of the overall system as suggested by the well-known model of Petschek (1964) and its successors. As observable in so many phenomena, the plasma seems to "like" the formation of small-scale structure, filaments. We must learn to understand the causes of this behavior in order to be able to make predictions for other situations. It is quite clear that the efficiency of a process is quite different when it is small scale and transient from what it would be when it is large scale and stationary.

3. AURORAL ACCELERATION PROCESS

There is an intimate relation between reconnection and the auroral acceleration process. Both processes violate the frozen-in condition of the magnetic field in a highly conducting plasma. We are used to call it reconnection or merging when it happens in a plasma with $\beta = \frac{8 \pi p}{B^2} \approx 1$. But magnetic lines of force can as well become reconnected when $\beta \ll 1$. It means that there is a shear of the transverse convection. Plasma that was distributed along a certain flux-tube at a certain time will be found on separate flux-tubes a moment later. The most striking visible expression of such an event was provided once by a barium ion jet experiment in the auroral magnetosphere (Wescott *et al.*,

1976). The upper part of a narrow barium ion streak which extended over a height range of many 1000 km broke suddenly into a sheet of parallel streaks which kept growing and dispersing, while the lower end ($\leqslant$ 5500 km) remained in its initial form. All this started when the flux-tube loaded with barium plasma came into contact with an auroral arc.

A quasi-stationary field-aligned shear of the transverse plasma motion, i.e. of the transverse electric field, can only exist in the presence of a parallel electric field. The occurrence of such fields with magnitudes much higher than what was to be expected on the grounds of the electrical conductivity was suspected long before the first direct measurements. The evidence came mainly from observations of the velocity distribution of primary auroral particles and anti-correlations of electrons and positive ions (protons) (e.g. Evans, 1975). Pronounced peaks in the energy spectrum at a few keV and field-aligned velocity distributions are typical for the electrons which generate structured auroral arcs. From their velocity dispersion one could also deduce that the electrostatic acceleration regions must be located at typically 1 R_E altitude, not in the distant magnetosphere. First direct experimental proof was provided by another barium ion jet experiment which showed clearly the upward field-aligned acceleration of barium ions by 5 keV at an altitude of 7500 km at the moment when a bright auroral arc developed at the projection points to the 100 km level (Haerendel et al., 1976).

It was only recently that extensive direct measurements inside the acceleration regions became available (Shelley et al., 1976; Mozer et al., 1977; Mizera and Fennell, 1977). They were obtained with the US Air Force satellite S3-3 on an elliptical polar orbit with 8000 km apogee. From all these measurements a heuristic model of the acceleration region was developed which is sketched in Figure 6. Strong transverse electric fields exist in thin sheets extending in E - W direction. The equipotential contours do not extend all the way to the ionosphere, but close at heights of several 1000 km. These regions are imbedded in sheets of intense field-aligned electric current emerging from the ionosphere. The current when directed upward is essentially carried by the energetic (few keV) primary auroral electrons which have undergone a linear acceleration in the region of $E_\parallel \neq 0$. Streams of keV protons, helium and oxygen ions are as well observed to emerge from these regions in the upward direction. Some have field-aligned distributions (protons), others (dominantly oxygen) have so-called conical distributions indicating the presence of additional strong transverse heating in the acceleration regions (Sharp et al., 1977). In addition, high amplitude plasma waves are present which have been identified as ion-cyclotron and lower hybrid waves and low frequency turbulence (Mozer et al., 1977; Kintner et al., 1978; Temerin, 1978). A coherent review of the observational facts and their physical interpretation has been given by Mozer et al. (1980).

The acceleration regions have been interpreted in terms of electro-

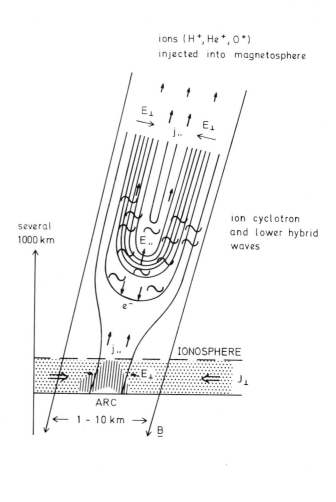

Figure 6. Phenomenological model of the auroral acceleration region above structured arcs at heights of several 1000 km. Strong transverse electric fields, $E_\perp$, are decoupled from the ionosphere by a region of $E_\parallel \neq 0$. The accompanying field-aligned current, $j_\parallel$, is carried mainly by accelerated electrons. Ions are injected upward and different plasma waves are excited.

static shocks (Swift, 1975; 1976), double layers (Shawhan et al., 1978) and regions of high anomalous resistivity (Hudson et al., 1978). It is fair to say that at this moment agreement on the basic physical process does not exist. Even the ultimate energy source feeding the acceleration region is a matter of debate. A convincing theory should provide answers to three basic questions:

(1) Why are the acceleration regions so narrow (as evidenced by the thinness of auroral arcs, typically 1 km)?

(2) What determines the strength of the electric current connected with the appearance of arcs?

(3) Why is the process located at several 1000 km altitude?

Already in 1971 Kindel and Kennel have shown that a constant field-aligned current meets the condition for two-stream ion-cycotron and ion-acoustic instabilities in the low density plasma above $\approx$ 1000 km. This argument taken from linear instability theory may provide the basic answer to question (3). However, an argument involving the state of fully developed plasma turbulence (consistent with the observed wavefields) would appear to be more satisfying. The exploration of various theoretical approaches to the understanding of the auroral acceleration process is at present in the center of magnetospheric research and presents one of the greatest challenges to the space plasma physicists.

I want to end this discussion with a brief outline of my own interpretation of the primary auroral process (Haerendel, 1980). The reader should be warned that a critical discussion of these ideas in the scientific community is just starting and that the elaboration of even the basic features of the model is still incomplete. The foundation of this theory is my conviction that the process is nothing else than another way of releasing magnetic energy stored in a magnetospheric-ionospheric current circuit. Under undisturbed conditions the energy is mostly dissipated by Ohmic losses in the ionosphere. Auroral arcs appear when at intermediate altitudes an effective resistance is building up which allows field-aligned potential drops, i.e. slippage of the magnetospheric plasma below and above the $E_\parallel \neq 0$ region with respect to each other. What is the nature of this slippage? It can only be in a sense as to carry the plasma and field configuration into a state of lower energy.

A simplified picture of the process is contained in Figure 7. The zero order magnetic field is taken as homogeneous. On either end of the system there are collision dominated regions representing the northern and southern ionospheres. In the center is a generator region representing the plasma sheet in the tail. Since the plasma is insulated from the central body by a non-conducting atmosphere it is usually in a state of slow convection (v_s), which is determined by the strength of the generated current and the Ohmic resistance of the ionospheres. The mechanical forces driving the generator are balanced by the viscous forces acting in the ionospheres.

This situation can suddenly be disturbed at intermediate altitudes when the acceleration region develops with a resistance that exceeds that of the ionosphere. This means the plasma is free to move in a sense as to relieve quickly the magnetic tension with velocities v_f exceeding by far the undisturbed convection velocity v_s. The knowledge of this event is propagated by elastic shear waves along $\bar{B}$. Within a short time, τ, the whole field line participates in the relief motion, and the magnetic energy reservoir becomes exhausted. However, all this happens only in thin shoots so that there is plenty of free energy left in the transverse dimension inside the gross current circuit. The process is rather similar to the breaking or tearing of an elastic medium. Once a crack has been formed it will propagate into the stressed medium.

Figure 7. Simplified situation representing the dragging of field lines by the mechanical forces in the generator region (plasma sheet) and opposed by the viscous forces provided by the ionosphere. The resulting slow convection (v_s) can become disturbed by sudden stress releases (velocity v_f) at intermediate altitude, the knowledge of which spreads along $\underline{B}$ by means of shear Alfvén-waves (Haerendel, 1980).

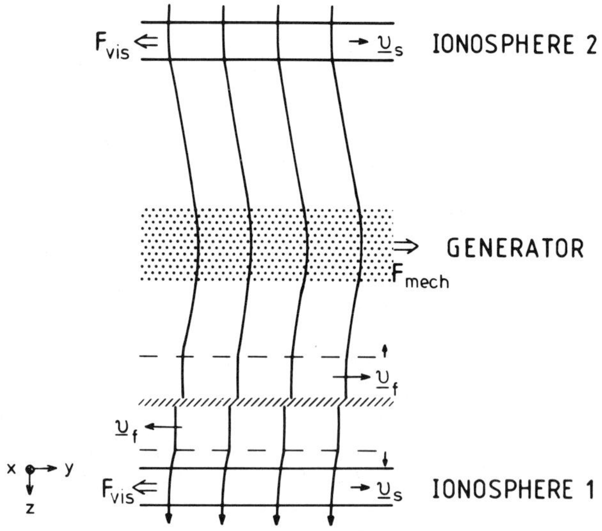

In Figure 8 we are looking sideways into the magnetospheric-ionospheric current loop. The "fracture zone" is enlarged. It is shown to propagate slowly transverse to $\underline{B}$ with velocity u_F into the reservoir of magnetic energy while the information about the breaking, i.e. the set up of anti-parallel convective motions $\underline{v}_\perp$, is propagating behind two wave fronts along the lines of force with a speed of the order of the Alfvén-velocity (v_A). These fronts form a pair of nearly transverse, slightly oblique waves ($u_F \ll v_A$) attached to the nose of the "fracture zone", the diffusion region. In a triangular region behind a second pair of somewhat slower standing waves, the opposing convective motions and their corresponding transverse electric fields are decoupled by an $E_\| \neq 0$.

For the quantitative analysis it is essential to determine the strength of the current, J_1, flowing inside the standing discontinuities and the propagation speed, u_F. For the first we take as a measure the field-aligned current density, $j_\|$, emerging from the ionosphere multiplied by the width, w, of the region. $w \cong u_F \cdot \tau$, where τ is the travel time of an Alfvén-wave to the ionosphere and back. It is easy to see that the speed of convection switched on behind the first pair of

Figure 8. Meridian cut through the magnetospheric-ionospheric current circuit with an enlarged view of the "fracture zone" at low altitudes. This region is structured by discontinuities (dashed lines) which are super- and sub-Alfvénic shear waves. They switch on and off a high transverse convection, $v_\perp$, and $E_\parallel \neq 0$ (Haerendel, 1980).

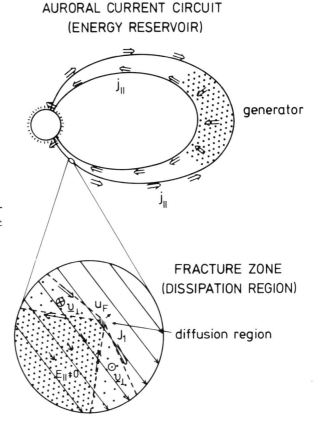

standing waves is:

$$v_\perp = \frac{c}{B} \cdot M_{n1}^{-1} \cdot \frac{4\pi v_A}{c^2} \cdot J_1 \qquad (4)$$

with $\quad J_1 = j_\parallel \cdot w$.

M_{n1} is the Alfvénic Mach-number of the first pair of waves. It has been shown by the author (Haerendel, 1979) that generalized Alfvén-waves exist which propagate with super- or sub-Alfvénic velocity if one allows for a jump of $E_\parallel$ across the discontinuity and conserves as usually mass, momentum, energy and magnetic flux. These waves are thought to structure the acceleration region in the manner sketched in Figure 8. In contrast to pure Alfvén discontinuities (intermediate waves) the thermodynamic properties of the plasma are strongly changing upon transition through the waves; the entropy increases.

The propagation of the "fracture zone" with speed u_F is regarded as a diffusion or rather heat conduction process caused by the pressure difference of the plasma in front of and behind the discontinuities. For this process, the transverse mobility is of great importance. It must be non-classical, i.e. of anomalous nature. We measure it by an effective transverse collision frequency, $v_\perp^*$, in units of the ion-cyclotron frequency, Ω_i:

$$v_\perp^* = \alpha_\perp \Omega_i \cdot (\alpha_\perp \approx 10^{-2}) \tag{5}$$

Taking all these elements together (Haerendel, 1980) one finds a characteristic current density, $j_\|$, for which the process can maintain itself:

$$j_c = M_{n1} M_{n2} \sqrt{\frac{e n c B}{2 \pi \alpha \tau}} \quad , \tag{6}$$

where $M_{n1,2}$ are the respective Alfvénic Mach-numbers. This current density is not related to a linear instability analysis as carried out by Kindel and Kennel (1971), but represents a condition for the stationary release of magnetic shear stresses by the described non-linear process.

The other relevant quantities like width, voltage drop etc. can be deduced as well. This exceeds the scope of this paper. The width, w, is related to the ability of the medium to support a parallel field. When introducing an anomalous resistivity by means of an effective parallel collision frequency, $v_\|^*$, one finds:

$$w \approx \sqrt{v_\|^* \tau} \cdot \frac{c}{\omega_{pe}} \quad , \tag{7}$$

where ω_{pe} is the plasma frequency. Thus, the thinness of auroral arcs appears to be related to the electron inertial length, c/ω_{pe}, and the effective collision frequency.

It is also easy to see why this "breaking" of field lines should occur above the topside ionosphere. It is here that the critical current defined by Equation 6 and projected to the ionosphere assumes its lowest value, i.e.

$$\frac{n}{\tau B} = \min \quad . \tag{8}$$

The sketched theory provides answers to the above formulated questions (Haerendel, 1980). In addition, it allows quantitative estimates of $j_\|$, $v_\perp$, w, field-aligned voltage, energy flux, etc. which turn out to be quite consistent with the observations, if one makes a reasonable guess on the microscopic parameters $\alpha_\perp$ and $\nu_\parallel^x$. The Mach-numbers turn out as $M_{n1} \gtrsim 1$, $M_{n2} \approx 10^{-1}$.

The great general interest in the auroral acceleration process lies in the suspicion that it may be rather universal. Sheared magnetic fields occur in a wide variety of cosmical situations, e.g. planetary magnetic fields which are distorted by satellites (Io in the Jovian magnetosphere), twisted magnetic fields in the solar corona, magnetic fields pervading the accretion column of binary neutron stars which are subject to distortion by the Rayleigh-Taylor instability. It is tempting to apply Equations 6 and 7 to such situations. However, an appropriate discussion of the assumptions and involved parameters cannot be given here. One can easily see though that rather high energies may be attainable by this process in systems of much higher magnetic field. So, it may play a non-negligible role for the origin of cosmic rays.

4. FINAL REMARKS

It is difficult to do justice to two so important and little understood processes in a short paper of this kind. My main aim is to draw attention to recent progress in their exploration by in situ measurements and their interpretations. Once we have developed theories that allow to derive, from first principles, the key features observed, we may be in the position to apply this knowledge with some confidence to distant objects. This is the exemplary nature of magnetospheric research in the wider context of astrophysics.

REFERENCES

Akasofu, S.-I., Hones, Jr., E.W., Bame, S.J., Asbridge, J.R., and Lui, Y.: 1973, J. Geophys. Res. 78, pp. 7257-7274.

Akasofu, S.-I.: 1977, Physics of Magnetospheric Substorms, Reidel Publ. Co., Dordrecht-Holland.

Anderson, K.A.: 1965, J. Geophys. Res. 70, pp. 4741-4763.

Baker, D.N., and Stone, E.C.: 1976, Geophys. Res. Lett. 3, pp. 557-560.

Belian, R.D., Baker. D.N., Hones, E.W.,Jr., Higbie, P.R., Bame, S.J., and Asbridge, J.R.: 1980, Timing of energetic proton enhancements relative to magnetospheric substorm activity and its implication for substorm theories, preprint, Los Alamos Scientific Lab.

Dungey, J.W.: 1961, Phys. Rev. Lett. 6, p. 47-48

Evans, D.S.: 1975, in "Physics of Hot Plasmas in the Magnetosphere",
 ed. by B. Hultqvist and L. Stenflo, Plenum, New York,
 pp. 319-340.

Haerendel, G., Rieger, E. Valenzuela, A., Föppl, H., Stenbaek-Nielsen,
H.C., and Wescott, E.M.:
 1976, European Space Agency, ESA-SP 115, pp. 203-211.

Haerendel, G.: 1978, J. Atmospheric Terr. Phys. $\underline{40}$, pp. 343-353.

Haerendel, G., Paschmann, G., Sckopke, N., Rosenbauer, H., and
Hedgecock, P.C.:
 1978, J. Geophys. Res. $\underline{83}$, pp. 3195-3216.

Haerendel, G.: 1979, Anomalous Shear Waves of Large Amplitude, preprint,
 Max-Planck-Institut f. extraterr. Physik.

Haerendel, G.: 1980, Proc. European Sounding Rocket and Balloon Progr.,
 ESA Symposium, Bournemouth, UK.

Heikkila, W.J.: 1975, Geophys. Res. Lett. $\underline{2}$, pp. 154-157.

Hones, E.W., Jr., Asbridge, J.R., Bame, S.J., Montgomery, M.D., Singer,
S., and Akasofu, S.-I.:
 1972, J. Geophys. Res. $\underline{77}$, pp. 5503-5522.

Hones, E.W., Jr., Palmer, I.D., and Higbie, P.R.:
 1976, J. Geophys. Res. $\underline{81}$, pp. 3866-3874.

Hudson, M.K., Lysak, R.L., and Mozer, F.S.:
 1978, Geophys. Res. Lett. $\underline{5}$, pp. 143-146.

Kindel, J.M. and Kennel, C.F.:
 1971, J. Geophys. Res. $\underline{76}$, pp. 3055-3078.

Kintner, P.M., Kelley, M.C., and Mozer, F.S.:
 1978, Geophys. Res. Lett. $\underline{5}$, pp. 139-142.

Mizera, P.F., and Fennell, J.F.:
 1977, Geophys. Res. Lett. $\underline{4}$, pp. 311-314.

Mozer, F.S., Carlson, C.W., Hudson, M.K., Torbert, R.B., Parady, B.,
Yatteau, J., and Kelley, M.C.:
 1977, Phys. Rev. Lett. $\underline{38}$, pp. 292-295.

Mozer, F.S., Cattell, C.A., Hudson, M.K., Lysak. R.L., Temerin, M.,
and Torbert, R.B.:
 1980, Satellite Measurements and Theories of Low Altitude
 Auroral Particle Accelerations, Space Sci. Rev. $\underline{27}$,
 pp. 155-213.

Parker, E.N.: 1957, J. Geophys. Res. 62, pp. 509-520.

Paschmann, G., Haerendel, G., Sckopke, N., Rosenbauer, H., and Hedgecock, P.C.: 1976, J. Geophys. Res. 81, 2883-2899.

Paschmann, G., Sonnerup, B.U.Ö., Papamastorakis, I., Sckopke, N., Haerendel, G., Bame, S.J., Asbridge, J.R., Gosling, J.T., Russell, C.T., and Elphic, R.C.:
1979, Nature 282, pp. 243-246.

Petschek, H.E.: 1964, AAS-NASA Symposium on the Physics of Solar Flares, NASA Spec. Publ. SP-50, pp. 425-439.

Rosenbauer, H., Grünwaldt, H., Montgomery, M.D., Paschmann, G., and Sckopke, N.: 1975, J. Geophys. Res. 80, pp. 2723-2737.

Russell, C.T., and Elphic, R.C.:
1979, Geophys. Res. Lett., 6, pp. 33-36.

Sarris, E.T., Krimigis, S.M., and Armstrong, T.P.:
1976, J. Geophys. Res. 81, pp. 2341-2355.

Sckopke, N., Paschmann, G., Haerendel, G., Sonnerup, B.U.Ö., Bame, S.J., Forbes, T.G., Hones, E.W., Jr., Russell, C.T.:
1980, Structure of the Low Latitude Boundary Layer, submitted to J. Geophys. Res.

Sharp, R.D., Johnson, R.G., and Shelley, E.G.:
1977, J. Geophys. Res. 82, pp. 3324-3328.

Shawhan, S.D., Fälthammar, C.-G., and Block, L.P.:
1978, J. Geophys. Res. 83, pp. 1049-1054.

Shelley, E.G., Sharp, R.D., and Johnson, R.G.:
1976, Geophys. Res. Lett. 3, pp. 654-656.

Sonnerup, B.U.Ö.: 1971, J. Geophys. Res. 76, pp. 6717-6735.

Sweet, P.A.: 1958, Electromagnetic Phenomena in Cosmical Physics, ed. by B. Lehnert, Cambridge Univ. Press, pp. 123-134.

Swift, D.W.: 1975, J. Geophys. Res. 80, pp. 2096-2108.

Swift, D.W.: 1976, J. Geophys. Res. 81, pp. 3935-3943.

Temerin, M.A.: 1978, J. Geophys. Res. 83, pp. 2609-2615.

Vasyliunas, V.M.: 1975, Rev. Geophys. Space Phys. 13, pp. 303-336.

Wescott, E.M., Stenbaek-Nielsen, H.C., Hallinan, T.J., and Davis, T.N.:
1976, J. Geophys. Res. 81, pp. 4495-4502.

ON THE MECHANISM OF GENERATION OF SOLAR COSMIC RAYS ENRICHED
BY HELIUM-3 AND HEAVY ELEMENTS

L.G. Kocharov
Leningrad Polytechnical Institute, Leningrad 195251, USSR

G.E. Kocharov
Physico-Technical Institute of the USSR, Academy of Sciences,
Leningrad 194021, USSR

The possible mechanism of helium isotopes separation in the solar atmosphere due to plasma effects have been proposed by G.E. Kocharov (1977), I. Ibragimov and G. Kocharov (1977). The cycle of papers: G. Kocharov and L. Kocharov (1978); L. Kocharov (1979 a,b,c,d, 1980); G. Kocharov, L. Kocharov and Yu. Charikov (1980) is devoted to the development of the theory and to the search of concrete model of corresponding physical processes in the solar plasma. Here we consider briefly the modern state of this problem.

According to our consideration to generate Solar Cosmic Rays with large enrichment of He^3 the following conditions are required. At first high non-isothermality $T_e/T_i \simeq 100$ has been formed over the acceleration region. Then for a short time ion-acoustic turbulence has been excited with the energy density $\sim 10^{-4} \div 10^{-3}$ of thermal energy density. During the existence of ion-acoustic turbulence, H and the main part of He are fully ionized and the degree of ionisation C, N, O does not exceed the equilibrium one at the temperature $T_e = 8 \cdot 10^4 K$, so $Z^{*4} \cdot A^{-2} \leqslant 1$, where Z^* is the ion's charge and A is its mass number. The total number of thermal protons in the acceleration region have to be rather large to provide the observed He^3 nuclei flux. Sometimes it may attain $10^{39} n^{-1}$, where n is the number of acceleration cycles for one event.

The joint analysis of these conditions and the experimental data on the X-ray radiation and the shape of the energy spectrum of protons, He^3, He^4 and other isotopes shows, that the acceleration occurs at the region with the characteristics of upper chromosphere.

According to the conventional notion if enough number of electrons accelerated in the corona with energy of about 10 keV penetrates the cold plasma of the upper chromosphere, the electron shock has to be formed. And ion-acoustic turbulence in the front of wave exists. More over the parameters of the generated turbulence may be in accordance with the parameters required to the preferentially acceleration of He^3 (L.G. Kocharov, 1980). High-energy electrons (~ 15 keV) will shoot through ion-acoustic front, generating before the front of the dense beam, which is unstable. As a result, strong Langmuir turbulence have been generated. It heats quickly thermal electrons and forms the required nonisothermality. The above considered ion-acoustic front have been propagated through this non-isothermality medium and provides the preferentially pre-acceleration of He^3 and also the injection of small part of protons, He^4, C, N, O to the mode of further acceleration. It is significant that the degree of ionisation of all elements except H in the front is nonequilibrium, so that the high non-isothermality is connected here with very quick heating of electrons. So, the enrichment of elements heavier than He as large as the He^3 enrichment has not been observed. It is essential that in the corona, where the degree of ionisation of heavy elements is high, the mechanism proposed by I. Ibragimov and G. Kocharov (1977) may provide the enrichment of solar cosmic rays by elements of Fe-group. The smaller enrichment of heavier elements as compared with He^3 is connected with the fact that in the corona the possible degree of non-isothermality is lower than in cold region of He^3 acceleration.

The natural consequence of our model is the generation of X-ray radiation. Really, the analysis of available data shows that He^3 rich events are accompanied by hard X-rays and a sufficient agreement between the intensity of X-ray and He^3 fluxes is available (G. Kocharov, L. Kocharov and Yu.Charikov). A very important peculiarity is the occurence of X-ray precursors of the flares (2-10 keV) in all He^3 rich events, for which the measurements of soft X-rays have been carried out. A mention should be made that 50% of He^3 rich events are accompanied by II-type radiobursts (L. Kocharov, 1979 b,c).

In the frame of the model discussed, all of these accompanying phenomena have a quality or quantity explanation. At last mention should be made that the heating of chromosphere should be accompanied by emission of UV radiation in He II lines. So it should be very important to carry out correlated experiments on SCR isotope composition, ultraviolet and X-ray solar radiation investigations especially with good spatial resolution.

References

Fisk, L.A.: 1978, Ap.J. 224, 1048.
Ibragimov, I.A. and Kocharov, G.E.: 1977, Proc. Int. Cosmic Ray Conf. Plovdiv, 12, 221.
Kocharov, G.E.: 1977, Proc. Int. Cosmic Ray Conf., Plovdiv, 12, 216.
Kocharov, G.E. and Kocharov, L.G.: 1978, Proc. X Leningrad Seminar, 37.
Kocharov, L.G.: 1979, a) Izv. AN SSSR.ser.fiz., 43, 730; b) Proc. Int. Cosmic Ray Conf., Kyoto, 12, 277; c) Izv. AN SSSR, ser.fiz., 43, 2529; d) Proc. XI Leningrad Seminar, 43, 1980.
Kocharov, L.G.: 1980, Izv. AN SSSR (in press).
Kocharov, G.E., Kocharov, L.G. and Charikov, Yu.E.: 1980, Preprint, Ioffe Phys. Tech. Inst. Leningrad.

ON ^{3}-HE RICH SOLAR PARTICLE EVENTS

E. Möbius and D. Hovestadt
Max-Planck-Institut für Physik und Astrophysik
Institut für extraterrestrische Physik
8046 Garching, W-Germany

Extensive measurements of the composition of energetic solar flare particles have revealed the frequent occurrence of solar flares with large enrichments of ^{3}He and heavy ions compared to the composition of the solar atmosphere. The basic characteristics of these events can be summarized as follows:

- ^{3}He/^{4}He-ratios up to more than 1 are observed
- the ^{3}He/^{4}He-ratio increases with energy
- no comparable enhancement of ^{2}H or ^{3}H is found
- there is a strong correlation of ^{3}He and heavy ion enrichment.

These main features of ^{3}He-rich flares can most reasonably be explained in terms of a two-stage process which is based on selective heating in a plasma instability followed by an acceleration process. A model proposed by Ibragimov et al. (1978) is based on an ion sound instability. In the following discussion, however, an alternative model proposed by Fisk (1978) is preferred which is based on resonant heating by an ion-cyclotron instability because of the higher selectivity for the He isotopes. ^{3}He is effectively heated at its gyroresonance and heavy ions with a charge to mass ratio of $Q/A \approx 1/3$ at the second harmonic. Our discussion is confined to the acceleration process which was not specified by Fisk. The acceleration threshold which is defined by the specific acceleration process determines the number of particles to be accelerated out of the suprathermal tail of the distributions. Following the idea of stochastic Fermi acceleration in the second-stage process (e.g. Sturrock, 1974) the threshold is set by a minimum rigidity condition. According to Sturrock (1974) the minimum rigidity is found to be equivalent to that of 1.5 - 15 keV protons for reasonable plasma-parameters in a coronal activity region ($n \approx 3 \cdot 10^9$ cm^{-3}, B = 10-100 Gauss). In Fig. 1 the enrichment factors for several ions are plotted normalized to ^{4}He during the May 7-12, 1974, ^{3}He-rich event (Hovestadt et al., 1975) in comparison with the following model calculation: Using the heating rates given by Fisk (1978) (assumption of a Maxwellian distribution) the threshold rigidity is treated as a free parameter to match the measured ^{3}He/^{4}He-ratio. For the heavy ion abundances the initial temperature of the flare material acts as a second free

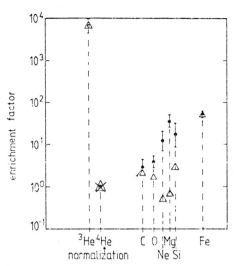
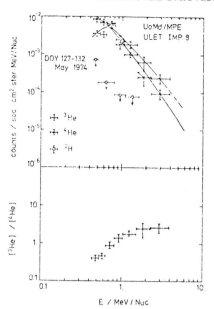

Fig. 1: Comparison of enrichment factors with a model calculation.
▮ Exp. results, May 7-12, 1974
△ Model: see text

Fig. 2: Spectra of ^{3}He and ^{4}He and upper limits of the D flux lower panel: ^{3}He/^{4}He ratio vs. energy.

parameter which determines the fraction of ions in the ionization state with Q/A ≳ 1/3 (taken from Jordan, 1969). The temperature is fitted to the abundance of Fe. A minimum rigidity equivalent to 5 keV protons and a temperature of $3 \cdot 10^6$K result which are compatible with the values in a coronal activity region.

The increase of the ^{3}He/^{4}He-ratio towards higher energies is mainly due to a harder ^{3}He spectrum compared to that of ^{4}He (Hovestadt et al., 1979; Möbius et al., 1980, example in Fig. 2). The spectral slope, however, is governed by the ratio of the acceleration time and the confinement time in the acceleration region τ_a/τ_c. The confinement time τ_c can be rigidity dependent as discussed by Scholer and Morfill (1975) in the context of the acceleration of energetic storm particles: $\tau_c \sim T^{-(3-n)/2} \cdot (A/Q)^{-(2-n)}$ where n is the exponent of the power spectrum of magnetic field fluctuations. If n is smaller than 2, τ_c increases for a larger charge to mass ratio. Thus, ^{3}He stays longer in the acceleration region than ^{4}He which results in a harder ^{3}He spectrum.

References

Fisk, L.A., *Ap. J.*, 224, 1048, 1978
Hovestadt, D. et al., *Proc. 14th Int. Cosmic Ray Conf.*, 5, 1613, 1975
Hovestadt, D. et al., *16th Int. Cosmic Ray Conf.*, Kyoto, 1979
Ibragimov, J.A. et al., *Dokl. Acad. Nauk USSR*, Nr. 588, 1978
Jordan, C., *Mon. Not. R. Astr. Soc.*, 142, 501, 1969
Möbius, E. et al., *Ap. J.*, June 1980 (in press)
Scholer, M. and G. Morfill, *J. Geophys. Res.*, 81, 5027, 1976
Sturrock, P.A., Coronal Disturbances, *IAU Symposium*, 57, 437, 1974

COSMIC RAY EVIDENCE FOR THE MAGNETIC
CONFIGURATION OF THE HELIOSPHERE

H. S. Ahluwalia
Department of Physics and Astronomy
The University of New Mexico
Albuquerque, New Mexico, 87131, U.S.A.

Sekido and Murakami (1958) proposed the existence of the heliosphere to explain the scattered component of the solar cosmic rays. The heliosphere of their conception is a spherical shell around the sun. The shell contains a highly-irregular magnetic field and serves to scatter the cosmic rays emitted by the sun. It thereby gives rise to an isotropic component of solar cosmic rays, following the maximum in the ground level enhancement (GLE). Meyer et al. (1956) showed that a similar picture applies to the GLE of 23 February 1956. They conclude that the inner and outer radii of the shell should be 1.4 AU and 5 AU respectively. They suggest that a shell is formed by the "pile-up" of the solar wind under pressure exerted by the interstellar magnetic field, as suggested by Davis (1955).

Simpson (1963) presented a digest of the results of the 11-year variation of cosmic rays studied by means of balloons, shielded ion chambers, and neutron monitors. He concludes that the observed characteristics of the variation may be explained by a change in the inner radius of the heliosphere from $\sim$5 AU at solar minimum to $\sim$30 AU at solar maximum; the variation in the size of the heliosphere is brought about by a change in solar wind pressure and the frequency of the occurrence of shock waves in the interplanetary medium, with the solar activity cycle. We now know that the mean solar wind velocity does not change significantly over the period 1964-73.

Axford et al. (1963) considered the problem of the termination of the solar wind and the solar magnetic field. They point out that interstellar atomic hydrogen has an important bearing on the termination problem. Dessler (1965) estimates the inner and outer radii of the shell to be 30 AU and 70 AU, respectively.

Ahluwalia and Escobar (1963) noted that if solar wind velocity decreases with increasing heliolatitudes then the heliosphere should be ellipsoidal, rather than spherical. Their argument is still valid if the radial flux of the solar wind decreases systematically with increasing heliolatitude. Kumar and Broadfoot (1979) show that the radial flux of the wind does indeed decrease with increase in heliolatitude. However, Ahluwalia-Escobar model is not general enough; it does not take into account the relative motion of the solar system with respect to the local interstellar medium. Ahluwalia (1980) shows that this modification results in two model magnetic field configurations for the heliosphere shown in Fig. 1a,b provided that the local interstellar magnetic field has a significant component aligned normal to the direction of motion of the solar system. A "closed" heliosphere is obtainable during odd solar activity cycles (e.g., #17, 19) and an "open" heliosphere during even cycles (e.g., #18, 20). Among other things these models help us understand (a) large-scale changes in the properties of the gross interplanetary magnetic field; (b) "anomalous" recovery of the 11-year variation of the cosmic ray intensity during even solar activity cycles, when the magnetic moment of the sun is directed from the south to north; and (c) existence of anomalous components of energetic particles and their large heliolatitudinal gradients, observed by McKibben, et al. (1979).

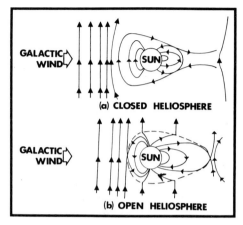

Figure 1a,b

References

Ahluwalia, H.S.: 1980, Solar and Interplanetary Dynamics; Proc. IAU Symposium #91, 27-31 August 1979. Published by D. Reidel Publ. Co., Amsterdam, Holland, p. 79.
Ahluwalia, H.S. and Escobar, I.: 1963, Geofisica Internacional 3, 21. Revista de la Union Geofisica Mexicana.
Axford, W.I., Dessler, A.J. and Gottlieb, B.: 1963, Astrophys. J. 137 1268.
Davis, L.: 1955, Phys. Rev. 100, 1440.
Dessler, A.J.: 1967, Rev. Geophs. 5, 1.
Kumar, S. and Broadfoot, A.L.: 1979, Astrophys. J. 228, 302.
McKibben, R.B., Pyle, K.R. and Simpson, J.A.: 1979, Astrophys. J. 227 L147.
Meyer, P., Parker, E.N. and Simpson, J.A.: 1956, Phys. Rev. 104, 768.
Sekido, Y., and Murakami, K.: 1958, Memoria del V Congreso Internacional de Radiacion Cosmica, Guanajuato, 5 al 13 de septiembre de 1955. Published by Instituto Nacional de Investigacion Cientifica, Mexico, pp. 253-261.
Simpson, J.A.: 1963, Semaine d'etude sur le problem du Rayonnement Cosmique dans l'space interplanetaire, 1-6 Octubre, 1962, Pontificiae Academiae Scientiarum Scripta Varia, pp. 323-343.

CONCLUDING REMARKS

Giancarlo Setti
Istituto di Radioastronomia CNR
University of Bologna, Italy

It is, of course, difficult to attempt to reach any general conclusion at the end of this Symposium where we have dealt with the physics of cosmic rays in so many different astrophysical conditions. As Prof. Ginzburg mentioned in his introductory talk, a much more appropriate title of the symposium would have been "Cosmic Rays Astrophysics". Alternatively, one may propose "Origins of Cosmic Rays" adding an s̲ to the original title. In this respect the most spectacular sites in the universe, whereby large amounts of energy are released in the form of relativistic particles, are to be found in active nuclei of galaxies, such as radiogalaxies, Seyferts' nuclei, etc. We still don't know what the ultimate source of energy is in these active nuclei, nor do we know how particles are so effectively accelerated, both in these compact objects or in the outer regions, such as the radio lobes of extended radio sources. However, whatever the detailed acceleration mechanisms may be, and whatever the precise nature of the ultimate energy source is, at least in this case it is clear where the "origin" of cosmic rays must be located. So, it is perhaps better to focus our attention on some definite subject, such as the problem of the origin of the "classical" cosmic rays, which, it seems to me, is still a largely unsolved problem. Here, I am listing a number of relevant topics, which, I feel, deserve further consideration and which may form the basis for the general discussion to follow. Since the time at our disposal is rather short, I will go very quickly in illustrating them.

The first point to be mentioned is the confinement region of cosmic rays in the Galaxy. During the meeting there was some discussion about the existence of a halo, or a "thick disk", as the trapping region of cosmic ray particles. It must be recalled at the start that in the framework of the galactic theory for the origin of cosmic rays the question of the existence of the halo has nothing to do with the energy requirements. The overall energy balance ultimately depends on the total amount of gas present in the Galaxy, which is mainly confined to the galactic disk. If I remember correctly, originally one of the main supporters of the existence of a quasi-spherical halo was the high degree of isotropy of the cosmic rays: a large volume surrounding the Galaxy in which the relativistic particles would be scattered many times back and forth. Now we know that there are several mechanisms by which particle trajectories can be effectively isotropized even within relatively small volumes in the galactic disk. As has been shown by Dr. Sancisi, radio observations of external galaxies have led to the discovery of a few spiral galaxies which possess more or less "flat" haloes. Although these findings are of great importance, there are a number of questions which are left unanswered. In particular, on the morphological side, we would like to know how this property correlates with the galaxy type and, even more important, how it correlates with the total power emitted at radio wavelengths. As far as I can judge, the existence of an external halo in M31, which is similar to our galaxy in its radio output, is still very much controversial.

In our own galaxy the distribution of relativistic electrons, which is inferred from the radio observations, appears to be thicker than the interstellar gas layer. However, a closer inspection of the galactic non-thermal radio emission reveals a rather complex distribution characterized by a number of features sticking out of the galactic plane, while in between these features, very little emission, if any, is sometimes recorded. Although the interpretation of the data is not straightforward, it is clear that the observations do not provide direct evidence of the existence of a large trapping volume in which the cosmic rays can be effectively confined for the required length of time. Moreover, to keep the cosmic rays down, the full weight of the interstellar gas is needed via the coupling provided by a rather weak magnetic field, and this is certainly difficult to achieve at great distances from the galactic plane. An alternative picture is the one in which cosmic rays are produced and remain confined essentially within the gaseous galactic disk for a few tens of million years and then propagate outward forming some sort of a galactic wind. Unfortunately, the position of the Sun close to the plane of symmetry of the Galaxy is not suitable to detect the effects of a gradual flux of cosmic rays out of the Galaxy. The more or less flat radio haloes one observes in some spiral galaxies may be just the result of this outward diffusion, or convection, of relativistic particles. It is to be hoped that future observations, and, in particular, high-resolution, high-sensitivity radio observations at different wavelengths of a suitable sample of spiral galaxies, will ultimately permit a better understanding of how relativistic particles and fields get out of the parent galaxies and of the physical parameters which control their propagation. This, in the end, is the relevant physical problem one wishes to understand, quite independent of the implications it may have on the economy of the cosmic rays in our galaxy.

The second important point which was discussed in this conference was the question of the uniqueness of the sources of cosmic rays. Various arguments based on the isotopic composition and on the electron component seems to indicate that perhaps supernovae are not "*the sources*" of galactic cosmic rays, or at least that they may not be the dominant contributors. A further argument which points toward the same conclusion is provided by the ratio between the energy densities in the proton and in the electron components. This ratio is ≈ 100 in the cosmic rays. We may now ask, "What is this ratio in the SN remnants?" In the case of the Crab Nebula it is well known that the ratio between protons and electrons energy densities cannot be very much different from unity. Let me quickly summarize the main arguments here. The first argument relates to the acceleration of particles presently taking place in this object and presumably due to the activity of the pulsar: If one takes canonical parameters for the neutron star and the observed slowing down of the pulsar period, one finds that about 40% of the rotational energy loss goes into the acceleration of the relativistic electrons to compensate for the radiation and adiabatic expansion losses. The second argument relates to the past history of the Nebula: One can show that the total kinetic energy involved in the acceleration of the shell to its present expansion velocity, presumably due to the pressure exerted by the relativistic gas, is about twice the total energy of the reservoir of relativistic electrons present in the Nebula. All together these two arguments show that the acceleration processes taking place inside the Nebula, and near the pulsar, are such as to channel about the same amount of energy in protons and electrons.

By inference one can assume that a similar situation exists for the other SN remnants, although the situation there is admittedly much less clear.

It may be pointed out here that the total kinetic energy in the shell amounts to $\sim 2 \times 10^{49}$ erg, while at least 10^{50} erg must be supplied in cosmic rays by a typical supernova to satisfy the energy requirements. Of course, the Crab Nebula may not be representative of a typical SN remnant. It is, however, rather unsatisfactory to find that the only object about which we can arrive at some definite conclusions does not appear to meet the general requirements.

One may note that if the acceleration would take place in a hydrodynamical shock at the moment of the SN explosion, whereby particles would be accelerated to the same speed, then one would expect a proton to electron energy ratio of $\sim 10^3$.

In summary, it appears likely that the cosmic rays are produced by a variety of sources. Perhaps different types of sources may also contribute in a diversified way to the various components of the cosmic ray flux. The old arguments which stemmed from the similarity of the spectra between various cosmic ray components as an indication of a common origin may not be as strong as they seemed in the past, since we now know that very similar spectra of particles are produced under completely different astrophysical conditions. At the moment it is not clear whether the cosmic rays can be accelerated in, or near, the sources or whether one has to invoke efficient acceleration mechanisms taking place in the I.S.M., somewhat along the lines which have been discussed this morning in the excellent review paper by Prof. Axford.

As a third point I would like to mention the question of the cosmic ray distribution in the Galaxy. First, let us consider the proton to electron ratio. It was usually assumed that this ratio is constant throughout the Galaxy and equal to ~ 100, the measured value close to the solar system. However, there is at least one argument which casts doubt on this assumption. The argument is as follows: To explain the observed radio emissivity at some properly chosen frequency out to a distance of ~ 1 Kpc from the Sun with the density of electrons present close to the solar system, one needs an average magnetic field of $\gtrsim 10\mu G$.

However, different types of observations indicate that the galactic magnetic field may be only a few µG. Since the galactic synchrotron emission depends on the square of the magnetic field, the density of relativistic electrons in the region we are considering must be about 10 times the measured value close to the solar system (Setti and Woltjer, Astrophys. Lett., 8, 125, 1971). Since it appears impossible to confine cosmic rays with a mean energy density much larger than that close to the Sun, the inference is that the proton to electron ratio is not constant. This, in turn, goes back to the question of the sources discussed above, because it would indicate that the proton and the electron components may have a different origin.

What about the distribution of protons? Are protons essentially produced and accelerated in spiral arms and then diffuse outward in the z-direction much more effectively than along the galactic disk? Is there any difference between the arm and the interarm regions? Arguments based on the study of meteoritic material seem to indicate that there hasn't been much variability (say a factor $\lesssim 2$) in the cosmic ray flux during the past 10^9y, or so. Since in that period of time the Sun moved in and out of the spiral arms several times, this indicates a rather uniform distribution of the protons, at least in the outer regions of the Galaxy. Some evidence in favour of a cosmic ray gradient has been provided by Prof. Wolfendale (this conference), who has shown that perhaps there is an increase of a factor four, or so, going from the outer parts of the galactic disk toward the galactic center. It seems to me that on the basis of the results which have been presented, this conclusion is still premature, although in view of the far reaching consequences the proof of the existence of such gradients would have, one should appreciate every effort spent in trying to establish its reality. The answer to these kinds of questions, of course, relies heavily on the utilization of γ-ray observations, much in the same way as radio observations play a key rôle in understanding the large scale distribution of the electrons. As has been shown during the conference, the interpretation of the galactic γ-ray flux is somewhat controversial. Before one can really use the γ-rays for the diagnosis of the cosmic ray distribution, the rôle played by the discrete sources must be fully settled. These may contribute most of the galactic γ-ray background much in the same way as it happens in the X-ray domain. Observations in the low energy γ-ray region of the e.m. spectrum will also play an important rôle in the understanding of this complicated matter.

As a final point I would like to make a few comments about the hypothesis of an extra-galactic origin of cosmic rays. Although it is quite clear that the cosmic ray nuclei and the electrons are produced and accelerated in the Galaxy, it seems to me that one cannot yet exclude the possibility that protons, and perhaps α particles, which convey most of the cosmic ray energy flux are of extra-galactic origin. The main argument against the extragalactic hypothesis has always been an argument of plausibility, since the energy requirements involved in a universal distribution of cosmic rays appeared too severe to most of the workers in the field. However, on the basis of our present knowledge of the universe and, in particular, on the frequency and strength of the active phases of galactic nuclei, we cannot exclude that the energy requirements for a universal cosmic ray flux can be met. Of course, the electrons which would be accelerated together with the protons in these hypothetical extra-galactic cosmic ray sources must give up all their energy in radiation. If one considers any plausible acceleration mechanisms, one can see that the relativistic electrons must radiate their energy in the far infrared part of the e.m. spectrum where part of the energy density, corresponding to a fraction of 1 eV/cm^3, could still be of non-thermal origin.

Again, a definite answer to this question will be provided by observations in the γ-ray domain. For instance, the limit imposed by the isotropic γ-ray background already tells us that a universal flux of cosmic rays, which most likely would be produced at high redshifts, would imply a low density Universe (present density $\rho_o \lesssim 10^{-7} \text{ g/cm}^3$). Also, as has been recalled by Prof. Ginzburg in his introductory talk, the interaction of universal cosmic ray protons with the interstellar gas in the Magellanic Clouds may produce an observable flux of γ-rays. Similar observational situations may be envisaged in the direction of other selected objects. However, the interpretation of the observational data may not be so clearcut, due to the presence of localized γ-ray sources.

Since time is getting short I think I will stop here. The considerations I have made, sometimes in an intentionally provocative way, concern some problems which, I believe, are worth our attention in this very last part of the meeting. Of course, I do not pretend to have made an exhaustive list of the important problems which have been discussed in these past few days, and anyone should feel free to add new topics to the list in the course of the general discussion.

GENERAL DISCUSSION

A.W. WOLFENDALE

This question of the contribution of discrete sources to the gamma ray flux is an important one, of course. Now, it seems to me that is dangerous to compare the situation in gamma rays with that in X-rays. You (Prof. Setti) are quite right that in the X-ray region the vast majority of the flux is due to discrete sources, but, I think, that in the gamma rays you should really compare more with the electron situation and there, of course, we know from synchroton data, which tell us something about the way electrons are distributed, that the bulk of the radiation comes from the continuum of the electrons moving in the interstellar magnetic field so that the contribution from discrete sources there is small. However, to look at perhaps the more important point of gradients, I may, perhaps, be allowed just to show this transparency again (Fig. 4, pg. 313, this volume). I think that what you want to do is to put a horizontal line through here (fig. on the left, $E_\gamma > 100$ MeV). Now, the point is that the local value has an uncertainty on it. This uncertainty, as shown, comes from the high-latitude gamma ray data, where one tries to determine the flux in that way. There's a point of, I think, better accuracy that one could draw there from our knowledge of the local proton spectrum, the local electron spectrum aligned with an enhancement as mentioned by you to fit in with radio, and our knowledge of nuclear physics. In other words, I think you have got to make your line go through there without any movement up and down, and, I think, it's stretching things a bit to get it quite flat, to have no gradient at all in that region. Coming to the other diagram (E_γ: 35-100 MeV), I think you would agree that it is very hard to get that to be flat, and I think one is really forced to the conclusion, with which you may agree, that the electrons show a gradient. But, I think it is true to say that within perhaps a matter of months there will be sufficient data from COS B to firm up these points, and particularly if we can get them at higher energies where the proton contribution is bigger and, if then, we really do see points with fairly high precision falling down like that, I think we will be safe, because the only flaw that you would make then, I think, is that we have underestimated the contribution from the discrete sources. If that is so, the point relative to the local value won't move, while those toward the galactic anticenter will come down even lower, so the gradient out there would be steeper. Here you have a problem because the point toward the galactic center will come down as well, but then, I think, it would mean that the target material is uncertain, the point being that in this region we are at the mercy of what happens in molecular clouds, how many there are, what their densities are, and so on. In the outer galaxy region the contribution of molecular clouds is very, very small. It's all nice neutral hydrogen, decently distributed, so I really do think that one would be able to say something in that region.

V.L. GINZBURG

There are four principal aspects related to the problem of the origin of cosmic rays, I mean the origin of the main part of cosmic rays observed near the Earth.
  1) Galactic versus metegalactic origin. It has been clear for about 30 years that the galactic model is correct. The corresponding energetic and dynamical arguments are quite strong and are becoming even stronger. γ-ray observations can provide direct proof and to some extent this proof already exists. New observations are needed (see Wolfendale, this symposium). But I do not see any real danger of a change here.
  2) Halo versus disk galactic origin models. I wrote about this so many times that I shall not discuss it again (see references in my paper at this Conference). My opinion is that the problem is clear enough in favour of the halo model, especially due to radioastronomical observations of edge-on galaxies (see Sancisi's paper at this Conference).
  3) Mixing problem. The question is whether, after subtraction of local sources, such as SN envelopes, etc., the energy density of cosmic rays, and particularly of the electronic component, is rather uniform or not over a large region of quasi-ellipsoidal shape and typical semi-major axis 10 to 14 Kpc and semi-minor axis 3 to 10 Kpc. I used to think, and still do, that the answer is affirmative. But the question is not clear. It must be analyzed carefully, and mainly by means of radio and gamma-ray astronomical methods.
  4) Sources of cosmic rays, places of acceleration, mechanisms of acceleration. These are old problems, but they are the hottest at the moment. The main competing possibilities are: a) acceleration inside young supernovae envelopes, b) acceleration by shocks from supernovæ in low-density interstellar medium, c) acceleration by other stars. I believe in a), but b) is now the most controversial possibility. From the talks and remarks by Axford, by Forman and by Völk we can conclude, as far as I understand, that acceleration in the interstellar medium can be effective enough only if in the shock region a sheet automatically appears with a diffusion coefficient D much smaller than the average for the halo model ($D_{eff} \sim 10^{28} - 10^{29} cm^2 s^{-1}$). In such a case, as Axford emphasized in his talk, acceleration can be effective even in the halo when the shock wave reaches there. But this means "in situ" accleration in the halo. In the past this possibility has been usually disregarded. For instance when I discussed the diffusive or "convective" propagation in the halo (see Proceedings of the Kyoto Conference, Vol. 2, page 148, 1979) I also neglected the "in situ" acceleration. The presence of such "in situ" acceleration would change the distribution of the brightness of radio continuum radiation with distance from the galactic plane at different frequencies. Therefore, from this point of view, the continuation of the radio observations of galaxies seen edge-on as

described by Sancisi in his paper at this symposium is especially needed.

R. COWSIK

I want to make a few points. Number one, regarding acceleration. All the mechanisms of acceleration we have so far discussed need an injection mechanism, that is, need a source of particles which are of higher energy. Then they will collide with matter, that is, they will collide with bulk motions, and this problem has not been understood at all. It is one of the most important problems to understand in terms of totally understanding the acceleration process. The second point I would like to make is that when we talk about the origin of cosmic rays we should also try to understand the origin of the cosmic ray electrons including the highest-energy ones, like those measured by Nishimura and those by Meyer and co-workers at Chicago. If we put the sources of cosmic rays very far away, we would not be able to get the highest energy electrons. The typical distances that are compared by the present measurements are of the order of a couple of hundred parsecs, purely from content losses on the $3°K$ background and the synchroton radiation in the galaxy. The third point that I would like to make is that Fermi-type or any stochastic type of acceleration in the region where the secondaries are also propagating, is precluded by the observation of the relative spectrum of secondaries and primaries. That is, it is precluded by the measurement of lithium-to-carbon ratio as a function of energy. These are the only three points that I wanted to make.

J. LINSLEY

When I was speaking a couple of days ago, I tried to make a point that the highest-energy cosmic rays, those with energies extending up to 10^{20} eV are certainly entitled to be considered cosmic, they are the most cosmic, in fact, of all. They have, to the highest degree, the properties that drew attention to cosmic rays in the first place and I have to say that we now have definite evidence on the properties of these experimental results, and these results are not explained, so that highest energy cosmic rays are still a mystery. Their acceleration is not explained and their propagation is not explained and it's not known for sure what part of the universe they fill and I want to comment that until this part of the cosmic rays is understood I think we can have no confidence in the explanations of the low energy. We can never be sure that the same discovery that explains the highest-energy cosmic rays will not overturn our provisional explanations and understanding of the properties of the low-energy cosmic rays.

G. SETTI

I feel I should intervene at this moment, because a number of arguments brought into the discussion are directly related to the remarks I made previously and, therefore, some clarification is in order. First of all, I wish to come back to this question of the gradient of cosmic rays illustrated again by Prof. Wolfendale. Quite generally I believe it is very dangerous to show diagrams with lines drawn on them because of very well known optical effects. More specifically, looking at your figure with $E_\gamma > 100$, if only the points were marked then I remain of the opinion that a horizontal line would fit them perfectly well, particularly in view of the fact that the last point is very uncertain. Concerning the diagram with $E_\gamma \simeq 35 - 100$ MeV, there I would agree that the points in the direction of the anticenter appear much lower than the average of the points in the direction of the galactic center. But, as you say, this diagram may be more relevant for the distribution of the electrons and it may, therefore, fit in more properly with the general question of the constancy of the proton to electron ratio. In any case, it seems to me, still, that before one can reach firm conclusions on this subject, the problem of the contribution of localized sources to the galactic γ-ray background must be fully settled. Coming now to some of the remarks made by Prof. Ginzburg, first of all, I think that the existence of a halo in our galaxy is fairly uncertain and that it is very doubtful if one can invoke some positive detections of haloes in external galaxies to prove anything with regard to our own galaxy. But I do not wish to repeat what I have already said in my concluding remarks. I think that the real important point, about which all of us may agree, is that particles must get out of galaxies and that one wishes to better understand how these particles diffuse outward, or are convected outward. One uses different names because, of course, one doesn't know the answer to this basic question. Concerning the question of the origin of cosmic rays I wish to make it clear that I am not advocating an extra-galactic origin for the bulk energy of the cosmic rays. The only thing I wanted to point out is that as yet, I do not see any compelling argument that makes this hypothesis unacceptable and that, therefore, this matter must be settled by observations rather than by beliefs.

J. WDOWCZYK

I was expecting that Prof. Ginzburg would cover this topic in his remarks, but he did not. I have taken the liberty to make a few remarks on the point of the Magellanic Clouds and I should be glad to be corrected if I am wrong: it concerns Prof. Setti's remark about whether the test of the Magellanic Clouds is conclusive for the origin, galactic or extragalactic, of the cosmic rays. Well, if I have not misunderstood the logic behind that topic, it seems that it is an argument that works in one way and not in two ways, in the sense that if we do see with the next generation gamma ray experiments that there is a finite gamma ray flux of the order of 10^{-7} ph/cm^{2}s, this does not necessarily mean that the cosmic rays are extragalactic because there can be sources in the Magellanic Clouds. However, if we do not see them then this is a clear indication that the cosmic rays are not extragalactic. If we take into consideration what the boundary conditions of this argument are, I think that this test is, in a certain sense, conclusive about the origin.

J. LINSLEY

The question of the existence of a halo is certainly interesting in connection with highest-energy cosmic rays because it makes a difference for understanding anisotropy in this galaxy, and also for models that would involve scattering by galaxies. So, it is an interesting question.

J. NISHIMURA

May I make some remarks about the different ratios of electrons to protons. Prof. Setti mentioned that one expects a very large magnetic field if one takes the electron spectrum observed at the earth. But, in fact, the electrons concern the energy of several hundred MeV, and at this energy, solar modulation is very high; as Prof. Wolfendale mentioned, the intensity could increase two or three times more. Besides, the synchroton radiation depends on the square of the magnetic field. So, one would expect a magnetic field of several μ-gauss. Of course, the electron to proton ratio may change, but the above argument may not be taken to stress that there is a different electron to proton ratio. Another point, which I want to ask Prof. Ginzburg, is that you mentioned that in the halo there is a change of the spectral index, but in that case the magnetic field may change with altitude so that it may affect also . . .

V.L. GINZBURG

Yes, the problem would be more complicated.

P. KIRALY

Well, about the question of galactic vs. extra-galactic origin in the low energy region, I would say that anisotropy results between 10^{11} and 10^{14}eV give a fairly certain result, and one can be fairly confident that it is due to galactic origin. Now, there may be some question of galactic interstellar acceleration which might give rise to some anisotropy. But it is rather an unlikely scenario that one has an extra-galactic origin and a substantial galactic acceleration causing anisotropy in the low energy region.